# CHEMICAL SENSES

Volume 2

# CHEMICAL SENSES

## Series

**Volume 1:** Receptor Events and Transduction in Taste and Olfaction, *edited by Joseph G. Brand, John H. Teeter, Robert H. Cagan, and Morley R. Kare*

**Volume 2:** Irritation, *edited by Barry G. Green, J. Russell Mason, and Morley R. Kare*

**Volume 3:** Genetics of Perception and Communication, *edited by Charles J. Wysocki and Morley R. Kare*

**Volume 4:** Appetite and Nutrition, *edited by Mark I. Friedman, Michael G. Tordoff, and Morley R. Kare*

# CHEMICAL SENSES

## Volume 2
## Irritation

*edited by*

**Barry G. Green**
**J. Russell Mason**
**Morley R. Kare**

*Monell Chemical Senses Center*
*Philadelphia, Pennsylvania*

**Marcel Dekker, Inc.** **New York and Basel**

ISBN 0-8247-8323-9

This book is printed on acid-free paper.



MARCEL DEKKER, INC.
270 Madison Avenue, New York, New York 10016

Current printing (last digit):
10 9 8 7 6 5 4 3 2 1

PRINTED IN THE UNITED STATES OF AMERICA

# Series Introduction

The Monell Chemical Senses Center celebrated its twentieth anniversary in 1988. Founded as a multidisciplinary organization dedicated to research in all aspects of the chemical senses, the Monell Center is today the only such institution of its kind and has become the focal point for chemosensory research. While the center has always had strong programs of research in many of the traditional areas of chemosensory science, it has also nurtured and helped to expand the scope of less traditional approaches to the study of the chemical senses. With this in mind, the center commemorated its twentieth year by hosting four international conferences on specialized topics within the framework of the chemical senses.

The proceedings of the four conferences are published by Marcel Dekker, Inc. as the series *Chemical Senses*. The first volume, *Receptor Events and Transduction in Taste and Olfaction*, contains the proceedings of the first international symposium that highlighted recent advances in understanding of receptor mechanisms in taste and olfaction. The second volume, *Irritation*, contains the proceedings of the second international symposium, dedicated exclusively to the developing field of common chemical sensitivity. The impact of genetics on the chemical senses was the topic of the third conference, which is covered in the third volume, *Genetics of Perception and Communication*. The proceedings of the final conference, which focused on recent research into the interdependence of the chemical senses and nutrition, appear in the fourth volume, *Appetite and Nutrition*.

The editors of these volumes are confident that the next twenty years will be even more fruitful than the past twenty and that as our research disciplines mature, this anniversary year of the Monell Chemical Senses Center will be recognized as a major turning point in our understanding of the mechanisms, processes, and functions of the chemical senses.

Joseph G. Brand, Mark I. Friedman, Barry G. Green,
Charles J. Wysocki, and Morley R. Kare

# Preface

This book is based on the International Symposium on Chemical Irritation in the Nose and Mouth held at the Monell Chemical Senses Center in June of 1988. The symposium was the first scientific meeting devoted exclusively to discussion of the sensitivity of the oro-nasal region to chemical irritants. The contributors to this volume come from a variety of scientific disciplines, including neuroanatomy, neurophysiology, respiratory physiology, psychophysics, animal behavior and food science. The diversity of the contributors reflects the broad relevance this topic has to basic and applied issues associated with the sensory function of the nose and mouth.

Although the study of chemical irritation is in its infancy compared to the study of gustation and olfaction, investigation of the sensitivity of the skin and mucous membranes to chemical irritants has occurred sporadically throughout most of this century. The event most often cited as promoting modern interest in the topic was the publication in 1912 of a paper by G. H. Parker, in which the concept of the "common chemical sense" was introduced. As Parker described it, the common chemical sense was a mucocutaneous chemosensory system, separate from gustation and olfaction, which responded to chemicals that were "more in the nature of irritants." It is not entirely clear why Parker chose to call this modality the *common* chemical sense, except that he believed the sensitivity to chemicals derived from the free nerve endings of the "general cutaneous nerves."

As readers of this book will discover, the adequacy of Parker's terminology, particularly his use of the word "common," is still under debate. Despite the discontent expressed by many of the contributors, a different terminology has not been formally established. Two alternatives come to mind in reviewing the chapters: One is to make a modest change in Parker's wording by replacing "common" with "general." The *general* chemical sense (which is suggested by Cain in Chapter 3) conveys the ubiquitous nature of the sensitivity to chemical irritants while avoiding the confusion over whether "common" refers to the nature of the stimuli, or, as Parker apparently intended it, to the nature of the sensory system. Alternatively, it may be preferable to follow the lead of Keele and Armstrong, who in 1964 coined the term "chemalgia" to describe pain produced by chemical stimulation. Given that chemical sensitivity is a general property of the skin and is associated with other sensations in addition to pain, it would seem appropriate to call it "chemesthesis." An appealing attribute of the latter term is that it does not imply that chemical sensitivity constitutes a separate sense; chemesthesis refers only to a chemical *sensibility*. This is an important distinction: Although the possibility remains that trigeminal chemoreception is mediated at least in part by specific chemoreceptors, the preponderance of evidence offered in this volume indicates a close relationship between chemical irritation and the senses of pain and temperature.

In addition to addressing these basic yet long-neglected issues, each of the chapters in this book contains new and often challenging data. The topics covered range from the anatomy and pharmacology of nerve fibers hypothesized to be sensitive to nasal irritants to the psychology of how and why individuals develop preferences for eating spicy foods. The book also provides an opportunity for researchers working in the chemical senses to learn about the rapidly growing body of data on the role cutaneous nociceptors play in the detection of endogenous and exogenous chemicals.

The General Discussion (which appears at the end of this volume) makes it apparent that more data will have to be collected before agreement can be reached on some of the most basic features of oral and nasal chemoreception. There still is, for example, little agreement not only about how to define chemical sensitivity, but also about the terminology that should be used to describe the stimuli and sensations associated with it. Most notably, should we even be using the terms "irritant" and "irritation"? Wedded to the issue of perceptual terminology is the question of quality coding, i.e., whether "trigeminal" chemoreception conveys many qualities, a few qualities, or only a single quality of sensation. The ground work for answering this question is layed in the chapters of this volume that explore either the physiological issue of what types of sensory fibers mediate trigeminal chemoreception, or the psychophysical issue of how sensations of "irritation" are subjectively labeled.

That the symposium seemed to produce more questions than answers testifies to the freshness of the discipline and to the fertile ground that awaits further study. It is hoped that this volume will serve not only as a source of current knowledge about the chemical sensitivity of the oral and nasal mucosa, but also as a stimulus and guide for future research in this emerging area of the chemical senses.

This symposium was made possible by generous donations from McCormick & Company, Inc. and Takasago International.

The editors also thank Ms. Jodie Carr for her indispensable help in making most of the arrangements for the symposium, and Ms. Janice Blescia for her expert work in transcribing the discussions.

Barry G. Green
J. Russell Mason
Morley R. Kare

# Contents

*Series Introduction* iii

*Preface* v

*Contributors* xiii

1. Affector and Effector Functions of Peptidergic Innervation of the Nasal Cavity 1

   *Thomas E. Finger, Marilyn L. Getchell, Thomas V. Getchell, and John C. Kinnamon*

   Discussion 19

2. Physiological Factors in Nasal Trigeminal Chemoreception 21

   *Wayne L. Silver*

   Discussion 39

3. Perceptual Characteristics of Nasal Irritation 43

   *William S. Cain*

   Discussion 59

4. Evidence for Interactions Between Trigeminal Afferents and Olfactory Receptor Cells in the Amphibian Olfactory Mucosa 61

   *André Holley, Jean-Francois Bouvet, and Jean-Claude Delaleu*

   Discussion 69

5. Trigeminal vs. Olfactory Input for Laryngectomized Patients 71

*Maxwell M. Mozell, David N. Schwartz, Steven L. Youngentob, Donald A. Leopold, Paul R. Sheehe, and James A. Listman*

Discussion 93

6. Responses of Normal and Anosmic Subjects to Odorants 95

*James C. Walker, John H. Reynolds IV, Donald W. Warren, and James D. Sidman*

Discussion 119

7. Brain Responses to Chemical Stimulation of the Trigeminal Nerve in Man 123

*Gerd Kobal and Thomas Hummel*

Discussion 137

8. Capsaicin, Irritation, and Desensitization: Neurophysiological Basis and Future Perspectives 141

*J. Szolcsanyi*

Discussion 169

9. Effects of Thermal, Mechanical, and Chemical Stimulation on the Perception of Oral Irritation 171

*Barry G. Green*

Discussion 193

10. Differences Between and Interactions of Oral Irritants: Neurophysiological and Perceptual Implications 197

*Harry T. Lawless and David A. Stevens*

Discussion 213

11. Personality Variables in the Perception of Oral Irritation and Flavor 217

*David A. Stevens*

Discussion 229

12. Getting to Like the Burn of Chili Pepper: Biological, Psychological, and Cultural Perspectives 231

*Paul Rozin*

Discussion 271

13. Effects of Menthol on Nasal Sensation of Airflow 275

*Ronald Eccles*

Discussion 293

14. Trigeminal Nerve Stimulation: Practical Application for Industrial Workers and Consumers 297

*Yves Alarie*

Discussion 305

15. Effectiveness of Six Potential Irritants on Consumption by Red-Winged Blackbirds (*Agelaius phoeniceus*) and Starlings (*Sturnus vulgaris*) 309

*J. Russell Mason and David L. Otis*

Discussion 323

General Discussion 325

Index 341

# Contributors

**Yves Alarie** Department of Industrial Environmental Health Sciences, Graduate School of Public Health, University of Pittsburgh, Pittsburgh, Pennsylvania

**Jean-François Bouvet** Université Claude Bernard, Villeurbanne, France

**William S. Cain** John B. Pierce Foundation Laboratory and Department of Epidemiology and Public Health and Department of Psychology, Yale University, New Haven, Connecticut

**Jean-Claude Delaleu** Université Claude Bernard, Villeurbanne, France

**Ronald Eccles** Common Cold and Nasal Research Centre, University of Wales College of Cardiff, Cardiff, Wales

**Thomas E. Finger** Department of Cellular and Structural Biology, School of Medicine, University of Colorado Health Sciences Center, Denver, Colorado

**Marilyn L. Getchell** Department of Anatomy and Cellular Biology, Wayne State University Medical School, Detroit, Michigan

**Thomas V. Getchell** Department of Anatomy and Cellular Biology, Wayne State University Medical School, Detroit, Michigan

**Barry G. Green** Monell Chemical Senses Center, Philadelphia, Pennsylvania

**André Holley** Laboratoire de Physologie Neurosensorielle, Université Claude Bernard, Villeurbanne, France

**Thomas Hummel** Department of Pharmacology and Toxicology, University of Erlangen-Nürnberg, Erlangen, Federal Republic of Germany

**John C. Kinnamon** Department of Molecular, Cellular, and Developmental Biology, University of Colorado, Boulder, Colorado

**Gerd Kobal** Institute of Pharmacology and Toxicology, University of Erlangen-Nürnberg, Erlangen, Federal Republic of Germany

**Harry T. Lawless** Department of Food Science, Cornell University, Ithaca, New York

**Donald A. Leopold** Department of Otolaryngology and Communication Sciences, State University of New York Health Science Center at Syracuse, Syracuse, New York

**James A. Listman*** Department of Physiology, State University of New York Health Science Center at Syracuse, Syracuse, New York

**J. Russell Mason** U. S. Department of Agriculture, Animal and Plant Health Inspection Service/Science and Technology, and Monell Chemical Senses Center, Philadelphia, Pennsylvania

**Maxwell M. Mozell** Department of Physiology, State University of New York Health Science Center at Syracuse, Syracuse, New York

**David L. Otis** U.S. Department of Agriculture, Animal and Plant Health Inspection Service/Science and Technology, Denver Wildlife Research Center, Denver, Colorado

**John H. Reynolds IV** Department of Biobehavioral Research and Development, Bowman Gray Technical Center, R. J. Reynolds Tobacco Co., Winston-Salem, North Carolina

**Paul Rozin** Department of Psychology, University of Pennsylvania, Philadelphia, Pennsylvania

**David N. Schwartz** The SUNY Upstate Clinical Olfactory Research Center, State University of New York Health Science Center at Syracuse, Syracuse, New York

**Paul R. Sheehe** Department of Preventive Medicine, State University of New York Health Science Center at Syracuse, Syracuse, New York

**James D. Sidman** Dental Research Center, University of North Carolina, Chapel Hill, North Carolina

**Wayne L. Silver** Department of Biology, Wake Forest University, Winston-Salem, North Carolina

**Present affiliation*: Department of Pediatrics, State University of New York Health Science Center at Syracuse, Syracuse, New York

**David A. Stevens** Francis L. Hiatt School of Psychology, Clark University, Worcester, Massachusetts

**J. Szolcsanyi** Department of Pharmacology, University Medical School of Pécs, Pécs, Hungary

**James C. Walker** Department of Biobehavioral Research and Development, Bowman Gray Technical Center, R. J. Reynolds Tobacco Co., Winston-Salem, North Carolina

**Donald W. Warren** Department of Dental Ecology, School of Dentistry, and Department of Otolaryngology, School of Medicine, University of North Carolina, Chapel Hill, North Carolina.

**Steven L. Youngentob** Department of Physiology, State University of New York Health Science Center at Syracuse, Syracuse, New York

# CHEMICAL SENSES

## Volume 2

# 1
# Affector and Effector Functions of Peptidergic Innervation of the Nasal Cavity

**Thomas E. Finger**
University of Colorado Health Sciences Center
Denver, Colorado

**Marilyn L. Getchell and Thomas V. Getchell**
Wayne State University Medical School
Detroit, Michigan

**John C. Kinnamon**
University of Colorado
Boulder, Colorado

## I. INTRODUCTION

The classical view of nonolfactory sensory innervation of the nasal cavity includes the concept that sensory nerve fibers originating from cell bodies of the trigeminal (Gasserian) ganglion terminate peripherally as free nerve endings within the epithelium (e.g., see DeLond and Getchell, 1987). Sensory impulses generated in these fibers are transmitted via the central processes of these same cells to sites of termination within the brainstem, such as the spinal trigeminal nucleus or possibly the nucleus of the solitary tract. In fact, this system and its brainstem centers are most likely responsible for many aspects of our reaction to trigeminal stimuli (see Silver, 1987) including reflex sneezing or coughing as well as perception of such qualities as temperature and pungency. Several recent lines of evidence indicate, however, that responses to trigeminal nerve stimuli may be more complex than those accounted for by the traditional brainstem routes (e.g., Lundberg et al., 1987; Bouvet et al., 1987a, 1988). This chapter will explore many of these other avenues of action of the trigeminal nerve system and will attempt to draw together many disparate lines of investigation into a conceptual framework for further consideration.

## II. NASAL TRIGEMINAL UNMYELINATED NERVE FIBERS CONTAIN NEUROPEPTIDES

A family of tachykinins, including substance P, is present throughout the length of many sensory ganglion cells that give rise to fine, unmyelinated sensory nerve fibers (C fibers) (Hua et al., 1985). Physiological experiments indicate that these fine fibers are responsible for many of the reactions to chemical irritants of the skin (Jancso et al., 1967) as well as of the nasal and respiratory passages (Lundblad et al., 1984, 1985). Immunocytochemical studies demonstrate that substance P-immunoreactive nerve fibers are present in the nasal epithelium and that these peptidergic fibers arise from the trigeminal nerve (Lundblad et al., 1983a; Bouvet et al., 1987a). Calcitonin gene-related peptide (CGRP) has no structural similarity to substance P and yet is often colocalized with substance P in sensory processes (Skofitsch and Jacobwitz, 1985a; Lee et al., 1985; Franco-Ceredada et al., 1986; Finger, 1986; Gibbins et al., 1987). Our results from studies on rats and mice indicate that such colocalization is the case for much of the peptidergic innervation of the nasal epithelium also (St. Jeor et al., 1988). For convenience, fibers exhibiting immunoreactivity to both substance P and CGRP antisera will be referred to as CGRP-SP fibers, although the reader is cautioned that immunocytochemical studies do not permit exact identification of the chemical nature of the substance with which the antisera are reacting. Furthermore, some fibers appear to contain detectable levels of only CGRP. Other classes of peptidergic fibers, e.g., luteinizing hormone releasing hormone (LHRH)-containing (Wirsig and Getchell, 1986) and vasoactive intestinal peptide (VIP)-containing (M. Getchell et al., 1987; Uddman et al., 1978), have also been observed in the epithelium but will not be discussed in this chapter.

## III. STIMULUS ACCESS: PROXIMITY OF PEPTIDERGIC FIBERS TO THE EPITHELIAL SURFACE

Although several previous investigators, using a variety of techniques, have described the presence of nonolfactory nerve fibers within the epithelium (see M. Getchell et al., 1988 for review), no definitive conclusion has been reached regarding how close to the epithelial surface such fibers extend. Light microscopic immunochemical studies illustrate immunoreactive fibers extending virtually to the epithelial surface (e.g., figure 1b in Lundblad et al., 1983a). However, our electron microscope observations of free nerve endings within the nasal epithelium of rats indicate that these nerve fibers are not exposed directly to the mucosal compartment. In all cases in which a superficial nerve process has been followed to its distal terminus, the nerve fiber remains below the level of the tight junctions formed between the epithelial cells (Fig. 1). Thus in order for chemical stimuli to reach the free nerve endings, these stimuli must (a) enter the nasal cavity, (b) partition into and diffuse through the mucus layer, (c) cross the epithelial cell membranes and/or intercellular tight junctions, either by active or passive processes, and (d) activate the free nerve ending. Since most effective stimuli of the common chemical sense are lipid-soluble (Silver, 1987), the line of epithelial tight junctions should not provide a major barrier to penetration of the epithelial surface by the stimulus.

However, it may not be necessary for the chemical stimulus to contact the sensory nerve fiber directly. One hypothesis of pain mechanisms holds that painful stimuli do not directly activate peptidergic fibers in the epithelium but rather act by producing bradykinin in the damaged tissue; the bradykinin then is envisioned as binding to specific bradykinin receptors on the pain fibers, thereby activating them (Steranka et al., 1988). If this model is applicable to trigeminal nasal chemoreception, then the inhaled stimuli may activate the sensory fibers by causing the formation of bradykinin within the cells of the nasal epithelium. One way of testing this hypothesis is by testing stimulus effectiveness before and after administration of bradykinin antagonists (Steranka et al., 1987).

## IV. DISTRIBUTION OF PEPTIDERGIC SENSORY FIBERS WITHIN THE NASAL CAVITY

Although the presence of peptidergic trigeminal nerve fibers within the nasal mucosa has been noted by several investigators (Änggård et al., 1979; Lundblad et al., 1983a; Papka and Matulionis, 1983; Bouvet et al., 1987b; M. Getchell et al., 1987), no detailed description has been offered regarding the intranasal distribution of such fibers. Furthermore, it is unclear whether significant interspecies differences exist regarding the prevalence and distribution of peptidergic nerve endings within the epithelium. In order to resolve issues such as these, we have undertaken studies on the distribution of peptidergic nerve fibers in the nasal epithelium of several species.

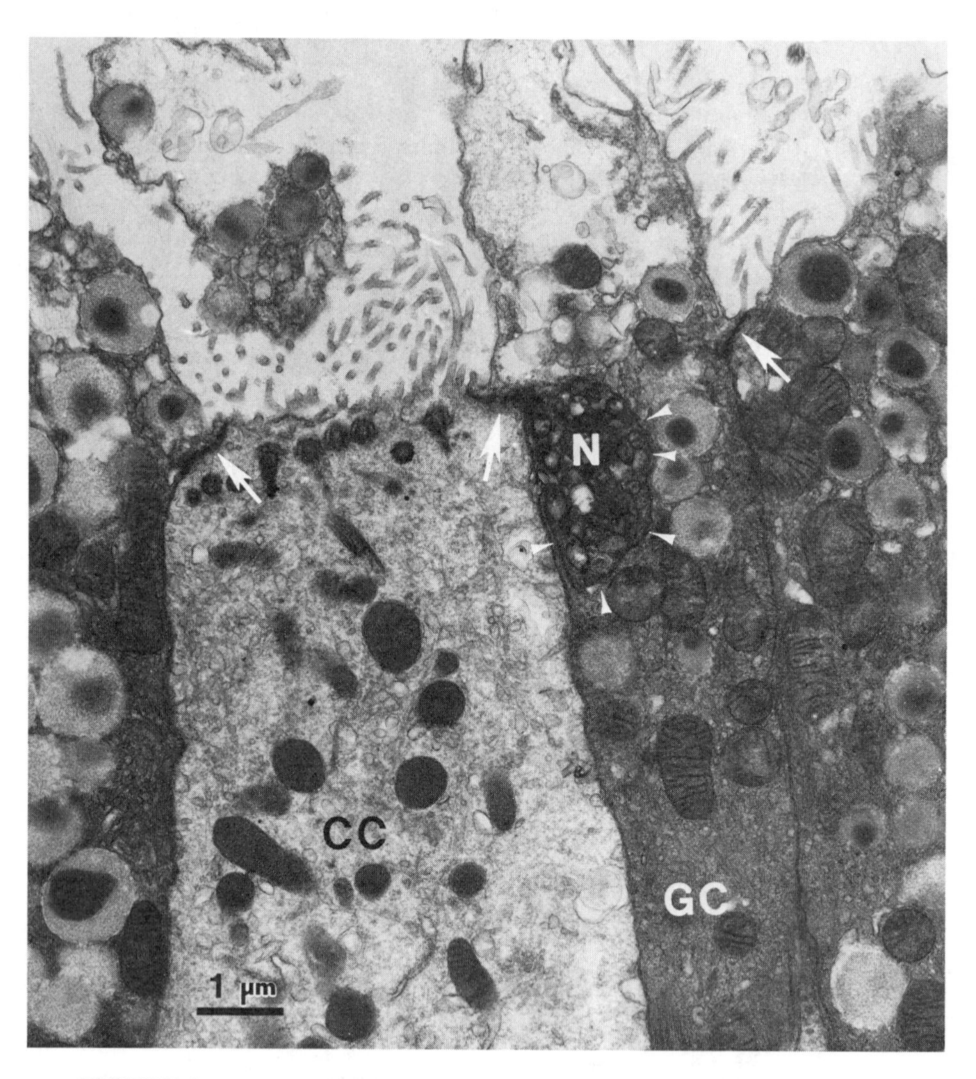


**FIGURE 1** Electron micrograph showing the closest approach of a free nerve ending to the surface of the nasal respiratory epithelium of a rat. The nerve process (N) lies between a goblet cell (GC) and a ciliated cell (CC), but below the level of the tight junctions (arrows). The plasma membrane of the nerve process is outlined by the arrow heads.

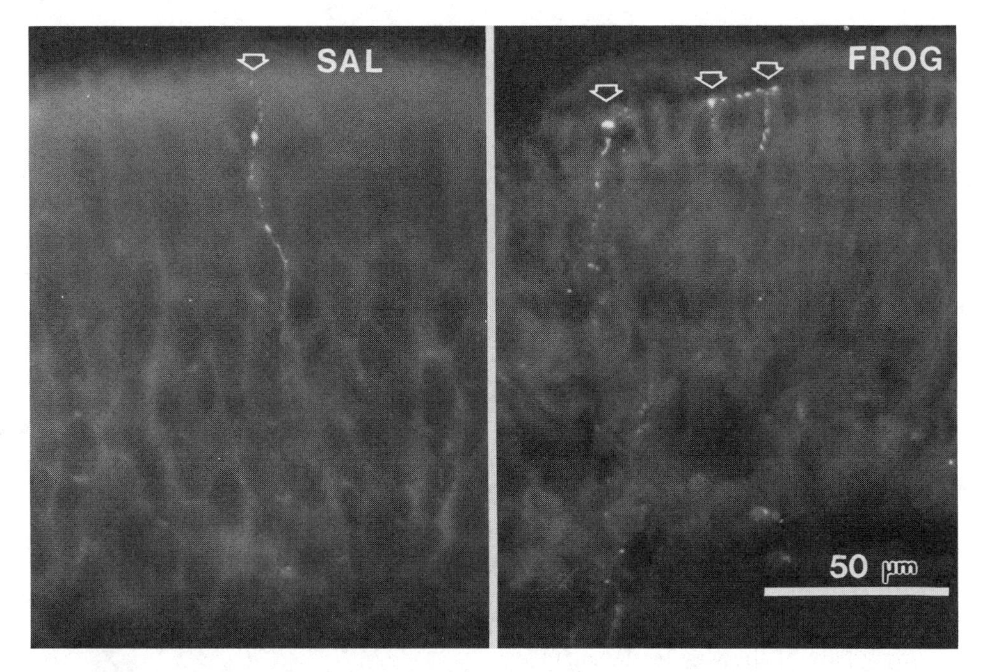


**FIGURE 2** Photomicrographs comparing the density of peptidergic innervation of the olfactory mucosa in the tiger salamander (SAL) and the grass frog (FROG). The fluorescent peptidergic fibers are indicated by the arrows.

## A. Amphibia

Immunohistochemical studies were undertaken on both urodelean and anuran species including tiger salamander, mudpuppy, bullfrog, grass frog, and marine toad. Even within this vertebrate class, profound differences were observed in the density of peptidergic innervation of the olfactory organs. In all anurans studied in the current series as well as those studied by Bouvet et al. (1987b), a relatively dense peptidergic innervation occurs throughout the olfactory epithelium. In contrast, in the two urodeles studied (and in all mammals examined to date) intraepithelial peptidergic fibers are comparatively sparse (Fig. 2). The CGRP-SP peptidergic fibers in salamanders are most prevalent in subepithelial layers. Peptidergic fibers occur between the cells of both the deep and superficial nasal glands. In addition, numerous peptidergic fibers course just below the basal lamina and are frequently associated with blood vessels. Occasional peptidergic nerve fibers can be seen within the olfactory epithelium. More often than not, these rare intraepithelial fibers are associated with the capillary loops that mark zones of epithelial maturation (M. Getchell et al., 1986; 1989).

### B. Rodents

We have begun to map the distribution of CGRP-SP fibers within the nasal cavities of mice and rats (St. Jeor et al., 1988). Preliminary findings, mostly obtained from complete sets of sections reacted with CGRP antisera, are presented below. Spot checks of these results utilizing double-label protocols indicate that most CGRP-immunoreactive fibers within the nasal epithelium also are immunoreactive to substance P antisera.

In virtually all areas of the nose, a rich plexus of CGRP-SP fibers can be observed within the lamina propria. These peptidergic fibers are most often associated with blood vessels, but scattered fibers also occur within glands or connective tissue elements within this layer. At many places within the nasal cavity, a few peptidergic fibers of the lamina propria will ascend slightly to form a plexus at the base of the epithelium. Occasionally, a single fiber can be seen emerging from this plexus to run perpendicularly outward toward the epithelial surface. These intrapeithelial CGRP-SP fibers usually traverse the entire height of the epithelium, although some fibers appear to extend only part of the way to the surface.

The distribution of intrapeithelial CGRP-SP fibers within the nasal cavity is not homogeneous, varying to some degree by region and by the nature of the overlying epithelium. The histological nature of the epithelium within the nasal cavity exhibits distinct regional variation. The epithelium in the anterior and ventral parts of the cavity is a medium-thick, goblet cell-rich columnar epithelium. As one proceeds caudally or dorsally within the nasal cavity, the epithelium changes to a taller, columnar epithelium virtually devoid of goblet cells. In other studies, e.g., Meisami et al. (1988), this change in epithelial structure was shown to coincide with the change from nonolfactory to olfactory epithelium. Whether this transition exactly corresponds to the transition from nonolfactory to olfactory epithelium has not been determined directly in our material.

Although the epithelium within the goblet cell-rich area is shorter than elsewhere in the nasal cavity, the incidence of epithelial CGRP-SP fibers also seems lower. Occasional transepithelial peptidergic fibers can, however, be observed, especially along the anteroventral portion of the septum near the vomeronasal organs.

Within the taller (presumed olfactory) epithelium, the incidence of epithelial peptidergic fibers is low. No quantitative estimate of the density of this innervation has been made, but a single 15 $\mu$m section through the entire nose might typically contain only a few such fibers. The epithelial CGRP-SP fibers appear most frequently in, but are not limited to, areas in which the nasal passage is narrowed, e.g., where a turbinate approaches the septum, or on sharply curved surfaces.

Within the vomeronasal organ, a very different picture emerges. Although CGRP-immunoreactive fibers are sparsely present within the vomeronasal sensory epithelium, they are comparatively abundant within the nonsensory epithelium of the organ (Fig. 3). At light microscopic levels, these fibers appear similar to the transepithelial fibers of the nasal cavity, i.e., fine varicose fibers extending to near the luminal surface. Their main distinguishing feature is their density. Not only are the CGRP-immunoreactive fibers dense in the vomeronasal nonsensory epithelium, but a dense subepithelial plexus exists as well. Numerous immunoreactive fibers course through the lamina propria to innervate blood vessels and the cavernous tissue within the organ.

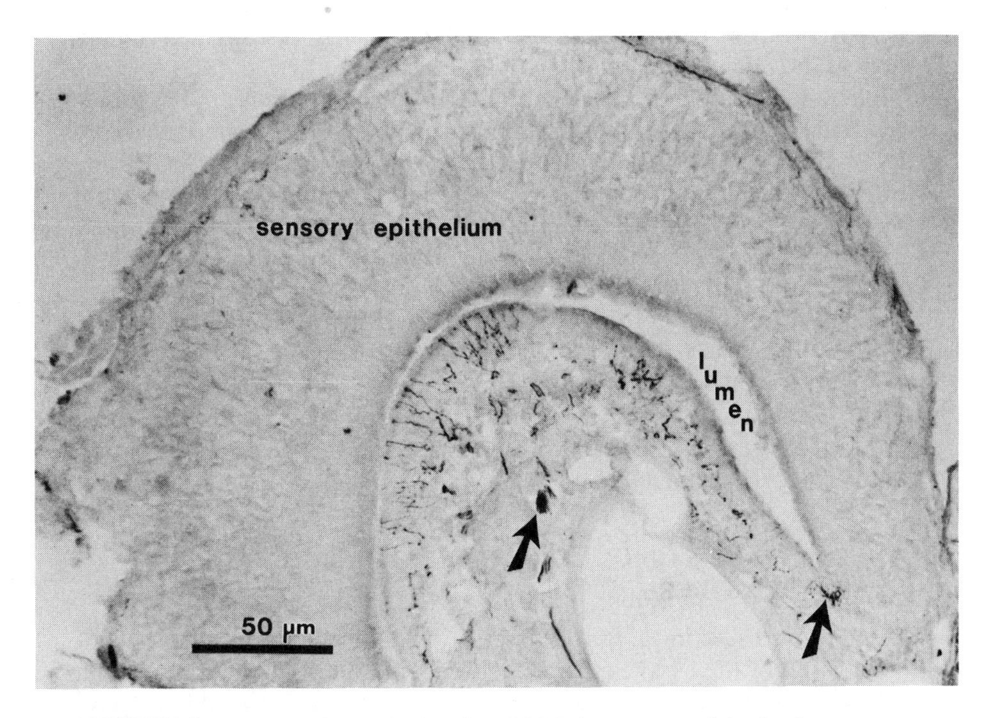


**FIGURE 3** Photomicrograph showing CGRP immunoreactivity in the vomeronasal organ of a hamster. The dark, immunoreactive fibers are plentiful in the nonsensory epithelium but sparse in the sensory epithelium of the organ. Numerous immunoreactive nerve bundles (arrows) can be seen deep to the intraepithelial fibers.

The prevalence of the CGRP-immunoreactive fibers in this specialized area may indicate a role for these fibers in the vomeronasal pump.

The subepithelial plexus beneath the olfactory epithelium includes CGRP-immunoreactive fibers which travel with the olfactory nerve fascicles. These peptidergic fibers do not appear to arise from olfactory or vomeronasal receptor cells since few CGRP- and substance P-like immunoreactive receptor cells were observed. CGRP-immunoreactive fibers are present not only within the olfactory nerve bundles in the nasal cavity but also within the olfactory nerve as it penetrates the cribriform plate to form the olfactory nerve layer of the olfactory bulb. Fine CGRP-immunoreactive fibers can be seen within the olfactory nerve layer as well as within many of the glomeruli within the bulb. Because of the involved and tortuous course of these fibers in rodents, tracing a single fiber from the periphery to a terminus in the bulb is a formidable undertaking. In salamanders, however, the olfactory nerve is much straighter. Because of this, we have been able to trace a single peptidergic fiber in serial light microscopic sections from its position in the olfactory nerve just below the epithelium to a terminus in a rostrally situated glomerulus of the olfactory bulb.

In order to account for these bulbar CGRP-immunoreactive fibers, Rosenfeld and coworkers (1983) suggested that a subset of olfactory receptors contain CGRP. Since few receptors seem to be immunoreactive for CGRP, we offer an alternative hypothesis to explain these findings. The peripheral, sensory fibers of the trigeminal nerve may have one branch in the olfactory epithelium with another branch entering the olfactory nerve to end in the olfactory bulb. Activation of these processes by a common chemical sensory stimulus then would not only result in propagation of an impulse back to the CNS but would antidromically activate the trigeminal nerve terminals within the olfactory bulb—similar in this respect to the peripheral axon reflex (Lembeck and Gamse, 1982). A similar type of mechanism has been postulated for trigeminal nerve collaterals reaching the sphenopalatine ganglion (Lundblad et al., 1983b).

## V. EFFECTS OF STIMULATION OF PEPTIDERGIC SENSORY FIBERS

Based on introspection as well as observation of other animals, one effect of stimulation of the nasal common chemical sense must be the transmission of sensory information to the central nervous system. We readily perceive noxious as well as nonnoxious stimuli via this mechanism (see Chapter 2 of this volume). The most likely route of transmission of this information within the CNS includes either the nucleus of the solitary tract or the spinal trigeminal nucleus. Both areas receive input from peptidergic fibers of the trigeminal nerve (South and Ritter, 1986). The higher order CNS areas involved in perception of trigeminal nasal stimuli are not clear but may involve components of the ventral posteromedial nucleus (VPM) of the thalamus as the site for integration of input from the trigeminal somatosensory nuclei as well as input from the visceral sensory systems. A cortical representation for trigeminal nasal chemoreception is unknown, but if the VPM is involved, then insular or suprainsular cortex would be the most likely site.

In addition, strong stimulation of trigeminal sensory fibers in the nose will elicit reflexes such as sneezing, coughing, or choking. All of these reflexes are probably mediated via intrinsic brainstem pattern generators and thus do not require the participation of higher neural centers. Furthermore, activation of the trigeminal nasal sensory fibers may modulate activity in preganglionic autonomic neurons via pathways involving brainstem and spinal reflex systems. However, increasing evidence suggests that peptidergic sensory fibers may participate in an "axon reflex" type of activity which serves to alter peripheral physiology without the intervention of the central nervous system. The classic example of this mechanism is neurogenic inflammation, in which activation of peripheral pain fibers results in generalized hyperemia and swelling within the innervated area. The mechanism responsible for this process is antidromic activation of collaterals of the sensory nerve fibers followed by release of the bioactive peptides into the surrounding tissue. Then various tissue components, such as blood vessel walls and mast cells, respond to the diffusely released peptide to produce significant alterations in the local tissue physiology.

The possibility that a type of axon reflex may also operate in the nasal trigeminal system must be considered (Lundblad et al., 1983b; M. Getchell et al., 1989). Before considering possible mechanisms, we must first consider what the possible peripheral distribution of a single peptidergic trigeminal fiber might be. Since current methodologies do not permit visualization of single trigeminal afferent fibers, it is necessary to extrapolate from the existing data available on the population as a whole. The reader is cautioned, however, that single fibers may not be as widely distributed as described in this hypothetical construct.

## VI. POSSIBLE COLLATERALIZATION OF SINGLE FIBERS

If we assume that the epithelial processes of the peptidergic fibers represent the site of transduction and depolarization, then peptide release will start at these sites. As described above, collaterals from these fibers ramify near the basal lamina of the epithelium, among secretory cells of nasal glands, and along blood vessel walls. The long conducting processes of these same fibers reach into the sphenopalatine ganglion (Lundblad et al., 1983b) while other

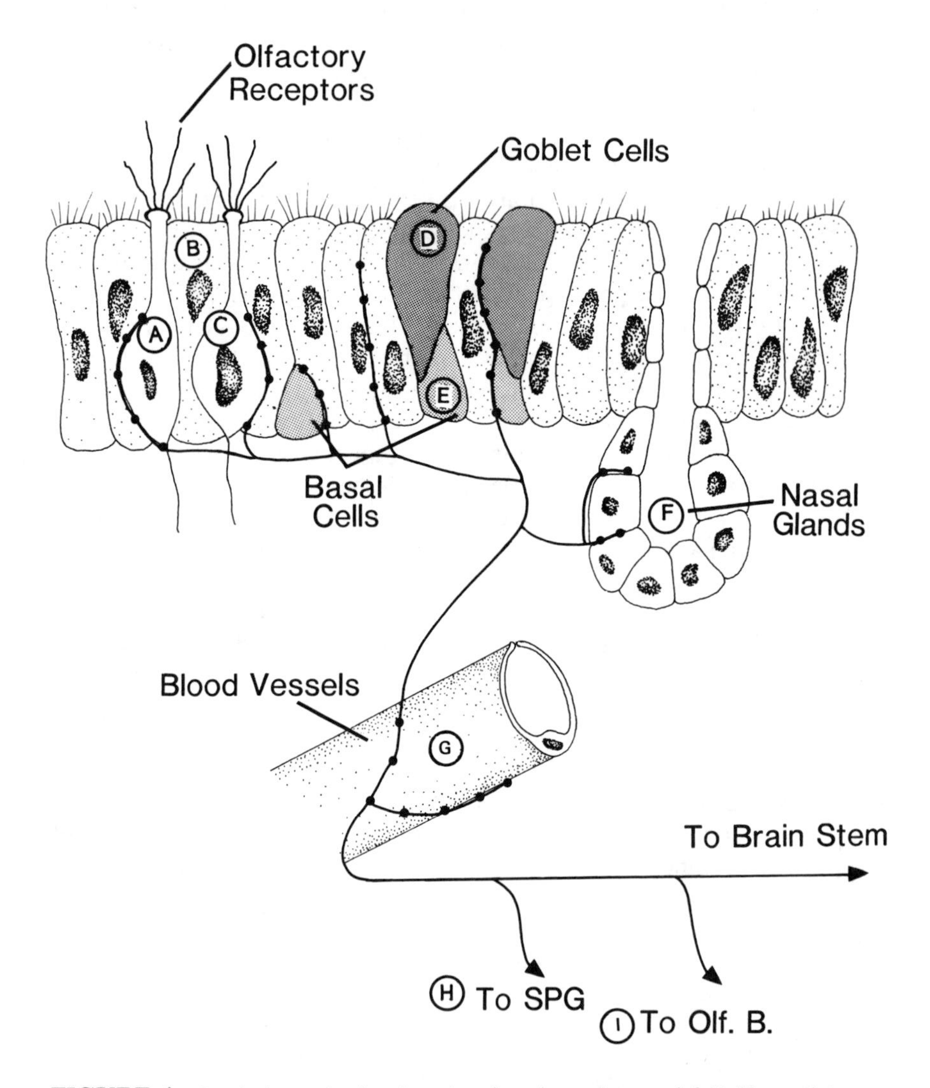


**FIGURE 4** Semischematic drawing showing sites of potential "effector" (axon reflex) actions of peptidergic fibers innervating the nasal epithelium. For the sake of illustration, a single peptidergic fiber is shown as affecting all sites. This need not be the case; various subsets of the peptidergic fibers may exist with each subset affecting the physiology at only a limited number of sites. The letters correspond to the letters in the text indicating possible sites of action for released peptides.

distal processes may even extend into the olfactory bulb. Thus activation of a single peptidergic trigeminal sensory fiber may cause the release of bioactive peptides into diverse tissues. At each of the aforementioned release sites, the peptides are likely to have different effects. Following is a discussion of the possible effects at each of the release sites; some of these effects are well documented, whereas others are purely speculative. We hope that this conceptualization will prove worthwhile as a framework for future research (Fig. 4).

### A. Olfactory Receptors

As described by Bouvet et al. (1987a, 1988), the release of neuropeptides in the vicinity of olfactory receptors may result in modulation of activity within these neurons. Substance P is known to affect ion channels in other neuronal systems, and there is no reason to believe that the peptide will not affect olfactory neurons similarly. The overall effect of modulation of the activity of such ion channels in olfactory receptor functioning is unknown.

### B. Ciliated Epithelial Cells

The release of bioactive peptides along the basal lamina and within the epithelium may affect the ciliated supporting cells as well as the receptors. Lindberg et al. (1987) reported the acceleration of mucociliary activity within the maxillary sinus in response to stimulation of the nasal cavity by ammonia vapor or by intravascular injection of capsaicin. These investigators noted that in the animal preparation used (rabbit), the passage between the maxillary sinus and nasal cavity is quite small. Thus the ammonia vapors may not have reached the epithelium of the maxillary sinus where the mucociliary acceleration was observed. One interpretation of these results is that both ammonia and capsaicin activate the widely branched intranasal peptidergic fibers, which then release their bioactive peptides within the epithelium of the sinus as well as that of the nose. The increase in ciliary activity may be due to activation of the ciliated epithelial cells by the peptides rather than by direct activation by the applied chemical stimuli.

### C. Paracellular Shunts

Paracellular current pathways are likely routes for passive NaCl flux to restore transmucosal ionic balance following active ion transport across the olfactory mucosa (Persaud et al., 1987). Substance P has been shown to increase the anion permeability of such paracellular pathways in the trachea (Rangachari and McWade, 1986). The overall effect of peptidergic modulation of shunt permeability on olfactory receptor cell activity is difficult to assess.

### D. Intraepithelial Secretory Cells

Both goblet cells and sustentacular cells are believed to contribute to nasal secretion. Both cell types reside in the epithelium in proximity to the peptidergic fibers. Electrical stimulation of the trigeminal nerve or local application of substance P to the mucosa results in morphological signs of secretory activity by sustentacular cells (M. Getchell et al., 1989). Changes in the mucus composition may affect prereceptor and perireceptor events in olfactory transduction.

### E. Basal Cells

A rich plexus of peptidergic fibers occurs on both the mucosal and serosal sides of the basal lamina. The release of peptides by these fibers is likely to impact on the basal portions of all epithelial cells including sustentacular cells and receptor cells. The proliferative basal cells will of course be most directly immersed in this milieu. Substance P itself has been shown or suggested to exhibit neurotrophic activity (Linder and Grosse, 1981; Jonsson and Hallman, 1982; Nishimoto et al., 1985); other kinins also exhibit growth factor-like activity. For example, bradykinin and bombesin are believed to act as growth factors in certain systems, e.g., bombesin specifically acts on bronchial epithelial cells (Willey et al., 1984). Conceivably the neuropeptides CGRP and substance P might act as growth factors for basal cells. Activation of the peptidergic fibers could then result in modification of the proliferative rate of basal cells or even of the most prevalent type of daughter cell that is produced.

### F. Nasal Glands

As described by several investigators (see M. Getchell et al., 1989), CGRP-SP fibers ramify among the secretory cells of both the superficial and deep nasal glands. Activation of the trigeminal nerve or topical administration of substance P significantly alters the secretory activity of the glandular elements (M. Getchell et al., 1989). Substantial alterations in the mucus layer overlying the epithelium are likely to result in profound changes in the prereceptor and perireceptor events preceding olfactory transduction (T. Getchell et al., 1984). Thus the secretory state of the nasal glands may be an important factor in the modulation of olfactory activity.

### G. Blood Vessels

As we and other investigators have shown (e.g., Lundblad et al., 1983a), peptidergic fibers contribute to the perivascular neural plexus. Lundblad et al. (1983b) demonstrated that antidromic activation of the trigeminal nerve results in long-duration vasodilation, apparently mediated by substance P. Vasodilation will produce increased blood flow within the local area supplied by the affected vessels. Based on studies by Lacroix et al. (1988), vasodilation should result in increased volume within the nasal cavity. However, substance P also causes plasma extravasation with resulting tissue edema. This is likely to result in long-term narrowing of the nasal passages. In any event, activation of trigeminal peptidergic fibers would be expected to alter air flow characteristics within the nasal cavity. This mechanism may operate globally, e.g., trigeminal activation may alter all nasal airways, or it may operate locally, e.g., local trigeminal stimulation may alter the passage between the stimulated areas. Also, although this mechanism may not be of vital importance to normal olfactory stimulus access, the abundance of peptidergic fibers within the vomeronasal organ suggests that peptides may play a large role in flow through the vomeronasal system.

### H. Collaterals to Parasympathetic Ganglia

Not only can peptidergic fibers directly influence vascular flow, but collaterals of the peptidergic fibers synapse within the sphenopalatine ganglion to modulate parasympathetic activity (Lundblad et al., 1983b). Thus a second, less direct mechanism exists whereby peptidergic fibers may influence glandular secretion and blood flow within the nasal cavity.

### I. Termination Within the Olfactory Bulb

Several investigators (Rosenfeld et al., 1985; Skofitsch and Jacobowitz, 1985b) noted the presence of CGRP-immunoreactive fibers within the olfactory nerve and glomeruli but interpreted these as processes arising from olfactory receptors. We offer the interpretation that these peptidergic fibers may include collaterals of trigeminal nerve fibers that innervate the nasal cavity. If this is the case, then yet another avenue exists whereby the trigeminal sensory fibers may influence olfactory sensory processes. The peptides CGRP and substance P are both neuroactive. Release of either or both of these substances into the primary processing layer of the olfactory bulb may be expected to modulate activity within these structures. Thus activation of trigeminal sensory fibers in the nasal cavity have a route by which they may directly influence olfactory processing without the necessity of synapsing within the brainstem primary trigeminal sensory nuclei.

## VII. CONCLUSION

The nasal epithelium is innervated by neuropeptide-containing sensory processes of the trigeminal nerve which we believe to participate in nonolfactory nasal chemosensitivity. The sensory processes not only relay information to the trigeminal nuclei of the brainstem but also serve as effector elements in a variety of peripheral axon reflexes. Many if not all of these axon reflexes will result in alterations in mucosal physiology that are likely to affect olfactory reception, transmission, or processing. Thus, trigeminal chemoreception should not be viewed as a simple sensory system that can be studied in isolation from the olfactory system. Stimulation of the trigeminal nasal system is likely to effect changes in perception of subsequent or concurrent olfactory stimuli.

## ACKNOWLEDGMENTS

This work was supported by NIH grants NS 23326 (TEF and JCK), NS 20486 (TEF), and NS 16340 (TVG), and by NSF grant BNS-85-07949 (MLG). The authors thank Var St. Jeor and Mary Womble for their excellent histological work. We also want to thank Wayne L. Silver whose enthusiasm and encouragement led to many of the original experiments described in this chapter.

## REFERENCES

Änggård, A., Lundberg, J. M., Hökfelt, T., Nilsson, G., Fahrenkrug, J., and Said, S. J. (1979). Innervation of the cat nasal mucosa with special reference to relations between peptidergic and cholinergic neurons. *Acta Physiol. Scand. Suppl.* 473: 50.

Bouvet, J. F., Delaleu, J. C., and Holley, A. (1987a). Olfactory receptor cell function is affected by trigeminal nerve activity. *Neurosci. Lett.* 77: 181-186.

Bouvet, J. F., Godinot, F., Croze, S., and Delaleu, J. C. (1987b). Trigeminal substance P-like immunoreactive fibers in the frog olfactory mucosa. *Chem. Senses* 12: 499-505.

Bouvet, J. F., Delaleu, J. C., and Holley, A. (1988). The activity of olfactory receptor cells is affected by acetylcholine and substance P. *Neurosci. Res.* 5: 214-223.

DeLong, R. E., and Getchell, T. V. (1987). Nasal respiratory function-vasomotor and secretory regulation. *Chem. Senses* 12: 3-36.

Finger, T. E. (1986). Peptide immunohistochemistry demonstrates multiple classes of perigemmal nerve fibers in the circumvallate papilla of the rat. *Chem. Senses* 11: 135-144.

Franco-Cereda, A., Henke, H., Lundberg, J. M., Petermann, J. B., Hökfelt, T., and Fischer, J. A. (1986). Calcitonin gene-related peptide (CGRP) in capsaicin-sensitive substance P-immunoreactive sensory neurons in animals and man: Distribution and release by capsaicin. *Peptides* 8: 399-341.

Getchell, M. L., Zielinski, B., and Getchell, T. V. (1986). Ontogeny of the secretory elements in the vertebrate olfactory mucosa. In: *Ontogeny of Olfaction*. W. Breipohl (Ed.) Springer-Verlag, Berlin, pp. 71-82.

Getchell, M. L., Finger, T. E., and Getchell, T. V. (1987). Localization of NGF like, VIP-like and CGRP-like immunoreactivity in the olfactory mucosa of the salamander, bullfrog and grass frog. *Chem. Senses* 12: 658.

Getchell, M. L., Zielinski, B., and Getchell, T. V. (1988). Odorant and autonomic regulation of secretion in the olfactory mucosa. In: *Molecular Biology of the Olfactory System.* F. L. Margolis and T. V. Getchell (Eds.). Plenum Press, New York, pp. 71-98.

Getchell, M. L., Bouvet, J.-F., Finger, T. E., Holley, A., and Getchell, T. V. (1989). Peptidergic Regulation of Secretory Activity in the Olfactory mucosa of the salamander: Immunohistochemistry and Pharmacology. *Cell Tiss. Res.* 256: 381-389.

Getchell, T. V., Margolis, F. L., and Getchell, M. L. (1984). Perireceptor and receptor events in vertebrate olfaction. *Prog. Neurobiol.* 23: 317-345.

Gibbins, I. L., Furness, J. B., and Costa, M. (1987). Pathway-specific patterns of the coexistence of substance P, calcitonin gene-related peptide, cholecystokinin and dynorphin in neurons of the dorsal root ganglia of the guinea-pig. *Cell Tissue Res.* 48: 417-437.

Hua, X.-Y., Theodorsson-Norheim, E., Brodin, E., Lundberg, J. M., and Hökfelt, T. (1985). Multiple tachykinins (neurokinin A, neuropeptide K and substance P) in capsaicin-sensitive sensory neurons in the guinea-pig. *Regul. Pept.* 13: 1-19.

Jancso, N., Jancso-Gabor, A., and Szolcsanyi, J. (1967). Direct evidence for neurogenic inflammation and its prevention by denervation and by pretreatment with capsaicin. *Br. J. Pharmacol.* 31: 138-151.

Jonsson, G., and Hallman, H. (1982). Substance P counteracts neurotoxin damage on norepinephrine neurons in rat brain during ontogeny. *Science* 215: 75-77.

Lacroix, J. S., St. Jarne, P., Änggård, A., and Lundberg, J. M. (1988). Sympathetic vascular control of the pig nasal mucosa: (1) increased resistance and capacitance vessel responses upon stimulation with irregular bursts compared to continuous impulses. *Acta Physiol. Scand.* 132: 83-90.

Lee, Y., Takami, K., Kawai, Y., Girgis, S., Hillyard, C. J., MacIntyre, I., Emson, P. C., and Tokyama, M. (1985). Distribution of calcitonin gene-related peptide in the rat peripheral nervous system with reference to its co-existence with substance P. *Neuroscience* 15: 1227-1237.

Lembeck, F., and Gamse, R. (1982). Substance P in peripheral sensory processes. In: *Substance P in the Nervous System.* R. Porter and M. O'Connor (Eds.). Ciba Foundation Symp. 91, Pitman, London, pp. 35-48.

Lindberg, S., Dolata, J., and Mercke, U. (1987). Stimulation of C fibers by ammonia vapor triggers a mucociliary defense reflex. *Am. Rev. Resp. Dis.* 135: 1093-1098.

Lindner, G., and Grosse, G. (1981). [The effect of substance P on the regeneration of nerve fibers *in vitro*.] *Z. Mikrosk. Anat. Forsch* 95: 390-394.

Lundberg, J. M., Lundblad, L., Martling, C.-R., Saria, A., St. Jarne, P., and Änggård, A. (1987). Coexistence of multiple peptides and classic transmitters in airway neurons: Functional and pathophysiologic aspects. *Am. Rev. Resp. Dis.* 136: 516-522.

Lundblad, L., Lundberg, J. M., Brodin, E., and Änggård, A. (1983a). Origin and distribution of capsaicin-sensitive substance P-immunoreactive nerves in the nasal mucosa. *Acta Otolaryngol.* 96: 485-493.

Lundblad, L., Änggård, A., and Lundberg, J. M. (1983b). Effects of antidromic trigeminal nerve stimulation in relation to parasympathetic vasodilation in cat nasal mucosa. *Acta Physiol. Scand.* 119: 7-13.

Lundblad, L., Lundberg, J. M., and Änggård, A. (1984). Local and systemic capsaicin pretreatment inhibits sneezing and the increase in nasal vascular permeability induced by certain chemical irritants. *Arch. Pharmacol.* 326: 254-261.

Lundblad, L., Brodin, E., Lundberg, J. M., and Änggård, A. (1985). Effects of nasal capsaicin pretreatment and cryosurgery on sneezing reflexes, neurogenic plasma extravasation, sensory and sympathetic neurons. *Acta Otolaryngol.* (Stockholm) 100: 117-127.

Meisami, E., Keating, T., Paternostro, M., and Tran, L.-H. (1988). Postnatal development of nasal olfactory structures in the rat and rabbit as seen in reconstructed 3-D models. *Chem. Senses* 13: 716.

Nishimoto, T., Ichikawa, H., Wakisaka, S., Matsuo, S., Yamamoto, K., Nakata, T., and Akai, M. (1985). Immunohistochemical observation on substance P in regenerating taste buds of the rat. *Anat. Rec.* 212: 430-436.

Papka, R. E., and Natulionis, D. H. (1983). Association of substance P-immunoreactive nerves with the murine olfactory mucosa. *Cell Tiss. Res.* 230: 517-525.

Persaud, K. C., DeSimone, J. A., Getchell, M. L., Heck, G. L., and Getchell, T. V. (1987). Ion transport across the frog olfactory mucosa: The basal and odorant-stimulated states. *Biochim. Biophys. Acta* 902: 65-79.

Rangachari, P. K., and McWade, D. (1986). Peptides increase anion conductance of canine trachea: An effect on tight junctions. *Biochim. Biophys. Acta* 863: 305-308.

Rosenfeld, M. G., Mermod, J.-J., Amara, S. G., Swanson, L. W., Sawchenko, T. E., Rivier, J., Vale, W. W., and Evans, R. M. (1983). Production of a novel neuropeptide encoded by the calcitonin gene via tissue-specific RNA processing. *Nature* 304: 129-135.

Silver, W. L. (1987). The Common Chemical Sense. In: *The Neurobiology of Taste and Smell.* T. E. Finger and Silver, W. L. (Eds.). John Wiley and Sons, New York, pp. 65-88.

Skofitsch, G., and Jacobowitz, D. M. (1985a). Calcitonin gene-related peptide coexists with substance P in capsaicin sensitive neurons and sensory ganglia of the rat. *Peptides* 6: 747-754.

Skofitsch, G., and Jacobowitz, D. M. (1985b). Calcitonin gene-related peptide: Detailed immunohistochemical distribution in the central nervous system. *Peptides* 6: 721-745.

South, E. H., and Ritter, R. C. (1986). Substance P-containing trigeminal sensory neurons project to the nucleus of the solitary tract. *Brain Res.* 372: 283-289.

St. Jeor, V. R., Kinnamon, J. C., and Finger, T. E. (1988). Neuropeptide immunoreactivity in nerve fibers of the nose of mice and rats. *Chem. Senses* 13: 731-732.

Steranka, L. R., DeHaas, C. J., Vavrek, R. J., Stewart, J. M., Enna, S. J., and Snyder, S. H. (1987). Antinociceptive effects of bradykinin antagonists. *Eur. J. Pharmacol.* 136: 261-262.

Steranka, L. R., Manning, D. C., DeHaas, C. J., Ferkany, J. W., Borosky, S. A., Connor, J. R., Vavrek, R. J., Stewart, J. M., and Snyder, S. H. (1988). Bradykinin as a pain mediator: Receptors are localized to sensory neurons, and antagonists have analgesic actions. *Proc. Natl. Acad. Sci. USA* 85: 3245-3249.

Uddman, R., Alumets, J., Densert, O., Hakanson, R., and Sundler, F. (1978). Occurrence and distribution of VIP nerves in the nasal mucosa and tracheobronchial wall. *Acta Otolaryngol.* 86: 443-448.

Willey, J. C., Lechner, J. F., and Harris, C. C. (1984). Bombesin and the C-terminal tetradecapeptide of gastrin-releasing peptide are growth factors for normal human bronchial epithelial cells. *Exp. Cell Res.* 153: 245-247.

Wirsig, C. R., and Getchell, T. V. (1986). Amphibian terminal nerve: Distribution revealed by LHRH and AChE markers. *Brain Res.* 385: 10-21.

# Chapter 1 Discussion

**Dr. Cain:** I'm intrigued by the sparsity of the peptidergic fibers in the nasal cavity. Don't you believe that those fibers also carry other information like pain, and that there is probably no spot in the nasal cavity that does not have pain sensitivity? I'd also like to ask two or three questions. First, do you believe that there is a kind of mixture of peptidergic and nonpeptidergic fibers and that the peptidergic ones give signature for chemosensation? Is there any evidence for that? Second, what do you think substance P and CGRP do? Do they play any role in creating afferent information or are they simply involved in things like the spread of information from one terminal to another?

**Dr. Finger:** I'll try to answer those; they are not exactly related. I tried to be very careful not to claim that I was showing the entire trigeminal innervation of the nasal cavity by use of these antibodies. One certainly can't rule out the possibility—in fact, I would put it as a probability—that there are nonpeptidergic trigeminal sensory fibers also innervating the epithelium that we can't detect at all. We're looking, then, at one class of trigeminal sensory fibers within the epithelia.

**Dr. Cain:** But if you were doing heat pain research, I think you would assume that all the fibers that are responding are of that sort. I don't hear those people talk about more than one kind of fiber. When they talk about responding, say, to heat-evoked pain, they talk about substance P-containing fibers. What I'm saying is that the mosaic would look so sparse it would seem like you'd only be sensitive in very localized regions. That seems somewhat unrealistic to me.

**Dr. Finger:** If you look at the cutaneous mosaic there is not a tremendous distribution of substance P fibers in skin either; they're not as extensive as one would imagine. Also, you have to remember what scale of magnification we are looking at here. That is, I'm not sure that if you could do heat stimulation in the nasal cavity that it would not be uniformally sensitive to pain. Of course, in skin there are several different types of pain fibers, only one of which is immunoactive for substance P and CGRP. To get to the second question, which was do these peptides play a role in sensory transmission, they probably do in the same sense that they play a role in transmission of pain information in the spinal cord. They probably coexist with a second neurotransmitter, a more classical neurotransmitter such as an excitatory amino acid, and they have more prolonged effects in terms of sensory transmission than do the short-lived classical neurotransmitters.

**Dr. Szolcsanyi:** It is rather common that all of these kinds of effects of substance P-immunoreactive fibers are related to the axon reflex. I want to stress again what we already published several years ago, that the only evidence that can be used to discriminate between a sensory ending and an effector ending in the axon reflex relies on Bruce's experiments that by means of local anesthesia an effective response could be abolished completely. But this is not true. We repeated the experiment with many different local anesthetics and the response could not be abolished. What does this mean? It means that the same ending could serve both sensory and effector functions. This is why we call these "sensory-effector" or efferent functions of the sensory endings. In this particular system, in the nasal cavity for example, Lundberg and coworkers showed that there is also a mucous ciliary movement induced by antidromic stimulation. It is again a piece of evidence that the peptidergic fibers could serve as effector endings. This is my question, because you did not list it: what is your opinion of the Lundberg experiment?

**Dr. Finger:** I didn't list it by oversight more than anything else. I was already at seven and running out of space. I have thought of at least three more since, and I think that's an entirely reasonable one to be added, along with activation of mast cells. There are mast cells in the epithelial tissue and I didn't discuss those. One thing I would hope to get out of this conference is adding other factors to this list, which certainly can be extended.

# 2
# Physiological Factors in Nasal Trigeminal Chemoreception

**Wayne L. Silver**
Wake Forest University
Winston-Salem, North Carolina

## I. INTRODUCTION

Virtually all animals possess the ability to detect chemical stimuli in their external environment. Animals use their chemical senses to locate and test food, to recognize and attract mates to identify conspecifics and predators, and to maintain a nutritional balance. Another important function of chemoreception in animals is to detect irritating and potentially harmful compounds and to initiate procedures to avoid them. Although the irritating nature of certain chemical stimuli had long been known, Parker (1912) was one of the first to recognize that the sense of irritation produced by chemicals was distinct from olfaction and gustation. He coined the term "common chemical sense" to describe the sense of irritation aroused by the action of noxious chemicals on exposed or semiexposed mucous membranes. Parker had observed that fish turned away from noxious chemicals presented to their flanks even when olfaction and gustation had been eliminated, and concluded that a common chemical sense mediated by free nerve endings in the epithelium and served by spinal nerves was responsible for the avoidance behavior. Parker suggested that the primary function of the common chemical sense was to protect the animal from potentially harmful chemicals.

The entire outer surface of many aquatic animals (including fish and amphibians) can be considered to be an exposed mucous membrane which is susceptible to the action of chemical stimuli. However, in most terrestrial vertebrates the development of a horny epidermis has effectively restricted the common chemical sense to the mucous membranes of the nose, mouth, eye, respiratory tract, and the anal and genital orifices. One of the major components of the common chemical sense in terrestrial vertebrates is the trigeminal (Vth cranial) nerve which innervates the mucosae of the eyes, nose, and mouth. Although trigeminal nerve endings can be stimulated by chemicals, they are usually considered as primary mediators of the sensations of pain, touch, temperature, and proprioception. Whether or not common chemical sensation involves the stimulation of receptors different from pain or temperature receptors has not been shown conclusively.

By far, the majority of the research conducted in the chemical senses has been on the olfactory and gustatory systems. Relatively little attention has been paid to trigeminal chemoreception. This is especially true for physiological studies. The majority of these studies examined the responses of trigeminal fibers in the nasal cavity. The nasal cavity is innervated by two branches of the trigeminal nerve, the ethmoid (ophthalmic division) and the nasopalatine (maxillary division), whose cell bodies, located in the trigeminal (Gasserian) ganglion, send their axons to the anterior dorsal and posterior ventral regions of the nasal cavity, respectively.

One of the first reports of electrophysiological recordings from trigeminal (ethmoid) nerve fibers in the nasal cavity in response to chemical stimuli was by Beidler and Tucker (1956). Subsequently, Tucker (1963, 1971) investigated nasal trigeminal responses to chemical stimuli in the rabbit and tortoise. In the freely respiring anesthetized rabbit, trigeminal amyl acetate thresholds were approximately 2 log units higher than olfactory thresholds ($\approx$ 400 to 4 ppm). The threshold concentration for phenethyl alcohol was usually lower than the olfactory threshold although responses were highly dependent on the depth and frequency of inspiration. For the gopher tortoise, olfactory, amyl acetate thresholds were 4 log units lower than trigeminal thresholds ($\approx$ 400 to 0.4 ppm) while the trigeminal response to benzyl amine and phenethyl alcohol appeared at concentrations below which olfactory responses appeared. Responses to chemical stimuli appeared to be mediated by much smaller, slower conducting fibers ($\approx$ 1 m/sec) than responses to mechanical stimuli.

Other physiological studies of nasal trigeminal chemoreception include those by (a) Cooper (1970) in which single ethmoid nerve fibers in the rat that responded to tactile stimulation did not appear to respond to chemical stimuli and vice versa; (b) Kulle and Cooper (1975) in which thresholds were determined from rat nasopalatine single-fiber responses to formaldehyde (0.25 ppm), ozone (5.0 ppm), and pentanol (300 ppt); (c) Ulrich et al. (1972) in which rat single-fiber responses were obtained in response to a 0.5% solution of chlorobenzylidene malonitrile in polyethylene glycol 200 placed directly in the nares and to cigarette smoke; and (d) Ito (1968) in which strong single-fiber responses from guinea pig trigeminal nerves were elicited by saturated acetic acid, formic acid, cineole, cyclohexanone, ether, peppermint, and camphor; weak responses were elicited by saturated amyl acetate, cyclopentanone, and methyl acetate; and no responses were elicited by saturated limonene, musk, ionine, and diphenyl ether.

My own laboratory has been examining the physiology of trigeminal chemoreceptors in the nasal cavity for the past several years. We have determined the response characteristics of the trigeminal nerve to a number of compounds, tested for the properties of a chemical stimulus that make it an effective trigeminal stimulus, investigated which fibers in the trigeminal nerve may be responsible for the response to chemical stimuli, and examined the nasal epithelium to determine where the nerve endings might be located. The remainder of this chapter will deal primarily with reviewing these findings. For other reviews of the physiological aspects of nasal trigeminal chemoreception, see Tucker (1971), Silver and Maruniak (1981), and Silver (1987).

## II. METHODS

Male Sprague-Dawley rats were used in these studies. The preparation is as follows. The rats are anesthetized with urethane (ethyl carbamate: (1-2.5 g/kg IP) and a tracheotomy is performed. One cannula is placed in the caudal end of the trachea to allow the rat to breathe room air. Another cannula is inserted into the rostral end of the trachea up to the nasopharynx. This cannula is connected through a flowmeter to a vacuum line, allowing for the precise control of air flow through the nasal cavity. The rat is placed in a head holder which allows the head to be rotated and moved up and down. The skin overlying the parasagittal ridge of the frontal bone is removed on one side and the contents of the orbit retracted by means of a hook inserted in the tissue. A cavity is thus formed exposing the ethmoid nerve as it leaves its foramen. The nerve is freed from the surrounding tissue for several millimeters distal to the foramen and gently stripped of its connective sheath.

Electrical activity is recorded by placing the nerve on a pair of platinum-iridium electrodes, with the electrode at the cut end serving as the indifferent lead. The rat is grounded through the head holder. Mineral oil is pipetted into the cavity, covers the nerve, prevents it from drying out, and ensures electrical insulation. The electrodes are connected to the high-impedance probe of an AC amplifier. The amplifier output is monitored with a storage oscilloscope and audio monitor and stored on magnetic tape. Amplified activity from the whole nerve is also passed through an integrator (see e.g., Kiyohara and Tucker, 1978) and displayed on a chart recorder to obtain a population response. Respiration is monitored by placing a thermocouple wire, connected to a DC amplifier, into the cannula allowing the rat to breathe room air. The output of the amplifier is displayed on the chart recorder and stored on magnetic tape.

Stimuli used in these studies are reagent grade and purchased from commercial suppliers. They are presented to the rat via an air dilution olfactometer (see, e.g., Walker et al., 1979). A clean air stream flowing at a known rate through a tube containing the stimulus becomes saturated with that stimulus and is mixed with a dilution stream of filtered air. The stimulus is delivered at a final flow rate of 2 L/min. Dilution is controlled by flowmeters which regulate the ratio of flow between stimulus-saturated and dilution streams. A clean 2 L/min air stream serves as the background and is presented to the rat's external nares throughout the experiment. Solenoid valves are used to switch from the background air stream to the test stimulus. For stimulus presentation, the vacuum line connected to the nasopharyngeal cannula is turned on for 30 sec. The flow rate through the nasal cavity is between 250 ml and 1 L. In the middle of this 30 sec period the test stimulus is delivered for 10 sec.

The multiunit neural responses obtained from the nerve and displayed on the chart recorder are a representation of the moving average of nerve impulse traffic. Response magnitude is measured as the height from baseline (i.e., the background neural activity which often increases after the vacuum

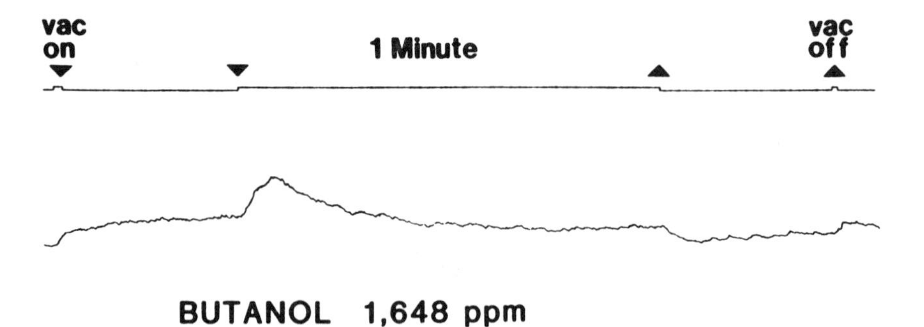


**FIGURE 1** Summated, multiunit response to a prolonged (1 min) presentation of butanol at vapor saturation. Vac on and vac off on the top trace correspond with the onset and offset of the vacuum (when air is drawn through the nose). ▼ corresponds with the onset and ▲ offset of the stimulus.

is turned on, due for the most part to mechanoreceptor stimulation) to the peak of the positive displacement. Because the magnitude of the integrated response is determined by several factors, most importantly the position of the nerve on the electrodes (e.g., if those fibers which are most responsive to the stimulus are farthest away from the electrode, the response will be small), the response is not calibrated absolutely but is instead measured in arbitrary

**TABLE 1** Concentration-Response Data for 29 Compounds

| Compound | Perceived intensity[a] | Maximum response[b] | Slope[c] | Threshold[d] |
|---|---|---|---|---|
| Phenethyl alcohol | 0.13 | 28 | 0.81 | 24 |
| α-Terpineol | 0.53 | 28 | 0.57 | 19 |
| Heptanoic acid | 0.87 | 5 | — | 11 |
| Hexanoic acid | 0.93 | 35 | 1.06 | 16 |
| Limonene | 0.93 | 15 | 2.38 | 497 |
| Benzyl acetate | 1.40 | 37 | 1.76 | 132 |
| Heptanol | 2.80 | 28 | 0.60 | 7 |
| Linalool | 4.00 | 48 | 1.07 | 45 |
| Valeric acid | 5.00 | 55 | 1.46 | 139 |
| Menthol | 6.14 | 27 | 1.07 | 27 |
| Amyl acetate | 6.67 | 77 | 1.53 | 252 |
| Butanol | 6.67 | 64 | 0.57 | 162 |
| Butyl acetate | 7.33 | 87 | 1.08 | 945 |
| Methanol | 7.67 | 121 | 0.66 | 3020 |
| Cyclohexanone | 7.80 | 205 | 0.73 | 112 |
| Butyric acid | 7.87 | 160 | 0.62 | 50 |
| Toluene | 7.87 | 146 | 1.16 | 2404 |
| Propionic acid | 8.73 | 148 | 0.57 | 27 |
| Acetic acid | — | 145 | 1.01 | 82 |
| *d*-Carvone | — | 64 | 1.15 | 45 |
| Ethanol | — | 108 | 0.72 | 1380 |
| Formic acid | — | 129 | 1.51 | 145 |
| Hexanol | — | 30 | 0.50 | 8 |
| l-Carvone | — | 88 | 1.56 | 23 |
| Nicotine | — | 109 | 0.73 | 5 |
| Octanoic acid | — | 18 | — | 9 |
| Octanol | — | 23 | 0.67 | 3 |
| Pentanol | — | 48 | 0.49 | 59 |
| Propanol | — | 82 | 0.65 | 457 |

[a]Human data from Doty et al., 1978.
[b]Rat data. Relative to the response to the standard ≈ 550 ppm cyclohexanone.
[c]Rat data. Determined from log-log concentration-response plots.
[d]Rat data. Concentration in ppm first eliciting a response larger than baseline.

units. Therefore, in order to compare results across animals, response magnitudes are reported as a percentage of the response to a standard cyclohexanone stimulus. The percentage of the standard response is then averaged for all rats at each concentration and mean concentration-response curves are obtained.

Electrophysiological threshold is defined as the concentration which first elicits a response distinguishable from baseline activity. This measurement of threshold is somewhat conservative since increased activity must at times be detected out of a noisy baseline. In actuality, the threshold would be somewhere between the concentration which first elicits a response distinguishable from baseline activity and the preceding concentration.

## III. RESULTS AND DISCUSSION

### A. Response Characteristics

Mechanical stimulation of the naris with a probe elicits vigorous responses from the ethmoid branch of the trigeminal nerve. Responses are also often seen when air is pulled through the nose in the absence of a chemical stimulus. This probably represents mechanical, thermal, or chemical stimulation of the nasal cavity, although the response most resembles that to mechanical stimuli. Two main types of responses to chemical stimuli are seen. High concentrations of stimulatory compounds elicit responses with an initial phasic component followed by a decline to a steady-state tonic level usually back down to baseline. This is seen in Fig. 1. Low concentrations of some stimulatory compounds produce a gradual increase in activity that lasts for the duration of the stimulus. Responses at all concentrations return rapidly to baseline levels after removal of the stimulus.

Concentration-response curves and thresholds have been obtained for 29 compounds, as listed in Table 1. All compounds were tested on at least four animals. Figures 2 and 3 show a typical concentration series and averaged

---

**FIGURE 2** Response to increasing concentrations of toluene. Numbers under the records denote the concentration (in ppm) presented to the rat. The top trace in each record is a representation of respiration, obtained from a thermocouple wire placed in the rat's trachea. The middle trace is the integrated, multiunit response recorded from the ethmoid branch of the trigeminal nerve. The two outer single marks on the lower trace corresponds to the onset and offset of the vacuum (when air is drawn through the nose). The two double marks correspond with the onset and offset of the stimulus. The time between the two double marks is 10 sec. Test stimulus responses are reported as a percentage of the response to the standard stimulus, $\approx$ 550 ppm cyclohexanone.

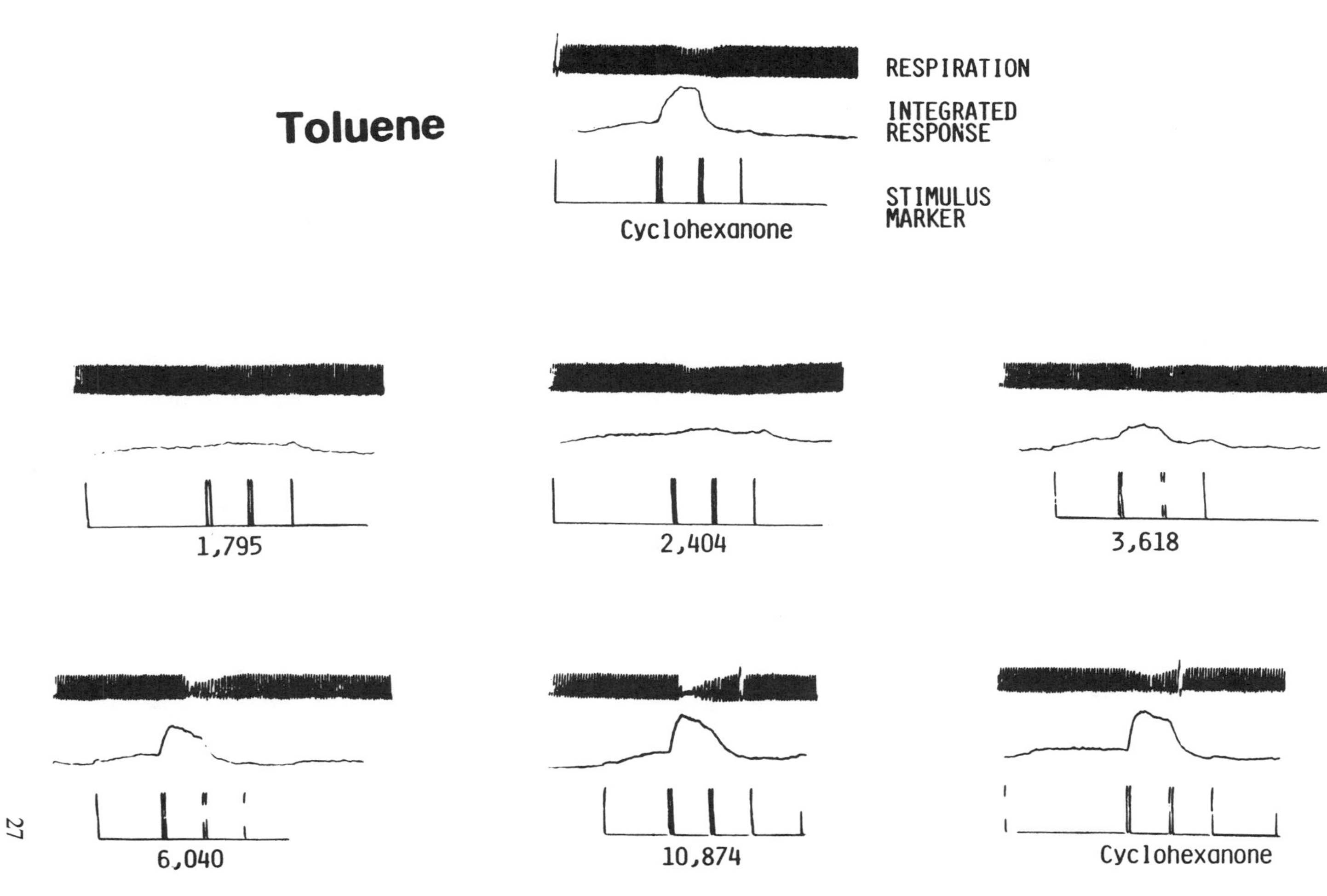
Toluene
RESPIRATION
INTEGRATED RESPONSE
STIMULUS MARKER
Cyclohexanone
1,795
2,404
3,618
6,040
10,874
Cyclohexanone

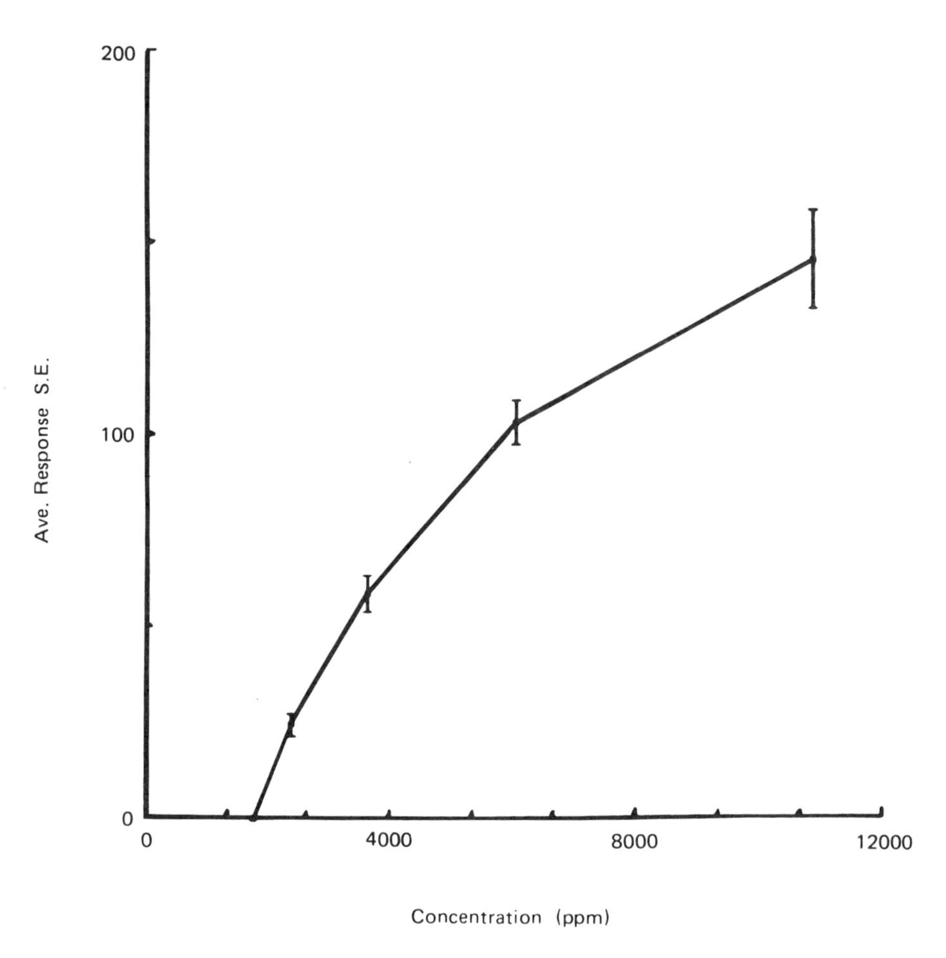


**FIGURE 3** Concentration-response curve for toluene. Data are averaged from seven rats (10 concentration series). Responses are reported as a percentage of the response to the standard stimulus, ≈ 550 ppm cyclohexanone. Capped vertical bars = the standard error of the mean.

concentration-response curve. For most compounds concentration-response curves continued to rise even at the highest concentrations tested (vapor saturation). However, for some compounds, especially those which elicited the greatest maximum responses (see Table 1), curves plateaued before vapor saturation was reached.

For human psychophysical studies, concentration-response curves are often described as power functions with the exponent of the function characterizing the growth of the response magnitude (perceived intensity) with increasing concentrations (Murphy, 1987). These functions are straight lines in log-log plots with the slope of the line equal to the exponent of the power function. For olfaction, response magnitude grows slowly as a function of concentration with exponents in the range of 0.2-0.7 (Murphy, 1987). Psychophysical functions for irritation (trigeminal stimulation) grow at a higher rate than that for odor (olfaction) (Cain, 1976). For example, an exponent of 1.2 was obtained for the irritant $CO_2$ (Cain and Murphy, 1980). The steeper the slope, the smaller the increase in concentration necessary to produce large increases in response magnitude. This is consistent with the trigeminal's possible role as a protection system. The concentration-response curves obtained from the rat trigeminal nerve also approximate straight lines when plotted on log-log coordinates. The slopes (exponents) for the compounds tested ranged from 0.49 to 2.38 (0.99 $\pm$ 0.09; x $\pm$ SE) (Table 1). Although a similar analysis has not been done for rat olfactory electrophysiological responses, the trigeminal results are consistent with the human psychophysical data.

The rat electrophysiological results also are consistent with another aspect of human trigeminal psychophysics. Doty et al. (1978) obtained perceived intensity ratings from human anosmics (i.e., lacking olfaction) for a number of compounds at vapor saturation. Eighteen of those compounds have now been used to stimulate the rat trigeminal nerve (Table 1). There is a significant correlation between the human perceived intensity ratings and the maximum rat summated response magnitude (Fig. 4). The maximum response (Table 1) was the response to a saturated stimulus or the response at which the concentration-response curve began to plateau. The human intensity ratings were not correlated with any other rat electrophysiological parameter including slopes or thresholds. Since anosmics judged intense stimuli as more unpleasant, the correlation between the rat and human data suggests that compounds which are unpleasant or irritating for humans may also be unpleasant or irritating for rats. Thus, the rat appears to be a good model for assessing the stimulatory effectiveness of chemicals on human nasal trigeminal receptors.

Trigeminal electrophysiological thresholds for 29 compounds are shown in Table 1. Thresholds ranged from 3 ppm for octanol to 3020 ppm for methanol. For the compounds tested, it appears that the greater the vapor pressure (at 20 °C), the lower the threshold. This is especially true for the aliphatic alcohols and acids (see B below). Some compounds which were rated as particularly irritating (e.g., methanol, toluene, and butyl acetate; Doty et al., 1978) exhibited high thresholds, while the thresholds for others (e.g., propionic acid menthol) were relatively low.

### B. Stimulus and Receptor Characteristics

A wide variety of chemicals can stimulate nasal trigeminal receptors. At least some human anosmics, for example, detected 45 of 47 compounds (Doty et

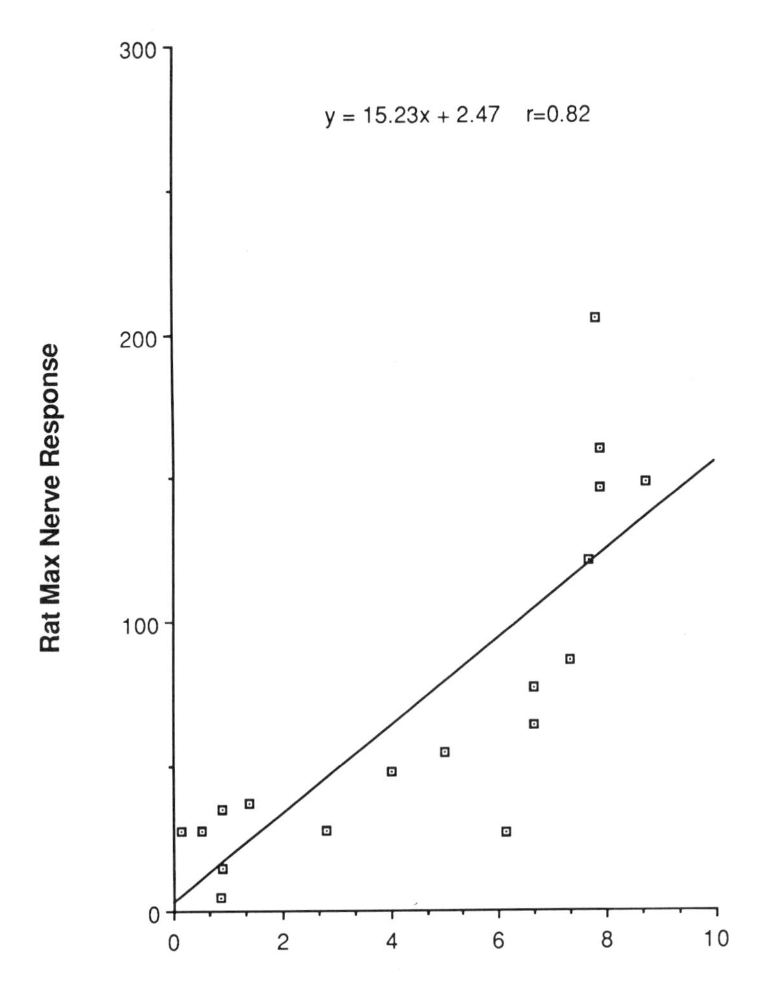


**FIGURE 4** Correlation between the maximum multiunit electrophysiological response magnitude from the rat ethmoid nerve and perceived intensity rating by human anosmics ($p < 0.001$). The human data are taken from Doty et al. (1978). The rat responses were averaged from at least four animals and standardized to the response to the standard, ≈ 550 ppm cyclohexanone. The compounds used for this analysis are shown in Table 1.

al., 1978). No compound systematically tested in our laboratory has failed to stimulate the trigeminal nerve in the nasal cavity if presented at high enough concentrations. One attempt (Moncreiff, 1951) to characterize chemical trigeminal stimuli was based on the sensations they aroused. Stimuli were classified as pungent spices, lacramatories, sternutatories, suffocants, or skin irritants.

Three classes of sensory irritants have been suggested by Alarie (1973). One group includes chemicals which react with nucleophilic groups (HO, $H_2N$, or HS groups) (Dixon and Needham, 1946; Moncreiff, 1951; and see Alarie, 1973). Compounds which react with SH groups fall into three categories (Alarie, 1973). These are thiol alkylating agents (e.g., chloro-acetophenone), dienophiles (e.g., acrolein), and mercaptide formations with trivalent arsenicals (e.g., diphenylcyanoarsine). A second class of irritants has been proposed which break S-S linkages (e.g., ammonia) (Alarie, 1973; Nielson and Babko, 1985). A third class of sensory irritants contain those which do not fit into the first two classes (Alarie, 1973). This class includes most of the chemicals listed in Table 1, such as ethanol. These compounds were suggested to function as protein precipitants, lipid solvents, or to possess some other action.

Based on these stimulus characteristics, a hypothetical receptor protein has been suggested to reside in the lipid bilayer of the trigeminal nerve ending and be similar to cholinergic receptors (Nielson and Alarie, 1982; Nielson and Babko, 1985). The proposed receptor protein would contain a site consisting of a nucleophilic group, a second site possessing a disulfide bond, and a third site with hydrogen donor ability, functioning as a physical adsorption site (Nielson et al., 1984; Nielson and Babko, 1985).

The ability of a molecule to activate the receptor by physical adsorption, i.e. by dissolving into the membrane (Nielson and Alarie, 1982) should be related to its oil-water partition coefficient. More effective stimuli should dissolve into the membrane easier and therefore have lower partition coefficients. $RD_{50}$ values (the concentration depressing the respiratory rate in mice to 50%) are one measure of effectiveness (Alarie, 1973). There is a strong correlation between log $RD_{50}$ for alkylbenzenes and alcohols and the log of their partition coefficients (Nielson and Babko, 1985) (Fig. 5). Similarly, there is a strong correlation between electrophysiological trigeminal thresholds for alcohols and the log of their partition coefficients (Silver et al., 1986) (Fig. 5). These results suggest that lipophilicity is an important determinant for sensory irritants.

The possibility that compounds such as the aliphatic alcohols may stimulate trigeminal nerve endings by a physical mechanism is also supported by the observation that a variety of systems respond similarly to these compounds. These systems include melanophores (Lerner et al., 1988), neuroblastoma cells (Kashiwayanagi and Kurihara, 1984), taste receptors (Kashiwagura et

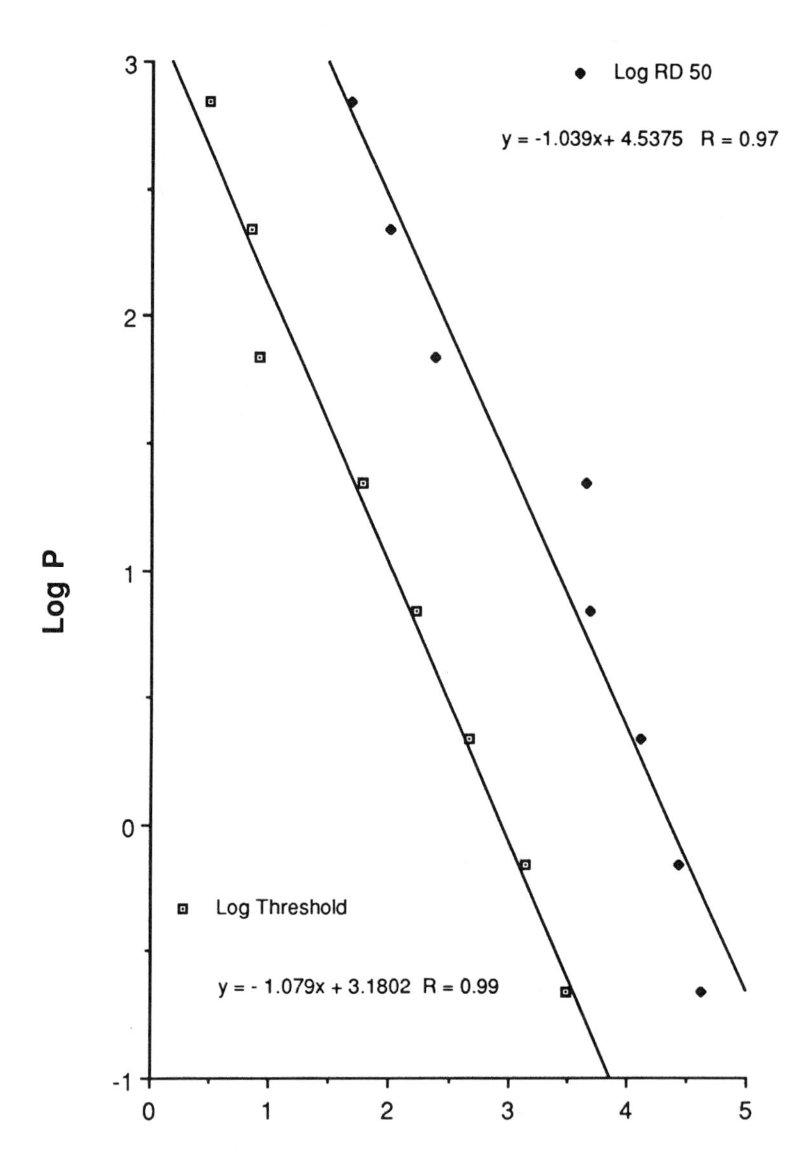


**FIGURE 5** Graph comparing the relationship between the log of the oil-water partition coefficient (from Lindenberg, 1951) and the log of rat trigeminal electrophysiological thresholds and the log of the mouse $RD_{50}$ values (the concentrations depressing the respiratory rate to 50%) (from Nielson and Babko, 1985) for *n*-aliphatic alcohols, $C_1$-$C_8$.

al., 1977), slime molds (Ueda and Kobatake, 1977), and olfactory receptors (Ottoson, 1958). As was suggested previously (Kurihara et al., 1986; Dionne, 1988), it is unlikely that specific receptors for these compounds would be found in such a diversity of systems. Indeed, these investigators as well as Price (1984) questioned whether specific protein receptors play a role in the detection of chemical stimuli in general. Price (1984) suggested several mechanisms based on general irritability rather than specific protein receptors which could account for stimulus-induced changes in the resting potentials of a chemoreceptive cell. These included alterations in the mucus composition, extracellular fluid composition, metabolic state, or membrane properties. One possibility is that stimulus molecules excite the receptor cell by changing the surface potential (phase boundary potential) at the external surface of the membrane (Kashiwayanagi and Kurihara, 1984; Price, 1984). Surface potentials are associated with the net fixed charge on all biological membranes. The stimulus would be adsorbed on the hydrophobic portion of the membrane causing a conformational change. This change, in turn, will alter the density of fixed charges on the membrane and thus the surface potential. Changes in surface potential would not significantly alter the membrane potential directly, but would alter the potential gradient within the membrane and could thus affect voltage-sensitive gating molecules in the membrane resulting in a depolarization of the membrane. These suggested mechanisms obviously do not preclude the possibility that chemical stimuli may interact with specific protein receptors in or on the cell membrane, but it would be interesting to test whether the sensory irritants described as reacting with nucleophilic groups or S-S linkages could stimulate a variety of systems including liposomes containing no proteins.

### C. Nerve Fiber Characteristics

The trigeminal (Vth cranial) provides sensory innervation to the epithelia of the head, including intranasal, some intraoral, and corneal surfaces. The sensations usually attributed to trigeminal nerve stimulation include pain, touch, temperature, and proprioception. Chemical stimuli also elicit responses from the trigeminal nerve but it appears that they may stimulate pain or temperature receptors rather than specific chemoreceptors.

The trigeminal nerve contains a wide variety of fiber sizes. For example, the infraorbital nerve, a major branch of the maxillary division, consists of both myelinated and unmyelinated fibers (Jacquin et al., 1984). The fiber distributions are unimodal with myelinated axons ranging from 0.8 to 14.9 $\mu$m and unmyelinated fibers from 0.3 to 1.5 $\mu$m. An analysis of the fibers in the ethmoid nerve also demonstrated a unimodal distribution of fiber sizes (Biedenbach et al., 1975). The ethmoid appears to contain more unmyelinated than myelinated axons although only a spectrum of myelinated fibers

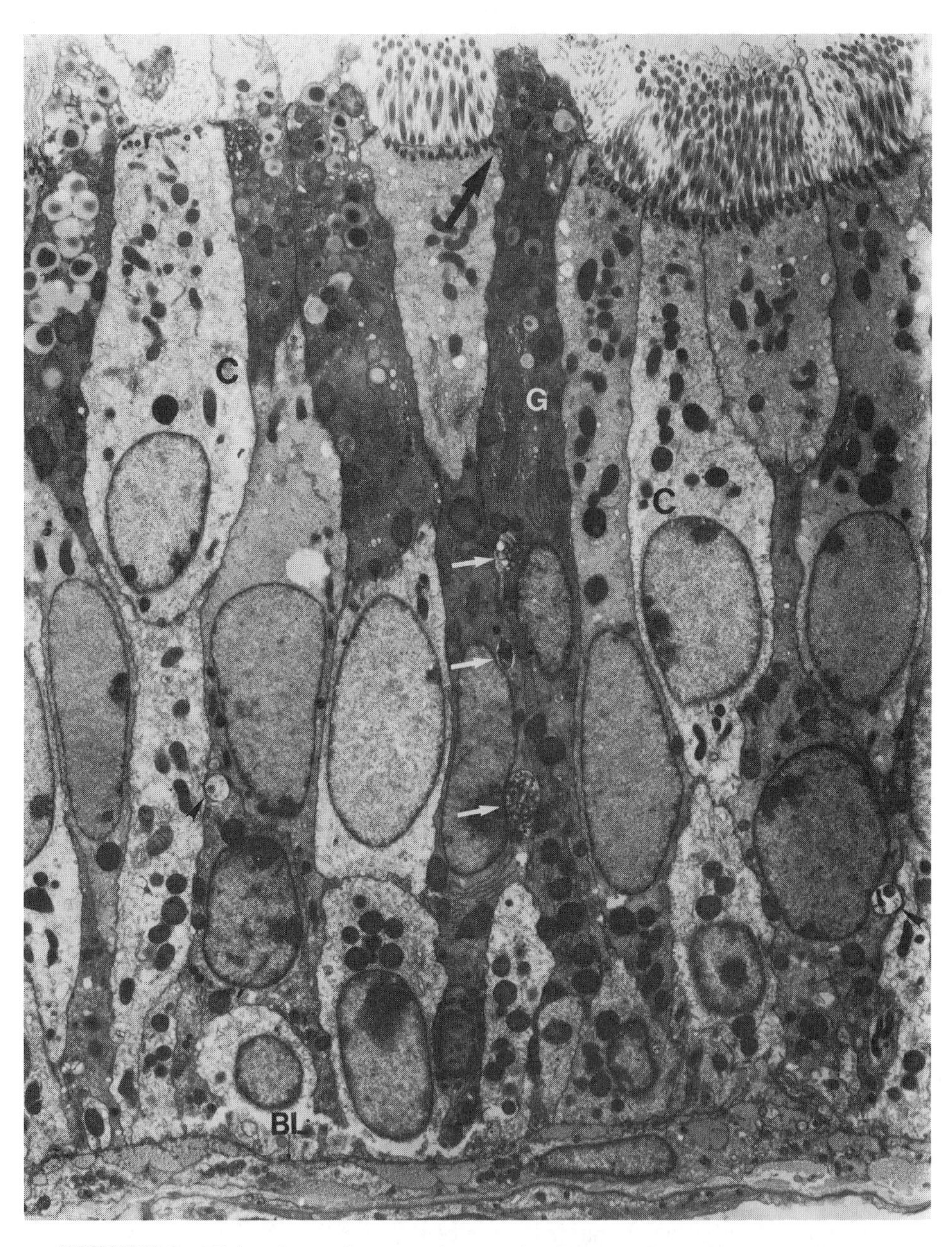


**FIGURE 6** High-voltage electron micrograph (HVEM) of a region of respiratory epithelium from a rat nose showing goblet cells (G) and ciliated cells (C) lying adjacent to the basal lamina (BL). Profiles of nerve fibers (arrow heads) are visible; one nerve fiber (white arrows) traverses nearly the entire depth of the epithelium, terminating near a tight junction (large arrow). The various segments of this nerve have been traced and connected using serial section analysis. ×4200. (Photo is courtesy of Var St. Jeor and John C. Kinnamon.)

was obtained in this light microscopic study. Fibers with diameters between 2 and 6 μm predominated in the ethmoid nerve. Which of the fiber types in the ethmoid nerve is responsible for the response to chemical stimuli has not been firmly established.

One attempt to address this question involved the use of capsaicin, the active ingredient in capsicum peppers (Silver et al., 1985). Chronic capsaicin administration desensitizes adult animals to chemically induced pain and may also raise thresholds for noxious mechanical and thermal stimulation (Nagy, 1982). The presumed mechanism for this desensitization is the depletion of substance P (or other neuropeptides) from unmyelinated primary afferent neurons including polymodal nociceptors. Trigeminal nerve responses to chemical stimuli in rats which have been injected with capsaicin are either eliminated or severely reduced (Silver et al., 1985). This suggests that the trigeminal nerve fibers responsible for the response to chemical stimuli may be small, unmyelinated fibers which subserve pain and that they may not represent a special class of trigeminal receptors.

An important question for the chemical stimulation of the trigeminal nerve is where the nerve endings are located in the epithelium. This will obviously have a bearing on how the stimulus reaches the receptor. Latency measurements for aliphatic alcohols were as short as 0.1 sec (Silver et al., 1986). Since diffusion of stimulus molecules through both the nonuniform mucous layer ranging up to 20 μm in depth and the respiratory epithelium (35-70 μm) would take considerably longer than 0.1 sec, trigeminal endings must lie fairly close to the surface of the epithelium if they are to be stimulated directly. Figure 6 is a high-voltage electron micrograph of a region of the respiratory epithelium of a rat nose. A nerve fiber can be seen to cross almost the entire epithelium, terminating near a tight junction just a few micrometers from the surface. A similar positioning of substance P-containing fibers also can be seen in the nasal respiratory epithelium (see Chapter 1 in this volume for figures and details).

## ACKNOWLEDGMENTS

I would like to thank T. Finger, R. Mason, C. Smeraski, A. Arzt, and D. Walker for their contributions to the data presented in this chapter. The work reported in this chapter was supported NIH grant NS-23326 and a grant from the R. J. Reynolds Tobacco Company.

## REFERENCES

Alarie, Y. (1973). Sensory irritation by airborne chemicals. *CRC Crit. Rev. Toxicol.* 2: 299-363.

Beidler, L. M., and Tucker, D. (1956). Olfactory and trigeminal responses to odors. *Fed. Proc.* 15: 14.

Biedenbach, M. A., Beuerman, R. W., and Brown, A. C. (1975). Graphic-digitizer analysis of axon spectra in ethmoidal and lingual branches of the trigeminal nerve. *Cell. Tiss. Res.* 157: 341-352.

Cain, W. S. (1976). Olfaction and the common chemical sense: Some psychophysical contrasts. *Sensory Proc.* 1: 57-67.

Cain, W. S., and Murphy, C. (1980). Interaction between chemoreceptive modalities of odour and irritation. *Nature* 284: 255-257.

Cooper, G. P. (1984). Responses of rat ethmoid nerve units to cutaneous and olfactory stimuli. *The Physiologist* 131: 171.

Dionne, V. E. (1988). How do you smell? Principle in question. *TINS* 11: 188-189.

Dixon, M., and Needham, D. M. (1946). Biochemical research on chemical warfare agents. *Nature* 158: 432-434.

Doty, R. L., Brugger, W. E., Jurs, P. C., Orndorff, M. A., Snyder, P. F., and Lowry, L. D. (1978). Intranasal trigeminal stimulation from odorous volatiles: Psychometric responses from anosmic and normal humans. *Physiol. Behav.* 20: 175-187.

Ito, K. (1968). Electrophysiological studies on the peripheral olfactory nervous system in mammalia. *Kitahanto J. Med. Sci.* 18: 405-417.

Jacquin, M. F., Hess, A., Yang, G., Adamo, P., Math, M. F., Brown, A., and Rhoades, R. W. (1984). Organization of the infraorbital nerve in the rat: A quantitative electron microscopic study. *Brain Res.* 290: 131-135.

Kiyohara, S., and Tucker, D. (1978). Activity of new receptors after transection of the primary olfactory nerve in pigeons. *Physiol. Behav.* 21: 987-994.

Kashiwagura, T., Kamo, N., Kurihara, K., and Kobatake, Y. Responses to frog gustatory receptors to various odorants. *Comp. Biochem. Physiol.* 56C: 105-108.

Kashiwayanagi, M., and Kurihara, K. (1984). Neuroblastoma cell as a model for olfactory cell: Mechanism of depolarization in response to various odorants. *Brain Res.* 293: 251-258.

Kulle, T. J. and Cooper, G. P. (1975). Effects of formaldehyde and ozone on the trigeminal nasal sensory system. *Arch. Environ. Health* 30: 237-243.

Kurihara, K., Yoshii, K., and Kashiwayanagi, M. (1986). Transduction mechanisms in chemoreception. *Comp. Biochem. Physiol.* 85A: 1-22.

Lerner, M. R., Reagan, J., Gyorgyi, T., and Roby, A. (1988). Olfaction by melanophores: What does it mean? *Proc. Natl. Acad. Sci. USA* 85: 261-264.

Lindenberg, B. A. (1951). Sur la sloubilite des substances organiques amphipatiques dans les glycerides neutres et hydroxyles. *J. Chim. Phys.* 48: 350-355.

Moncrieff, R. W. (1951). *The Chemical Senses.* Leonard Hill, London, pp. 172-193.

Murphy, C. (1987). Olfactory psychophysics. In: *Neurobiology of Taste and Smell.* T. E. Finger and W. L. Silver (Eds.). John Wiley and Sons, New York, pp. 251-273.

Nagy, J. I. (1982). Capsaicin: A chemical probe for sensory neuron mechanisms. In: *Handbook of Psychopharmacology*, vol. 15. S. D. Iverson, L. L. Iverson, and S. S. Snyder (Eds.). Plenum Press, New York, pp. 185-235.

Nielson, G. D., and Alarie, Y. (1982). Sensory irritation, pulmonary irritation, and respiratory stimulation by airborne benzene and alkylbenzenes: Prediction of safe industrial exposure levels and correlation with their thermodynamic properties. *Toxicol. Appl. Pharmacol.* 65: 459-477.

Nielson, G. D., and Babko, J. C. (1985). Exposure limits for irritants. *Ann. Am. Conf. Ind. Hyg.* 12: 119-133.

Nielson, G. D., Babko, J. C., and Holst, E. (1984). Sensory irritation and pulmonary irritation by airborne allyl acetate, allyl alcohol, and allyl ether compared to acrolein. *Acta Pharmacol. Toxicol.* 54: 292-298.

Ottoson, D. (1958). Studies on the relationship between olfactory stimulating effectiveness and physico-chemical properties of odorous compounds. *Acta Physiol. Scand.* 43: 167-181.

Parker, G. H. (1912). The reactions of smell, taste, and the common chemical sense in vertebrates. *J. Acad. Nat. Sci. Phila.* 15: 221-234.

Price, S. (1984). Mechanisms of stimulation of olfactory neurons: An essay. *Chem. Senses* 8: 341-354.

Silver, W. L. (1987). The common chemical sense. In: *Neurobiology of Taste and Smell.* T. E. Finger and W. L. Silver (Eds.). John Wiley and Sons, New York, pp. 65-87.

Silver, W. L., Mason, J. R., Marshall, D. A., and Maruniak, J. A. (1985). Rat trigeminal, olfactory, and taste responses after capsaicin desensitization. *Brain Res.* 333: 45-54.

Silver, W. L., Mason, J. R., Adams, M. A., and Smeraski, C. (1986). Trigeminal chemoreception in the nasal cavity: Responses to aliphatic alcohols. *Brain Res.* 376: 221-229.

Silver, W. L., and Maruniak, J. A. (1981). Trigeminal chemoreception in the nasal and oral cavities. *Chem. Senses* 6: 295-305.

Tucker, D. (1963). Olfactory, vomeronasal and trigeminal receptor responses to odorants. In: *Olfaction and Taste I.* Y. Zotterman (Ed.). Pergamon Press, New York, pp. 45-69.

Tucker, D. (1971). Nonolfactory responses from the nasal cavity: Jacobson's organ and the trigeminal system. In: *Handbook of Sensory Physiology, Vol IV, Chemical Senses, Part 1, Olfaction.* L. M. Beidler (Ed.). Springer-Verlag, New York, pp. 151-181.

Ueda, T., and Kobatake, Y. (1977). Changes in membrane potential, zeta potential and chemotaxis of Physdarum polycephalum in response to n-alcohols, n-aldehydes and n-fatty acids. *Cytobiology* 16: 16-26.

Ulrich, C. E., Haddock, M. P., and Alarie, Y. (1972). Airborne chemical irritants. Role of the trigeminal nerve. *Arch. Environ. Health* 24: 37-42.

Walker, J. C., Tucker, D., and Smith, J. C. (1979). Odor sensitivity mediated by trigeminal nerve in pigeon. *Chem. Senses Flav.* 4: 107-116.

# Chapter 2 Discussion

**Dr. Eccles:** I found your presentation very interesting. However, there are two points which I think make it difficult to interpret some of the results. The first point is when considering the trigeminal nerve, its distribution in the anterior part of the nose is going to include the nasal vestibule, where you have a squamous epithelia or skin-like epithelia, and also toward some part of the respiratory mucosa, where you have mucosal epithelium. Now the substances you are applying are going to get into these epithelia at different rates because one of them is covered by a mucous layer and one by a skin layer. In between these you may have a transition region which is perhaps a mixture of the two. So the odorants you are applying make the responses complicated as far as availability to the differing types of epithelia. The other point is when we consider that you are recording from a whole-nerve bundle, you are going to have mechanoreceptors, thermal receptors, nociceptors and, presumably, receptors responsive to chemical stimulation. The question I would pose is: When you start your air flow and draw the air through the nose, you get a response. Now is that related to some stimulation of, say, a temperature receptor or a flow receptor? And can you get a dose-response relationship to air flow if you are flowing air through at different rates?

**Dr. Silver:** Tucker did that experiment back in the 1960s and showed that there is a dose-response to flow rate. You can increase the flow rate and get a larger response when turning the vacuum on. Tucker demonstrated that for some compounds the flow rate was more important than others for trigeminal stimulation. In the tests that I have done, what we have decided to use as a flow rate now is pretty much the rate that appears to be normally drawn to the rat's nose from the work of Youngentaub. I agree and I think it is important that different stimuli will be drawn out and be taken up at different parts of the nasal cavity, and Dr. Mozell can probably talk about that.

**Dr. Mozell:** If the lipid solubility is increasing, the water solubility is decreasing. How does it get through the mucus?

**Dr. Eccles:** I would say it never goes through the aqueous phase and it could well be penetrating a thin, a very thin cornified layer in the nasal vestibule, therefore not actually penetrating a water or aqueous phase on a mucosal surface.

**Dr. Kobal:** I would like to make a comment on the contribution of the cornified skin to your responses. I don't think the stimuli excite the receptors in the skin because in human studies we see if we blow $CO_2$ or the other odorants onto the skin of the face you don't have any sensation at all. A question I would like to ask is, do we have any idea as to what extent C fibers and A delta fibers are involved in these responses?

**Dr. Silver:** The only evidence I have one way or another is, as Tom Finger suggested, we've done experiments where we have injected capsaicin and desensitized the animals and basically wiped out the response to stimuli. That suggests that peptidergic C fibers are involved in the response.

**Dr. Alarie:** A couple comments and a question. In your rat preparation, you have the eye right there. Have you ever just gently blown those irritants onto the cornea and recorded from your nerve?

**Dr. Silver:** No, I haven't.

**Dr. Alarie:** Well, we have done that. With the cornea you have a much simpler system; you have no vessels there, no mucous glands, no ciliary . . . you have nothing there; you have just a few epithelial cells and then the nerve endings. And what you see of course is the same nerve activity in the ethmoid nerve if you stimulate the cornea as if you stimulate the nasal mucosa. In humans, every single compound we have tested in man that is a nasal trigeminal stimulant is also a corneal stimulant. So these compounds will go through in the gas phase; they don't go through in the condensed phase. They will reach those nerve endings very easily. So I think that with your preparation you could look at the threshold to irritate the cornea at the same time you look at the threshold to irritate the nasal mucosa. You have a beautiful preparation here.

**Dr. Silver:** We actually cut the nerve that goes to the cornea to get to the other one. But Dawson showed many years ago, recording from the trigeminal nerve innervating the cornea of the frog, that the responses looked very similar to Tucker's responses in the turtle; so you are right.

**Dr. Kobal:** You should try $CO_2$ on the cornea; you will be surprised—there is almost no response in humans.

# 3
# Perceptual Characteristics of Nasal Irritation

**William S. Cain**
John B. Pierce Foundation Laboratory and Yale University
New Haven, Connecticut

## I. TWO PERSPECTIVES ON IRRITATION

Interest in the topic of sensory irritation from airborne chemicals arises from two sources. The first is toxicological, whereby interest derives from the tendency for certain chemicals to injure epithelial tissue during brief exposures. A chemical that can injure the epithelium can also evoke noxious sensations. Whether or not some damage must precede a sensation remains unclear. It may prove true for irritants of moderate or high potency, whereby the mechanism of action may entail making or breaking covalent bonds (Alarie, 1973). Conceivably, however, such chemical interactions could take place between irritants and neural receptors, putatively protein molecules, without epithelial damage (Neilsen and Alarie, 1982). The need for damage to precede the sensation seems less likely for irritants of low potency, which probably stimulate via mere physical rather than chemical interactions (see Ferguson, 1939).

The second source of interest in sensory irritation is more strictly sensory per se. Many odoriferous materials evoke pungency at concentrations well above their olfactory thresholds (Tucker, 1971). The materials involved would usually fall into the class of irritants of low potency that stimulate via physical interactions. Their ability to irritate might even have passed unnoticed except for occasional reports of pungency in experiments on the perception of their odors. Such materials include linalool (a major component of lavender), isoamyl acetate (banana smell), benzaldehyde (maraschino cherry smell), pyridine (gassy smell), and many others (Doty et al., 1978).

## II. PERCEIVED "ODOR" INTENSITY

From the standpoint of research on odor perception, irritant sensations pose both a liability and an opportunity. The liability stems from the possibility that various functional relations derived for olfaction contain significant perceptual "contamination." The psychophysical function for odor intensity provides an example. It is generally quite flat but varies from substance to substance, as does the odor threshold. (Indeed, these two parameters show a reasonable correlation greater than 0.5, which may have ultimate importance, though not to our example.) The psychophysical function for irritation or pungency, on the other hand, is relatively steep and also varies somewhat from substance to substance. When a substance possesses the ability to stimulate both olfaction and irritation, its psychophysical function may pos-

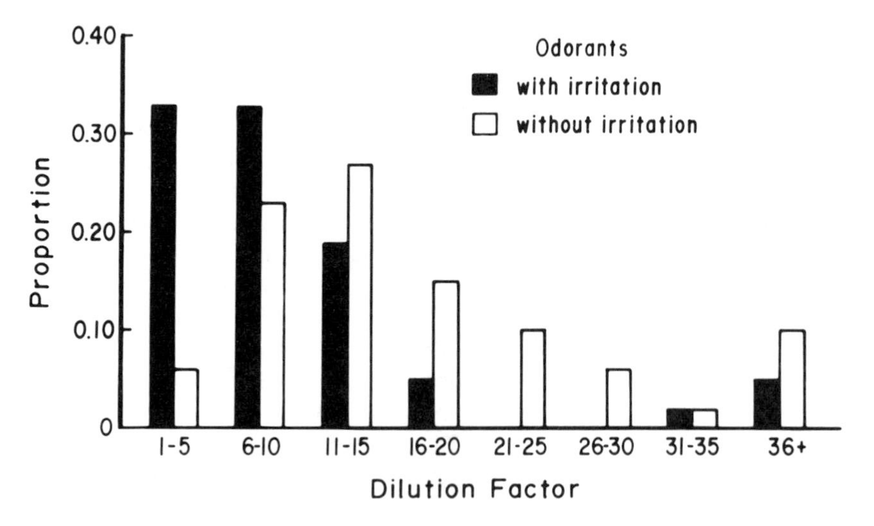


**FIGURE 1** In a study of how odor intensity varied with concentration, Katz and Talbert (1930) summarized the growth of 69 psychophysical functions in terms of dilution factors, the dilution necessary to reduce odor intensity by one point on a five-point category scale. A high dilution factor described a flat function and a low dilution factor a steep function. The range among the 69 functions exceeded 10:1. Odorants shown to have the ability to evoke irritation yielded a disproportionate share of low factors, i.e., steep functions, even when any irritating sensation was presumably consciously ignored. (From Cain, 1988.)

sibly reflect stimulation of both modalities, even when an investigator requests judgments of odor alone (Cain, 1974a).

Figure 1 shows that substances with the ability to arouse irritation typically also produced steep psychophysical functions for odor (Katz and Talbert, 1930; Cain, 1988). Whether determinants of the growth of the psychophysical function arise strictly from olfaction, or in part from olfaction and in part from irritation, has relevance to the possible mechanism for the variation in growth rate.

The admixture of olfactory with irritant sensations might also play a role in measurements of the time course of adaptation, where higher concentrations may exhibit more resistance to fading (Fig. 2) (Cain, 1974b). Whatever disadvantage this might engender for the study of olfaction, however, it may serve adaptively for the organism. It would seem adaptive to remain aware of high concentrations of airborne materials in the environment. Indeed,

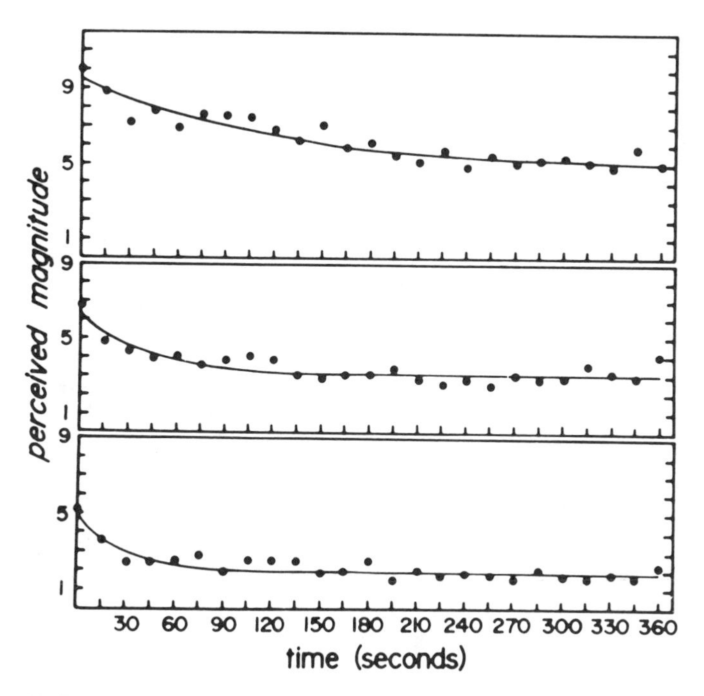


**FIGURE 2** Time course of adaptation to three levels of *n*-butyl acetate, the highest of which (top panel) evoked noticeable pungency and exhibited slower fading. (From Cain, 1974b.)

one might argue similarly regarding steepness of the psychophysical function; if irritation amplifies sensations at higher concentrations, it at least alerts regarding the presence of high levels of vapors.

## III. IRRITANT SENSE VS. COMMON CHEMICAL SENSE

The ability of a substance to evoke odor and irritant sensations simultaneously presents both practical and theoretical opportunities. On the practical side lies the ability of low-level irritation to signal certain desirable properties in foods and other commercial products. In order to discuss this benefit, we must diversity our vocabulary. Whereas someone who approaches the study of irritation from toxicology may feel comfortable with the term *irritant sensations* to describe all perceptual experiences within this realm, the person who approaches the topic from smell or taste will wish to expand the vocabulary to include such terms as pungency, piquancy, crispness, bite, tang, burn, fizz, and feel. Such diversity might seem merely euphemistic, but in fact it encourages viewing functioning over diverse experiences where irritation hardly seems to signal danger. Mexican chili burns. Hot and sour soup has piquancy, both in flavor and aroma. Lemonade has tang. Mainstream tobacco smoke produces a feel in the mouth. Air streaming down from Canada seems crisp and clean; so do appropriately formulated tile cleaners. Sushi with horseradish is pungent. Such terms can claim little scientific precision, but they imply benefits hardly implicit in the term irritant sensations.

The toxicologist and the chemical senses researcher may even differ in what they call the modality through which irritation arises. Whereas the former will feel comfortable with the term irritant sense, the latter will choose common chemical sense. The term *common* could just as well be replaced by *general.* The modality responds to almost any kind of chemical stimulus, volatile or involatile (e.g., in the mouth or nose, salt solutions of sufficient concentration will stimulate the modality), but, despite the perceptual nuances described above, lacks the qualitative range experienced from the stimulation of taste or smell. Qualitative differences in common chemical functioning seem much like those experienced via touch. Depending on intensity and the spatiotemporal distribution of stimulation, we may perceive a touch as simple pressure, as a punch, a tickle, rough, slippery, and so on. One might actually see the common chemical sense as part of somesthesis, where a chemical stimulus replaces mechanical or thermal stimuli.

## IV. CHEMICAL SOMESTHESIS

Like other somesthetic modalities, the common chemical sense exhibits spatial summation (Garcia Medina and Cain, 1982). For instance, inhaling the

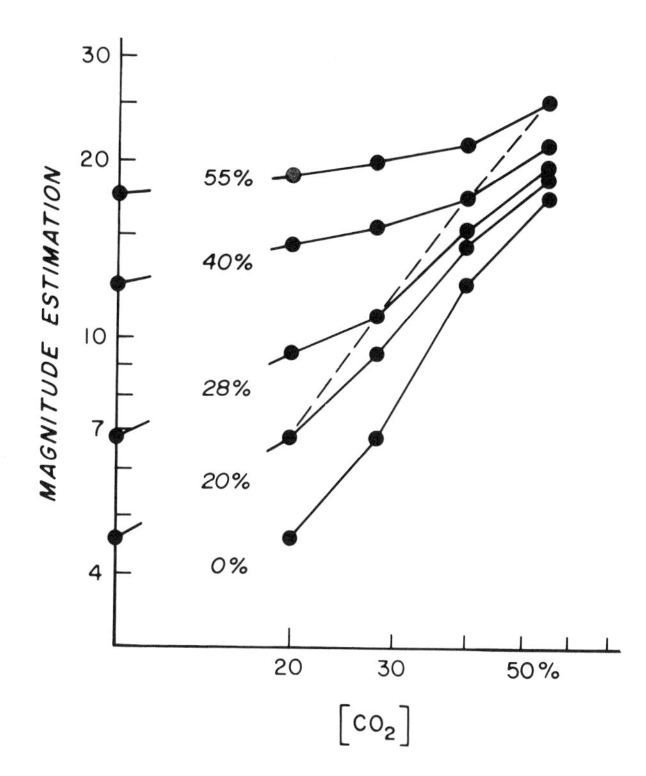


**FIGURE 3** Family of functions that show the pungency of various combinations of concentrations of carbon dioxide presented to the two nostrils. The abscissa shows concentration in one nostril and the parameter alongside the functions shows concentration in the other nostril. The dashed line connects points where both nostrils had the same concentration. (From Garcia Medina and Cain, 1982.)

pungent stimulus carbon dioxide via both nostrils produced a stronger sensation than inhaling it via only one (Fig. 3). The summation fell short of perfect with respect to stimulus mass per se, i.e., stimulation of two nostrils did not feel like stimulation of one at twice the concentration. Instead, summation followed a sum-of-squares rule between nostrils. Accordingly, the trading relation that described the distribution between nostrils for constant sensation followed the Pythagorean theorem (Fig. 4).

A trading relation can also describe the distribution of stimulus over *time* for constant sensation (Cometto-Muñiz and Cain, 1984). Figure 5 illustrates trading relations that imply almost perfect temporal summation of sensation for ammonia over single breaths lasting up to about 4 sec, i.e., time and

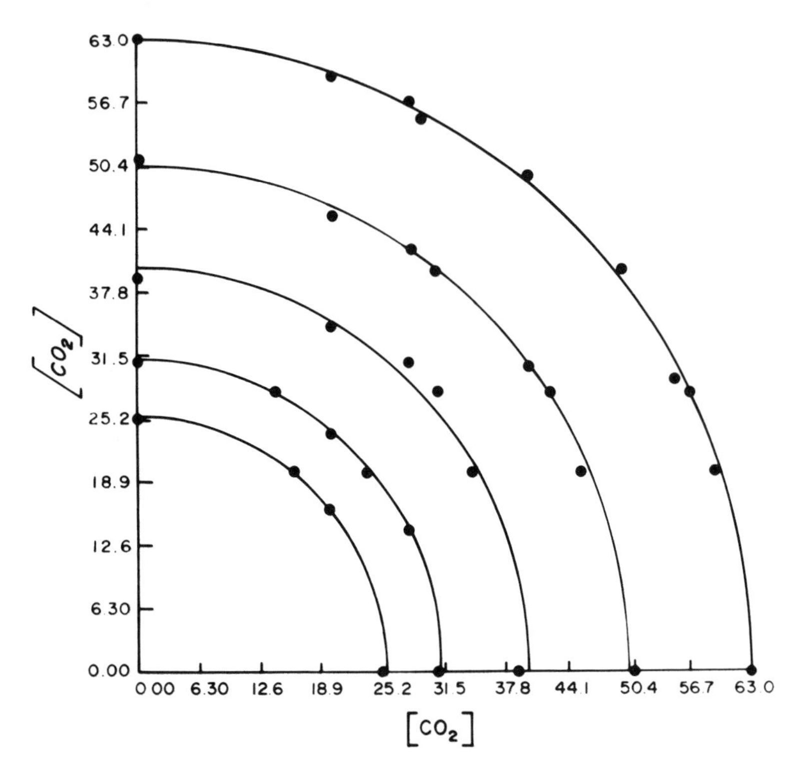


**FIGURE 4** Trading relations that depict bilateral (right nostril, left nostril) combinations of concentrations that led to equal perceived magnitude at various levels. (From Garcia Medina and Cain, 1982.)

concentration could be traced almost proportionally for constant sensation. Such temporal summation will occur over much shorter durations for olfaction. Hence, the summation shown in Fig. 5 comes from the pungency of the ammonia rather than from its odor (see Fig. 6). As one takes a long inhalation of ammonia, therefore, its odor will quickly (almost imperceptibly) reach a maximum but its pungency will grow. As the pungency grows, it may even suppress the odor (see Fig. 9). The difference in the time course between olfactory and the common chemical sensations suggest different modes of stimulation for the two modalities, even when stimulated with the same substance.

Sensations of the common chemical sense will grow not only during an inhalation, but also between inhalations and over relatively long durations. The time course of this behavior may vary with the particular mechanism of

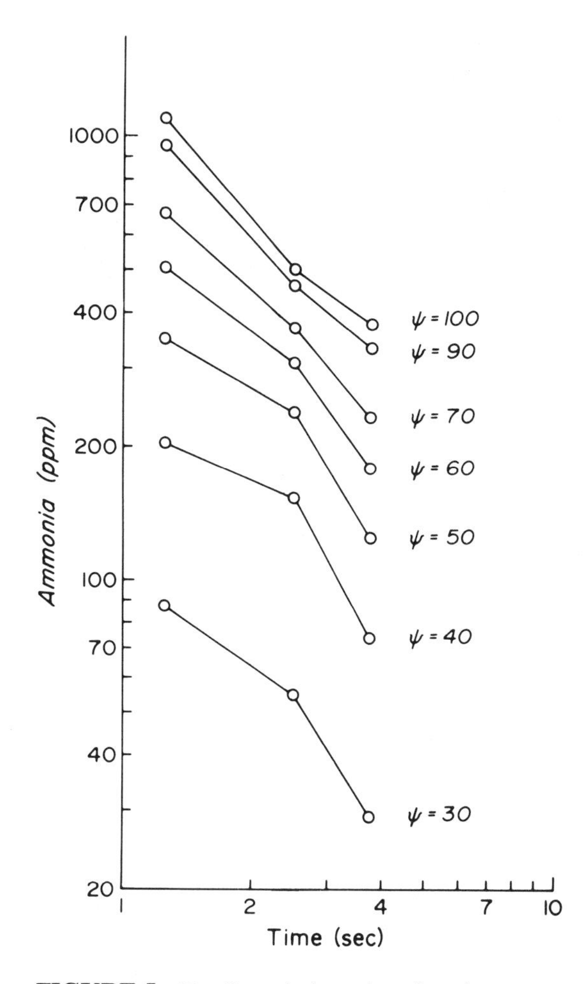


**FIGURE 5** Trading relations that show how concentration of ammonia could be traded with time in order to maintain perceived magnitude constant at various levels. (From Cometto-Muñiz and Cain, 1984.)

stimulation. In the case of formaldehyde, we found growth of perceived intensity over many minutes (Fig. 7) (Cain et al., 1986). Whereas growth alone characterized low levels of irritation, growth followed by some indication of adaptation characterized higher exposures. An exposure to 1 ppm for 90 min showed a clear picture of waxing followed by waning with the time constant of the former about 10-fold above that of the latter. The mechanism of adaptation in the common chemical sense may or may not relate to the mechanism for the buildup of sensation.

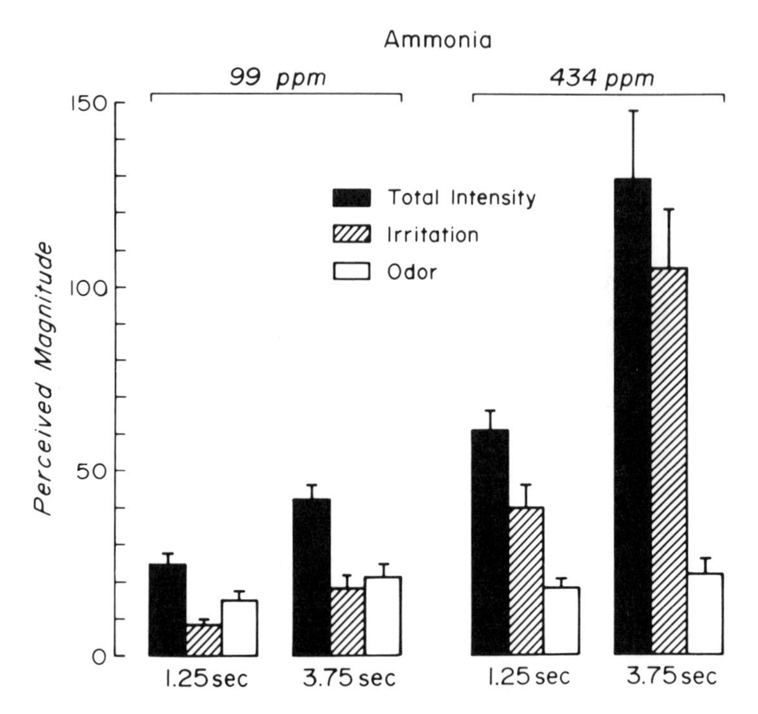


**FIGURE 6** Perceived odor, pungency, and total intensity for two concentrations of ammonia inhaled for two durations. Note how pungency increases much more dramatically than odor. (From Cometto-Muñiz and Cain, 1984.)

Does the initial buildup of sensation represent (a) a benign temporal integration of sensation of purely physiological, i.e., neural, origin, (b) accumulation of incident stimulus at the neural receptor, (c) repetitive damage to some structure, such as an epithelial cell, which might repetitively secrete endogenous chemicals that serve as the actual stimulus for the common chemical sense? The first possibility seems remote, though we can note that positive and negative afterimages in vision can alter visual perception long after cessation of the original stimulus (see also Kobal and Hummel, 1988).

Possibility b seems somewhat more reasonable since the mucosa does accumulate molecules with repeated inhalations (Hornung and Mozell, 1977). On the other hand, perhaps we should feel less surprise that common chemical sensations do build over time than that olfactory sensations do not. Even Aristotle noted that when a sniff stops, so does the odor sensation (Cain, 1978). When an irritant stops, however, the sensation may persist. Figure 7 gave evidence of such persistence in the interval between the half-hour expo-

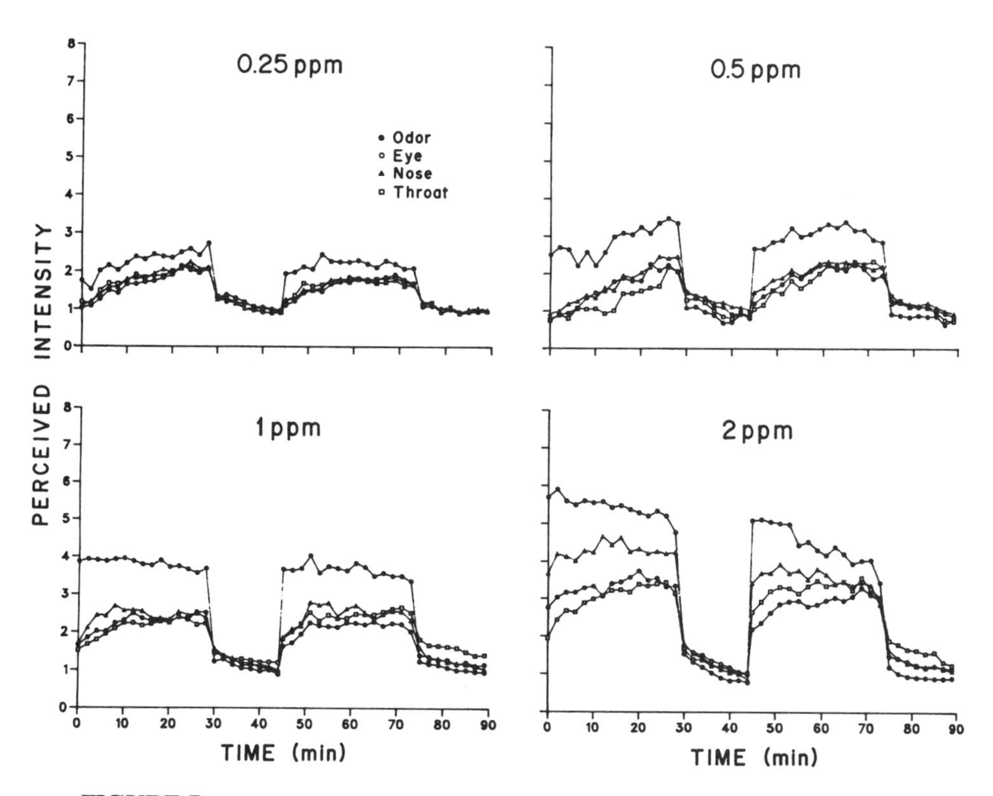


**FIGURE 7** Perceived odor intensity, and eye, nose, and throat irritaiton for various concentrations of formaldehyde. Participants received two 30 min exposures with a 15 min break in-between. (From Cain et al., 1986.)

sures to formaldehyde. Even though subjects left the chamber for a waiting room in that interval, they still reported relatively slowly declining irritation. This difference between modalities could in principle occur if surrounding tissue insulated common chemical receptors from the atmosphere and allowed accumulated stimulus to continue its action.

A thin layer of mucus forms the only barrier between the atmosphere and olfactory receptors. Hence, the concentration of stimulus at the active sites on the cilia of the receptors may follow that in the vapor phase relatively faithfully. Although some of the free nerve endings of the trigeminal nerve may lie near the surface of the epithelium, the majority probably lie somewhat below it. Hence, the epithelium may impede the progress of the molecules to the receptors in the manner of a leaky capacitor. Incident molecules

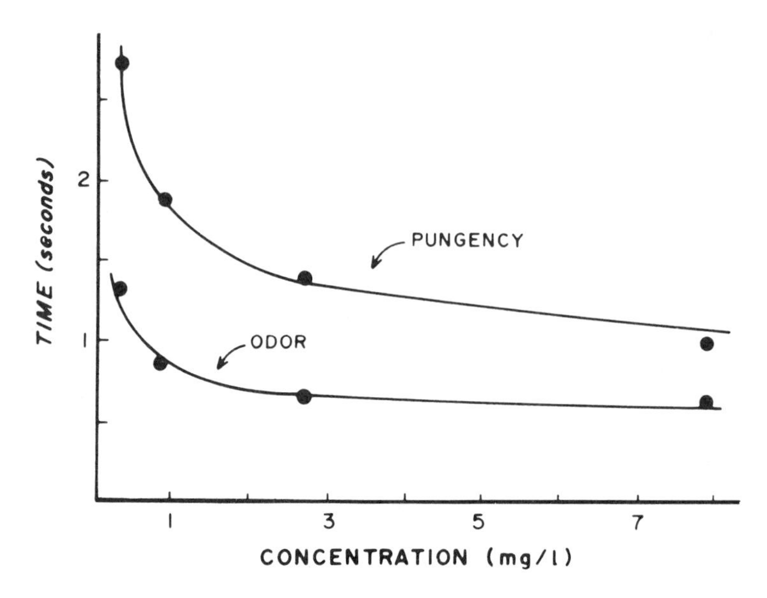


**FIGURE 8** Latency of the odor and pungency of *n*-butyl alcohol. The curves were fitted by a model that derived latency from diffusion time to receptors plus irreducible reaction time (Getchell et al., 1980). The latencies here were consistent with the conclusion that the free nerve endings of the trigeminal nerve by which pungency arose lay about 40 $\mu$m deeper than the cilia of olfactory receptors—an anatomically realistic difference. (From Cain, 1981.)

may "charge" the mucus and epithelium and diffusion may leak the stimulus to the subepithelial receptors. The stimulus may reach the receptors at a slowly increasing concentration before steady state occurs. If this were true, then the latency of pungency would lag well behind that of olfaction and indeed that does occur (Fig. 8).

Possibility c, though quite different from b in some ways, would prove difficult to distinguish from it with the data in hand. This possibility suggests that the receptors for common chemical sensations might respond to endogenous chemicals, such as histamine, released when cells suffer some temporary or permanent damage. If the amount released increased with cumulative damage, then sensation might increase over relatively long periods of time and might continue after cessation of the stimulus because of low-level inflammation. Furthermore, the latency of the sensations would probably prove longer because of the intermediate step.

The three possibilities to explain buildup could have analogs in the explanation of adaptation. That is, adaptation may represent a neural refractoriness per se (possibility a), or it may represent a buffered change in the rate at which molecules reach and more or less temporarily occupy the receptors (possibility b), or it may represent a change in the amount of endogenous chemical released from affected tissue and taken up by receptors (possibility c). The latter two possibilities again overlap. Let us assume that a person is exposed to an irritant that, like formaldehyde, largely stimulates by means of a chemical reaction rather than by some physical process, such as allosteric inhibition. Loss of sensitivity to the chemical might then last a relatively long time after exposure, either because its chemical action actually disables any affected neural receptors via covalent bonding or because affected epithelial tissue has already undergone its reaction and new epithelium must generate to replace the old. Possibilities b and c would suggest that loss of sensitivity to irritants could last a very long time after exposure and indeed this seems to happen (see Amoore, 1986). Workers exposed to formaldehyde, for example, claim to become inured to the irritating effects of relatively high airborne levels. Smokers also become less sensitive to airborne irritants. The change seems to have an acute as well as a chronic component in the sense that abstinence may bring back some, but not all, sensitivity almost immediately (Cometto-Muñiz and Cain, 1982).

Interest in the time course of common chemical sensations should not neglect what we might call temporal "microstructure." Variation in the temporal profile of a given sensation may alter its quality and acceptability. A sensation with a sharp onset may deserve the descriptor "irritating," whereas a sensation of similar overall magnitude but more muted onset may deserve the descriptor "mellow." Without the development of psychophysical techniques to explore differences on a time scale of tenths or hundredths of seconds, the subtleties of temporal variation will remain poorly explored. They nevertheless receive empirical attention in product formulation, where it might become known, for example, that the addition of an aroma chemical to a product may soften its sharpness or edge. Such an effect could occur for physical or physiological reasons.

## V. INTERACTION BETWEEN OLFACTION AND THE COMMON CHEMICAL SENSE

Although the addition of an aroma chemical to an irritating stimulus may soften irritation via subtle alteration in the time course of the sensation, it may also reduce its overall perceived intensity. The converse—a reduction of odor by irritation—also occurs and indeed is more pronounced. Figure 9 depicts how the addition of the pungent stimulus carbon dioxide to various concentrations of amyl butyrate depressed the odor of the latter (Cain and Murphy, 1980). The finding that the mutually suppressive interaction can

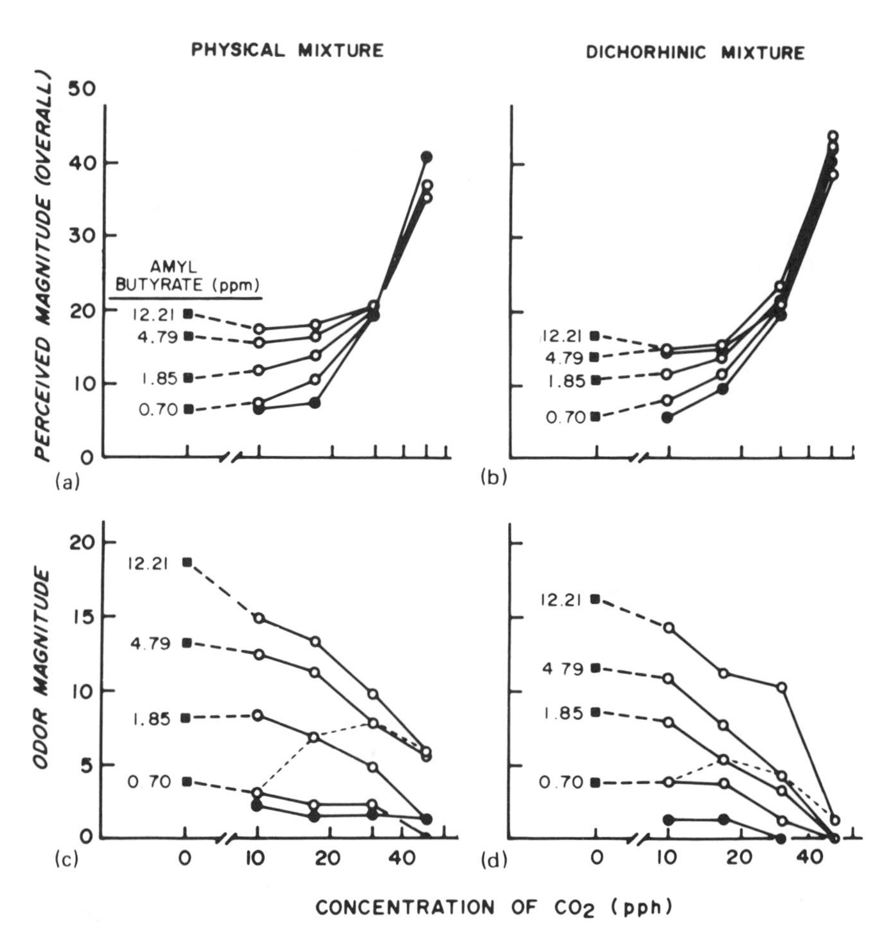


**FIGURE 9** (a) Perceived magnitude of mixtures of the pungent stimulus carbon dioxide and the benign, fruity smelling odorant amyl butyrate plotted as a function of the concentration of carbon dioxide. The numbers alongside the functions show the concentration of amyl butyrate in ppm. The filled circles show the perceived magnitude of carbon dioxide alone. The convergence of the family of functions implies that the components of the mixture do not show simple perceptual additivity. (b) Same as part a, except that the pungent and odoriferous stimuli were presented to separate nostrils, i.e., dichorhinically. Regarding perceptual additivity, the outcome was essentially the same as with the physical mixtures. (c) Perceived magnitude of the odor component of the mixtures. As pungency due to carbon dioxide increased, odor magnitude decreased. (d) Same as part c, except dichorhinic "mixtures." (From Cain and Murphy, 1980.)

occur in the dichorhinic case, where the odorant enters via one nostril and the irritant via the other, suggests a locus in the central nervous system. Behavioral and neurophysiological data in organisms ranging from fish to rabbits support an interaction between olfaction and the common chemical sense, with the preponderance of data in favor of a central neural locus (Stone, 1969; Belousova et al., 1983). Nevertheless, Bouvet et al. (1987) also demonstrated a possible peripheral mechanism.

An interaction between olfaction and the common chemical sense may have relevance for product formulation, as in the above example regarding the reduction of sharpness. It may also have relevance in the domain of indoor air quality. Complaints of low-level irritation from office and home furnishings have increased markedly over the last few years. Often the irritation seems to come on gradually over hours or even days. It seems quite possible for a product, such as a new carpet with its characteristic smell, that the smell may inhibit the perception of irritation at first, only to have it unmasked as olfactory sensitivity wanes. Just as anosmic persons served well to assess the immediate trigeminal effect of various odorants (Doty et al., 1978), such persons could also screen furnishings for duration-dependent irritation potential.

## VI. NASAL REFLEX AS A PSYCHOPHYSICAL ADJUNCT

Eccles (1982, p. 189) noted that "chemical irritation of the nasal mucosa [via branches of the trigeminal nerve] by substances such as tobacco smoke, chloroform, ammonia, sulphur dioxide, etc., may initiate potent respiratory and cardiovascular reflexes resulting in expiration with apnea, closure of the larynx, slowing of the heart and variable changes in blood pressure." Various investigations in this lab have addressed whether the irritation necessary to trigger a particular apneic reflex occurs at a criterion level of perceived intensity. Upon inhalation of an intense enough stimulus, the inhalation will be aborted by closing of the pharynx or larynx. A thermocouple placed at the nostril can readily record the interruption of breathing (Fig. 10).

We explored irritation-induced momentary apnea in males vs. females, in two nostrils vs. one, in smokers vs. nonsmokers, in elderly vs. young subjects, and at longer vs. shorter durations of stimulation (Dunn et al., 1982; Garcia Medina and Cain, 1982; Cometto-Muñiz and Cain, 1982, 1984; Stevens and Cain, 1986). In all cases, changes in the threshold for the reflex agreed quantitatively with companion psychophysical data. The threshold for the reflex showed the same degree of spatial summation across the nostrils as obtained from ratings of intensity; the latency for the reflex exhibited almost perfect trading between time and intensity; and the threshold for smokers

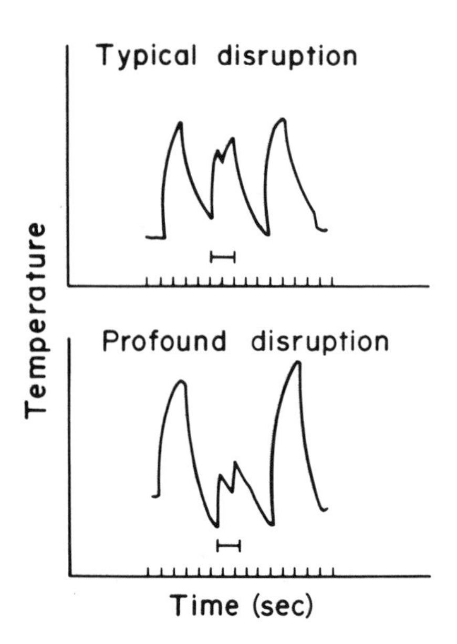


**FIGURE 10** A thermocouple placed at the rim of the nostril recorded, via temperature changes, the breathing pattern of a subject before, during, and after exposure to an irritant. At the threshold for reflex closing of the pharynx or larynx, the pattern of breathing showed a notch (top tracing). Well above the threshold, the pattern showed almost complete suppression of the inhalation. (From Cometto-Muñiz and Cain, 1982.)

fell above that of nonsmokers in a degree commensurate with their respective ratings of irritation, as did the thresholds for females vs. males and young vs. old. The data therefore endorse the conclusion that the reflex occurs at a criterion level of perceived intensity, which in turn suggests that it could serve as an objective measure of common chemical sense functioning.

## VII. CONCLUDING THOUGHT

The toxicologist and the chemical senses researcher bring somewhat different perspectives to the study of the irritant or common chemical sense. The toxicologist has shown the greater interest in mechanism, particularly as potentially explicable in structure-activity relations. The toxicologist has relatively little interest in such features as temporal and spatial summation, except as they might modify the risk of harm. The chemical senses researcher has focused more on characterization of the modality as a modality and has made little overt attempt to explicate mechanism. Insofar as the task of both researchers is to "define the stimulus" in as complete a fashion as possible, then the two approaches should begin to converge and even collaborate. I suspect that such collaboration could bring about a complete understanding of the modality, whichever name we choose for it, in our lifetime.

## ACKNOWLEDGEMENT

Preparation of this chapter was supported by NIH grant NS21644.

## REFERENCES

Alarie, Y. (1973). Sensory irritation by airborne chemicals. *CRC Crit. Rev. Toxicol.* 2: 299-366.

Alarie, Y., and Luo, J. E. (1986). Sensory irritation by airborne chemicals: A basis to establish acceptable levels of exposure. In: *Toxicology of the Nasal Passages.* C. S. Barrow (Ed.). Hemisphere, Washington, D.C., pp. 91-100.

Amoore, J. E. (1986). Effects of chemical exposure on olfaction in humans. In: *Toxicology of the Nasal Passages.* C. S. Barrow (Ed.). Hemisphere, Washington, D.C., pp. 155-190.

Belousova, T. A., Devitsina, G. V., and Malyukina, G. A. (1983). Functional peculiarities of fish trigeminal system. *Chem. Senses* 8: 121-130.

Bouvet, J. F., Delaleu, J. C., and Holley, A. (1987). Olfactory receptor cell function is affected by trigeminal nerve activity. *Neurosci. Lett.* 77: 181-186.

Cain, W. S. (1974a). Contribution of the trigeminal nerve to perceived odor magnitude. *Ann. NY Acad. Sci.* 237: 28-34.

Cain, W. S. (1974b). Perception of odor intensity and the time-course of olfactory adaptation. *ASHRAE Trans.* 80: 53-75.

Cain, W. S. (1976). Olfaction and the common chemical sense: Some psychophysical contrasts. *Sensory Processes* 1: 57-67.

Cain, W. S. (1978). History of research on smell. In: *Handbook of Perception, Vol. 6A, Tasting and Smelling.* E. C. Carterette and M. P. Friedman (Eds.). Academic Press, New York, pp. 197-229.

Cain, W. S. (1981). Olfaction and the common chemical sense: Similarities, differences, and interactions. In: *Odor Quality and Chemical Structure* (ACS Symposium Series No. 148). H. R. Moskowitz and C. B. Warren (Eds.). American Chemical Society, Washington, D.C., pp. 109-121.

Cain, W. S. (1988). Olfaction. In: *Stevens' Handbook of Experimental Psychology, Vol. 1: Perception and Motivation*, revised edition. R. C. Atkinson, R. J. Herrnstein, G. Lindzey, and R. D. Luce (Eds.). John Wiley and Sons, New York, pp. 409-459.

Cain, W. S., and Murphy, C. L. (1980). Interaction between chemoreceptive modalities of odour and irritation. *Nature* 284: 254-257.

Cain, W. S., See, L. C., and Tosun, T. (1986). Irritation and odor from formaldehyde: Chamber studies. In: *IAQ '86: Managing the Indoor Air for Health and Energy Conservation.* The American Society of Heating Refrigerating and Air-Conditioning Engineers, Atlanta, pp. 126-137.

Cometto-Muñiz, J. E., and Cain, W. S. (1982). Perception of nasal pungency in smokers and nonsmokers. *Physiol. Behav.* 29: 727-731.

Cometto-Muñiz, J. E., and Cain, W. S. (1984). Temporal integration of pungency. *Chem. Senses* 8: 315-327.

Dunn, J. D., Cometto-Muñiz, J. E., and Cain, W. S. (1982). Nasal reflexes: Reduced sensitivity to $CO_2$ irritation in cigarette smokers. *J. Appl. Toxicol.* 2: 176-187.

Doty, R. L., Brugger, W. E., Jurs, P. C., Orndorff, M. A., Snyder, P. J., and Lowry, L. D. (1978). Intranasal trigeminal stimulation from odorous volatiles: Psychometric responses from anosmic and normal humans. *Physiol. Behav.* 20: 175-185.

Eccles, R. (1982). Neurological and pharmacological considerations. In: *The Nose: Upper Airway Physiology and the Atmospheric Environment.* D. F. Proctor and I. Andersen (Eds.). Elsevier, Amsterdam, pp. 191-214.

Ferguson, J. (1939). The use of chemical potentials as indices of toxicity. *Proc. Roy. Soc. Lond., B: Biol. Sci.* 127: 387-404.

Garcia Medina, M. R., and Cain, W. S. (1982). Bilateral integration in the common chemical sense. *Physiol. Behav.* 29: 349-353.

Getchell, T. V., Heck, G. L., DeSimone, J. A., and Price, S. (1980). The location of olfactory receptor sites: Inferences from latency measurements. *Biophys. J.* 29: 397-412.

Hornung, D. E., and Mozell, M. M. (1977). Factors influencing the differential sorption of odorant molecules across the olfactory mucosa. *J. Gen. Physiol.* 69: 343-361.

Katz, S. H., and Talbert, E. J. (1930). *Intensities of Odors and Irritating Effects of Warning Agents for Inflammable and Poisonous Gases* (Technical Paper 480). U. S. Department of Commerce, Washington, D.C.

Kobal, G., and Hummel, C. (1988). Cerebral chemosensory evoked potentials elicited by chemical stimulation of the human olfactory and respiratory nasal mucosa. *Electroencephalogr. Clin. Neurophysiol.* 71: 241-250.

Neilsen, G. D., and Alarie, Y. (1982). Sensory irritation, pulmonary irritation, and respiratory stimulation by airborne benzene and alkylbenzenes: Prediction of safe industrial exposure levels and correlation with their thermodynamic properties. *Toxicol. Appl. Pharmacol.* 65: 459-477.

Stevens, J. C., and Cain, W. S. (1986). Aging and the perception of nasal irritation. *Physiol. Behav.* 37: 323-328.

Stone, H. (1969). Effect of ethmoidal nerve stimulation on olfactory bulbar electrical activity. In: *Olfaction and Taste: Proceedings of the Third International Symposium.* C. Pfaffmann (Ed.). Rockefeller University Press, New York, pp. 216-220.

Tucker, D. (1971). Nonolfactory responses from the nasal cavity: Jacobson's organ and the trigeminal system. In: *Handbook of Sensory Physiology, Vol. 4: Chemical Senses, Part 1, Olfaction.* L. M. Beidler (Ed.). Springer-Verlag, Berlin, pp. 151-181.

# Chapter 3 Discussion

**Dr. Alarie:** Maybe we can settle some of our differences on $CO_2$. Of course, you can call $CO_2$ an irritant and it is a nice tool; it is convenient, you can use a puff of it, and so on. But let's remember the mechanism of action by which you are getting irritation from $CO_2$. I don't think you can extrapolate your data on $CO_2$ to irritants in general. The way that you are getting irritation from $CO_2$ is that you are simply changing the pH of the cornea or changing the pH of the nasal mucosa. It is the same thing as if you put a drop of dilute sulfuric acid on the membrane. That's why I am saying that $CO_2$ is not an irritant, that is, it does not fit into one of the three categories that were presented this morning. In other words, the mechanism here is the change in pH.

**Dr. Cain:** I'm not addressing mechanism per se. I think that the properties—temporal, spatial properties and so on—are general characteristics of the common chemical sense. I don't think you dispute that.

**Dr. Alarie:** No, I don't dispute that. But when you are looking at the relationship of $CO_2$ to other irritants, I don't think you can extrapolate results from $CO_2$ to other irritants in general. That's what I am worried about.

**Dr. Eccles:** In the early part of your talk, you mentioned applying $CO_2$ to one nasal passage and then another nasal passage and later on you referred to dichorhinic studies. I presume from that you are assuming that both nasal passages are similar in air flow characteristics, vasomotor activity, secretory activity, etc. Now I would argue that this assumption is wrong in that in animals and man there is a regular nasal cycle of vasomotor and secretory activity and air flow, and this will perhaps complicate the interpretation of your stimuli when applying them to, say, a left nasal passage which may be congested and at a higher temperature.

**Dr. Cain:** I am well aware of this and everything is balanced. But I don't think that you can argue that the data we show aren't right; the fact that there is a nasal cycle and the fact that there are changes doesn't mean that these data are not right. They don't rely on the idea that the nostrils are absolutely identical. Everything is counterbalanced. I think that the nasal patency story has come up with a goose egg when it comes to explaining essentially anything in smell and irritation. We have studied nasal patency in hundreds of people and have come up with zero.

**Dr. Eccles:** So you would argue that the asymmetry in nasal air flow and secretory function has no effect at all?

**Dr. Cain:** No, I'm not saying that. As a limiting factor, sure, you close one nostril down, you're not going to get anything through it and in that respect patency matters. But throughout most of the nasal cycle you are only slightly more sensitive on one side than the other. I'm not arguing against that. What I am saying is that patency is an overrated variable.

# 4

# Evidence for Interactions Between Trigeminal Afferents and Olfactory Receptor Cells in the Amphibian Olfactory Mucosa

**André Holley, Jean-Francois Bouvet, and Jean-Claude Delaleu**
Université Claude Bernard
Villeurbanne, France

In the olfactory mucosa of amphibians, trigeminal nerve afferents coexist with olfactory cells. Several recent studies have provided immunohistological, pharmacological, and electrophysiological evidence that nerve fibers pertaining to the trigeminal system, or closely associated with it, participate in a local regulation of the olfactory receptor cell environment and possibly in a direct control of receptor cell activity. This chapter reviews the available findings and discusses their relevance to the physiological expression of olfactory function.

## I. IMMUNOHISTOLOGICAL EVIDENCE

In the frog and salamander, the olfactory epithelium and the lamina propria are innervated by fibers which present a substance P (SP)-like immunoreactivity (Bouvet et al., 1987a; Getchell et al., 1987, 1989). These fibers run parallel to the receptor cells in the epithelium and appear to terminate without profuse branching at or near the surface. Some of them travel parallel to the surface for a few micrometers. The immunoreactive fibers seem to emerge from a fluorescent plexus extending near the basal membrane. In the lamina propria the fibers contact blood vessels and Bowman glands. Some terminals penetrate between acinar cells of these glands. The fluorescent fibers are also present in some nerve bundles which seem to contain myelinated fibers.

Substance P-like immunoreactive fibers correspond to a component of the ophthalmic branch of the trigeminal nerve (NV ob). This can be inferred from the fact that all immunoreactivity disappears from the olfactory mucosa within 10 days following NV ob transection near the Gasser ganglion (Bouvet et al., 1987a). The relation of immunoreactive fibers to the trigeminal system was confirmed by the observation of SP-reactive cell bodies within the trigeminal ganglion (Getchell et al., 1989). There is therefore substantial evidence that the neural elements containing SP or a closely related compound which innervate the olfactory epithelium and the lamina propria are peripheral processes of bipolar ganglion cells pertaining to the trigeminal system.

## II. PHARMACOLOGICAL EVIDENCE

Exogeneous SP delivered in superfusion to the olfactory mucosa of amphibians affected the secretory activity of supporting cells and mucosal glands (Getchell et al., 1989). In supporting cells this activity was indicated by the formation of domes and blebs at the cell apical surface. In Bowman glands it was manifested by swelling and vacuolation of acinar cells. These effects appeared to be dose-dependent. Penta-SP, the C-terminal peptide, was more effective than SP 1-7, the N-terminal peptide.

Exogenous SP also affected several functional properties of the olfactory epithelium. When delivered in superfusion at a low concentration ($10^{-9}$ M), SP induced transepithelial slow potentials which presented some similarity to odor-evoked electroolfactograms (EOGs) (Bouvet et al., 1984). Some of these voltage transients were monophasic and surface-negative. Others were biphasic with a negative component first followed by a positive wave. A few examples of purely positive voltage changes have also been observed. These variations in time course and polarity could not be correlated with SP concentration. Acetylcholine (ACh) elicited a similar type of mucosal slow potential.

SP ($10^{-6}$-$10^{-5}$ M) altered the spontaneous activity of virtually all receptor cells investigated as single units with extracellular microelectrodes (Bouvet et al., 1988). The most usual receptor cell response was an increase in discharge frequency. Suppressive responses were also observed. Acetylcholine ($5.10^{-6}$-$5.10^{-4}$ M) affected about 50% of the investigated receptor cells, triggering excitatory responses almost exclusively. *d*-Tubocurarine (d-TC, $10^{-5}$ M) was found to strongly excite all receptor cells. Conversely, atropine (ATR, $10^{-4}$ M), exerted a suppressive action which was preceded in some cases by a brief excitation. This cholinergic antagonist was very effective in suppressing the excitatory effects of ACh. It also antagonized receptor cell responses to SP, but this action was partial and limited to the excitatory responses. Suppressive responses evoked by SP were not modified by ATR.

Finally, it is worth noting that SP ($10^{-7}$ M) affected the receptor cell responses to the olfactory stimulus isoamyl acetate (Bouvet et al., 1987b).

Since SP induced the secretory activity of supporting cells and Bowman glands, at least a part of the electrophysiological findings can be interpreted with reference to the secretory action of SP. Slow voltage changes could have originated from supporting cell secretory activity (Okano and Takagi, 1974) or resulted from variations in the extracellular electrolyte balance in connection with Bowman gland secretion. Another contribution to the voltage transient might be SP-induced vasodilation and increased vascular permeability (Lembeck and Holzer, 1979; Couture and Cuello, 1984).

It is not likely that all single-unit level observations can be explained without assuming some chemical action on receptor cells. This action could be either directly exerted by SP or indirectly elicited by a compound whose release was evoked by the peptide. Preliminary investigations aimed at identifying the location of SP receptors have failed because the olfactory epithelium presented a high level of unspecific binding to SP and other tachykinins. The duality of the responses elicited by SP could be explained as a direct effect on specific SP receptors together with the indirect cholinergic action of the peptide. Since SP has been shown to cause the release of ACh from certain parts of the CNS (Pepeu, 1974), the similarity of responses of olfactory receptor cells to ACh and SP might suggest that SP-evoked excitation was caused by the release of endogenous ACh. This is in agreement with the fact that SP-evoked excitatory responses were partly antagonized by ATR. Intracellular studies have shown that the muscarinic ACh-evoked depolarization of neurons was accompanied by an increase in membrane resistance which presumably reflected a decrease in $K^+$ conductance (Krnjević et al., 1971; Dodd et al., 1981). Similarly, SP was reported to induce a decrease in $K^+$ conductance in neurons (Krnjević, 1977; Nowak and MacDonald, 1981). The $K^+$ current involved in the action of ACh seems to be the M current (M for muscarinic inhibition) first observed in neurons of the frog sympathetic ganglion (Brown and Adams, 1980). This M current was also inhibited by SP (Adams et al., 1983). Pharmacological evidence therefore supports the notion that muscarinic receptors are present on receptor cells. Nevertheless the fact that SP was active on almost all receptor cells investigated whereas ACh affected only half of them suggests that the neuropeptide acts on several types of receptor cells. In connection with this, there is a possibility that olfactory receptor sites were involved in the generation of the recorded electrical responses.

## III. ELECTROPHYSIOLOGICAL EVIDENCE

There are experimental arguments that SP can be released from trigeminal afferents. Short trains of electrical pulses delivered to the NV ob reproduced several effects of the application of SP. In the salamander, 2 to 10 sec stimulations of the nerve branch resulted in morphological signs of secretion in supporting cells and in acinar cells of superficial and deep Bowman glands (Getchell et al., 1989).

The antidromic excitation of the NV ob was also found to affect the electrophysiological properties of the oflactory epithelium (Bouvet et al., 1987b; Getchell et al., 1989). The nerve stimulation evoked transepithelial slow potentials in the frog and the salamander. In most preparations, the slow potential was a single negative wave which started after a latency of 1-2 sec from the stimulation onset and lasted for 2-15 sec. The peak amplitude was 1-2.5 mV. In a few preparations, the first wave was followed by a second, negative component which lasted for 1-3 min, with a peak amplitude of 2-4 mV. These voltage transients rapidly decreased in amplitude and disappeared when the stimulation was repeated every 15 sec. Recovery occurred slowly over a period of several min. Capsaicin, which is known to deplete SP from primary sensory neurons, suppressed the slow potentials evoked by stimulating NV ob. When delivered in gas phase to the mucosa, the substance evoked a marked negative going voltage transient which started like an EOG, and presented ample fluctuations in later phases. Ten minutes after the onset of exposure to capsaicin, the olfactory epithelium no longer responded to NV ob stimulation. Atropine was found to be a powerful antagonist of nerve stimulation. A SP antagonist, (D-Pro$^{2}$, D-Trp$^{7,9}$)-SP (Novabiochem, $5.10^{6}$ M), reduced the amplitude of NV ob evoked potentials by no more than 20% (unpublished observations).

Simultaneous recording of slow potentials and single-unit responses showed that NV ob stimulation tended to evoke an increase of the spontaneous discharge during the first negative wave of the slow potential, which was followed by a decreased spike activity during the late component.

The trigeminal electrical stimulation interfered with epithelial responses to the olfactory stimulus. When a stimulation with isoamyl acetate was delivered during the second negative phase of a long-lasting, trigeminally evoked slow potential, the peak amplitude of the EOG was reduced by 15-50% with respect to that of the control. Similarly, in the same conditions single receptor cell responses were partly or even totally suppressed.

In many aspects the antidromic stimulation of the ophthalmic component of the trigeminal system reproduced the effects of SP application. The findings are therefore in agreement with the notion that the production of spike activity in the peripheral segments of the trigeminal nerve releases SP or a closely related peptide. Since SP itself was found to induce composite effects, there is no need to assume that nerve stimulation released several substances in order to account for the complex effects observed. However, this hypothesis cannot be ruled out.

## IV. PHYSIOLOGICAL IMPLICATIONS

The concept of the involvement of the trigeminal system in the local control of receptor cell activity and sensitivity to odors can be proposed to integrate the experimental findings considered above. However, the evidence that the trigeminal system is implicated in nonexperimental, more physiological conditions is not direct. In particular, there is no demonstration available of a significant release of SP during normal functioning of the olfactory system.

Among the findings in support of the hypothesis of physiological involvement, one may mention that (a) many odorants can stimulate trigeminal nerve endings in the nasal cavity, even at low, nonirritating concentrations (Beidler, 1965; Tucker, 1971; Silver and Moulton, 1982); (b) olfactory stimulation triggers secretory activity in supporting cells and acinar glands of the olfactory mucosa (Getchell et al., 1987). That trigeminal fibers, which are sensitive afferents, play a role of efferent effectors was assumed in other cases. The association of sensitive and secretory functions in the same type of nerve element has been proposed to explain the origin of the edema induced by local application of capsaicin in the rat nasal mucosa (Lundblad et al., 1983). Finger (1986) proposed a similar explanation for the function of a population of perigemmal fibers found in the taste buds.

One may assume that irritant or odorant stimulation of the olfactory mucosa triggers impulses in the sensitive terminals of the trigeminal nerve which lie close to the surface of the epithelium. Along their orthodromic way to the ganglion, the impulses antidromically invade collateral branches innervating the epithelium, glands, and blood vessels. They modify receptor cell properties and stimulate mucus secretion and plasma extravasation through the release of SP and possibly other chemicals.

The mechanism can be seen as a protective mechanism which ensures a feedback control of the receptor cell chemical environment. It could regulate the rate of secretion of the mucus and possibly the relative concentrations of its different components as a function of the nature and the quantity of the chemicals dissolved in the mucus. How can the modulation of receptor cell activity be integrated in this scheme? Is this modulation an essential component of the postulated protective mechanisms, or is it a side effect? The question remains open.

## REFERENCES

Adams, P. R., Brown, D. A., and Jones, S. N. (1983). Substance P inhibits the M-current in bullfrog sympathetic neurones. *Br. J. Pharmacol.* 79: 330-333.

Beidler, L. M. (1963). Comparison of gustatory receptors, olfactory receptors, and free nerve endings. *Cold Spring Harbor Symp. Quant. Biol.* 30: 191-200.

Bouvet, J. F., Delaleu, J. C., and Holley, A. (1984). Réponses électriques de la muqueuse olfactive de grenouille à l'application d'acétylcholine et de substance P. *C.R. Acad. Sci.,* 298 série 3 169-172.

Bouvet, J. F., Godinot, F., Croze, S., and Delaleu, J. C. (1987a). Trigeminal substance P-like immunoreactive fibres in the frog olfactory mucosa. *Chem. Senses* 12: 499-505.

Bouvet, J. C., Delaleu, J. C., and Holley, A. (1987b). Olfactory receptor cell function is affected by trigeminal nerve activity. *Neurosci. Lett.* 77: 181-186.

Bouvet, J. C., Delaleu, J. C., and Holley, A. (1987c). Does the trigeminal nerve control the activity of the olfactory receptors cells? In: *Olfaction and Taste IX.* S. D. Roper and J. Atema (Eds.). *Ann. N.Y. Acad. Sci.* 510: 187-189.

Bouvet, J. C., Delaleu, J. C., and Holley, A. (1988). The activity of olfactory receptor cells is affected by acetylcholine and substance P. *Neurosci. Res.* 5: 214-223.

Brown, D. A., and Adams, P. R. (1980). Muscarinic suppression of a novel voltage-sensitive K+ current in a vertebrate neuron. *Nature* 283: 673-676.

Couture, A. C., and Cuello, A. C. (1984). Studies on the trigeminal antidromic vasodilatation and plasma extravasation in the rat. *J. Physiol.* (London) 346: 273-285.

Dodd, J., Dingledine, R., and Kelly, J. S. (1981). The excitatory action of acetylcholine on hippocampal neurones of the guinea pig and rat maintained in vitro. *Brain Res.* 209: 109-127.

Finger, T. T. (1986). Peptide immunohistochemistry demonstrates multiple classes of perigemmal nerve fibers in the circumvallate papilla of the rat. *Chem. Senses* 11: 135-144.

Getchell, M. L., Finger, T. E., and Getchell, T. V. (1987). Localization of NGF-like, VIP-like, substance P-like and CGRP-like immunoreactivity in the olfactory mucosae of the salamander, bullfrog and grass frog. *Chem. Senses* 12: 658.

Getchell, M. L., Bouvet, J. F., Finger, T. E., Holley, A., and Getchell, T. V. (1989). Peptidergic regulation of secretory activity in the amphibian olfactory mucosa: Immunohistochemistry, neural stimulation and pharmacology. *Cell Tis. Res.* 256: 381-389.

Krnjević, K., Pumain, R., and Renaud, L. (1971). The mechanism of excitation by acetylcholine in the cerebral cortex. *J. Physiol.* (London) 215: 247-268.

Krnjević, K. (1977). Effects of substance P on central neurons in cats. In: *Substance P*, Nobel Symposium 37, U.S. Von Euler and B. Pernow (Eds.). Raven Press, New York, pp. 217-230.

Lembeck, F., and Holzer, P. (1979). Substance P as neurogenic mediator of antidromic vasodilatation and neurogenic plasma extravasation. *Neunyn-Schmiedeberg's Arch. Pharmacol.* 310: 175-183.

Lundblad, L., Saria, A., Lundberg, J. M., and Änggård, A. (1983). Increased vascular permeability in rat nasal mucosa induced by substance P and stimulation of capsaicin-sensitive trigeminal neurons. *Acta Otolaryngol.* 96: 479-484.

Nowak, L. M., and Mac Donald, R. L. (1981). Substance P decreases a potassium conductance of spinal cord neurons in cell culture. *Brain Res.* 214: 416-423.

Okano, M., and Takagi, S. F. (1974). Secretion and electrogenesis of the supporting cell in the olfactory epithelium. *J. Physiol.* (London) 242: 353-370.

Pepeu, G. (1974). The release of acetylcholine from the brain: An approach to the study of central cholinergic metabolism. *Prog. Neurobiol.* 2: 259-288.

Silver, W. L., and Moulton, D. G. (1982). Chemosensitivity of rat nasal trigeminal receptors. *Physiol. Behav.* 28: 927-931.

Tucker, D. (1971). Non olfactory responses from the nasal cavity: Jacobson's organ and the trigeminal system. In: *Handbook of Sensory Physiology*, Vol. 4. *Chemical Senses* Part 1, *Olfaction*. L. M. Beidler (Ed.). Springer-Verlag, Berlin, pp. 151-181.

# Chapter 4 Discussion

**Dr. Szolcsanyi:** Related to the blocking effect of atropine and capsaicin, I would like to stress that these agents are selective. However, as with most drugs, the selectivity holds only over a few orders of magnitude. In the range of $10^{-5}$-$10^{-6}$ both drugs start to have a local anesthetic effect. So in this case it is rather difficult, because you must apply capsaicin in the air, to be sure what the concentration is. I would propose that you do some tests to avoid this other effect of capsaicin that is not specific to substance P fibers and which in the range of $10^{-5}$-$10^{-6}$ could block every kind of neural response. My point is, I did preliminary experiments in which I put frogs in a rather concentrated capsaicin solution, and there was no sign of pungency. This is very interesting because the substance P fibers are even more superficial to the epithelial level in frogs than in the mammals. Have you tried applying capsaicin in awake animals?

**Dr. Holley:** No, but we have recorded the direct effect of capsaicin blown into the olfactory mucosa and there is a strong local response to capsaicin that continues for several minutes and then progressively the response disappears.

**Dr. Silver:** Regarding the EOG latencies when you put substance P on vs. when you stimulated the nerve, were they about the same or were the latencies longer when you put substance P on?

**Dr. Holley:** It is difficult to give you a good reply because when we perfuse the olfactory mucosa with substance P, it takes time for the substance to go from the injection site to the olfactory mucosa. So I cannot say anything about latency, except that there is a latency.

**Dr. Kobal:** Is the shape of the local, slow trigeminal response the same everywhere, or is it different at loci beside the olfactory mucosa?

**Dr. Holley:** We could not relate the variations in the responses that we measured to the location. It seemed to vary most depending on the state of the frog.

# 5
# Trigeminal vs. Olfactory Input for Laryngectomized Patients

**Maxwell M. Mozell, David N. Schwartz, Steven L. Youngentob, Donald A. Leopold, Paul R. Sheehe, and James A. Listman**
SUNY Health Science Center at Syracuse
Syracuse, New York

## I. LARYNGECTOMY-INDUCED OLFACTORY DECREMENTS

An association between a decrement in olfactory ability (hyposmia or even anosmia) and total laryngectomy has long been recognized both from clinical observation and from more formal investigations (Diedrich and Youngstrom, 1966; DeBeule and Damste, 1972; Gilchrist, 1973). Although our primary purpose in two recently completed studies (Mozell et al., 1986; Schwartz et al., 1987) was to investigate the mechanisms underlying this laryngectomy-induced olfactory decrement, the strategies developed for these studies have also made them relevant to this book. That is, these strategies involve a comparison of trigeminal and olfactory nerve inputs in a rather unusual context, namely, in the context of nasal air flow as generated by laryngectomized patients. However, to pursue this comparison we must first review the rationale and design of the two studies, which were primarily intended to explore how laryngectomy might lead to olfactory deficits.

Of course, the simplest mechanism for a laryngectomy-induced anosmia or hyposmia would be the loss of the normal ability to sniff air through the nose. That is, after laryngectomy, air and the odorants it carriers enter the trachea through a stoma in the neck, thus bypassing the nose. Although this

*Publisher's note*: Throughout this chapter, "air flow" is synonymous with "airflow."

might be the simplest mechanism, not all studies agreed. For instance, Henkin and his coworkers (Henkin et al., 1968; Hoye et al., 1970; Henkin and Larson, 1972) did not agree because in their hands the nasal air flow brought about by "gentle" insufflation of odorants into the nose did not lower the elevated detection and recognition thresholds of their laryngectomized patients. Furthermore, they argued that if, as in at least one study (Ritter, 1964), normal olfactory function results from odorant insufflation, it is due to the excitation of the accessory olfactory regions supplied by the trigeminal, glossopharyngeal, and vagus nerves rather than to the excitation of the primary olfactory region supplied by the olfactory nerve itself. Thus they pointed out that, without the proper precautions, subjects deprived of the primary olfactory input might still detect some odorants. At any rate, rather than ascribing the hyposmia of laryngectomy to the compromised nasal air flow, Henkin and his coworkers proposed that it was due to the disruption of certain neural networks dependent on an intact innervation of the larynx by the recurrent laryngeal nerve (Henkin and Larson, 1972). They based this proposal on their observation that hyposmia does not occur with traumatic or surgical alteration of the larynx unless the innervation of the larynx by the recurrent laryngeal nerve is severed.

There are, of course, other mechanisms that could possibly explain the olfactory deficit following laryngectomy, some of which may be secondary to a long-term lack of nasal air flow. These include: some sort of disuse atrophy of the receptors; changes in the secretion, movement, and composition of the mucus surrounding the receptors (Toppozada and Gaafar, 1974); changes in the mucosal characteristics [such as venous congestion (Dixon et al., 1949)] which, by affecting nasal patency, block the access of the molecules to the receptors.

Note, however, that even if there were mechanisms contributing to the hyposmia of laryngectomy other than the compromised nasal air flow, the effect of these other mechanisms would be masked by the compromised nasal air flow because without adequate odorant transport into the nose there can be no olfactory response. Thus, to evaluate the contribution of the compromised nasal air flow and to allow the other possible mechanisms to emerge unmasked, we developed what we call the "larynx bypass" (Mozell et al., 1983, 1985, 1986). This device assures an adequate transport of odorant molecules by permitting laryngectomized patients to sniff odorants into their noses with near-normal air flows.

## II. THE LARYNX BYPASS

In the larynx bypass (Fig. 1) a plastic tube passes from the tracheostoma into the mouth via a small plastic mouthpiece. The stomal connection is made

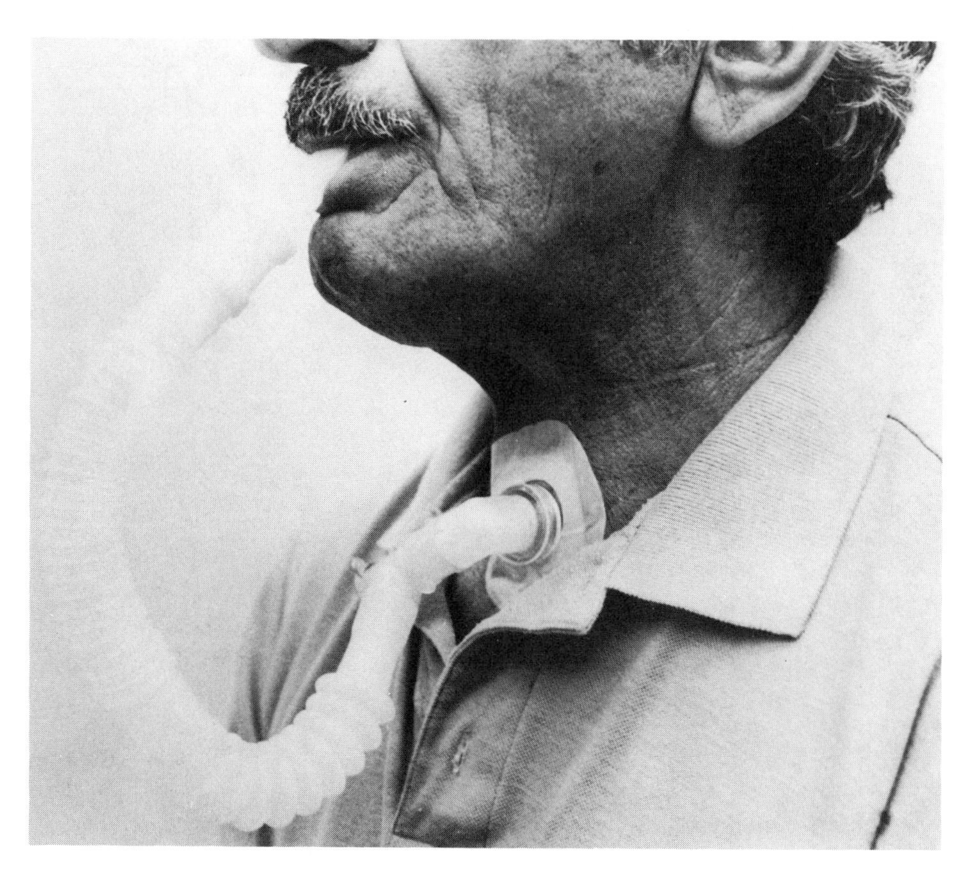

**FIGURE 1** The larynx bypass fitted to a laryngectomee. See text for description. (Schwartz et al., 1987; reproduced by permission of The Laryngoscope Journal.)

with a plastic elbow inserted into the flange of a rubber ring held securely around the stoma by a circle of water-resistant adhesive tape. Many of these parts come with a commercially available device designed to protect the stoma of the laryngectomee when he or she is showering (Stomaguard I, Medart, Glendora, CA). With lips sealed around the tube, the negative pressure developed in the lungs draws air through the nose, allowing the patient to sniff odorants from small glass "odorant bottles" (volume: 45 ml; length 85 mm; Teflon-lined screw-top caps) held at the nares.

In order to measure the air flow and sniff profiles generated by each patient when sampling odorants, we used a No. 2 Fleisch pneumotachometer connected to a plastic breathing mask held snugly against the patient's face

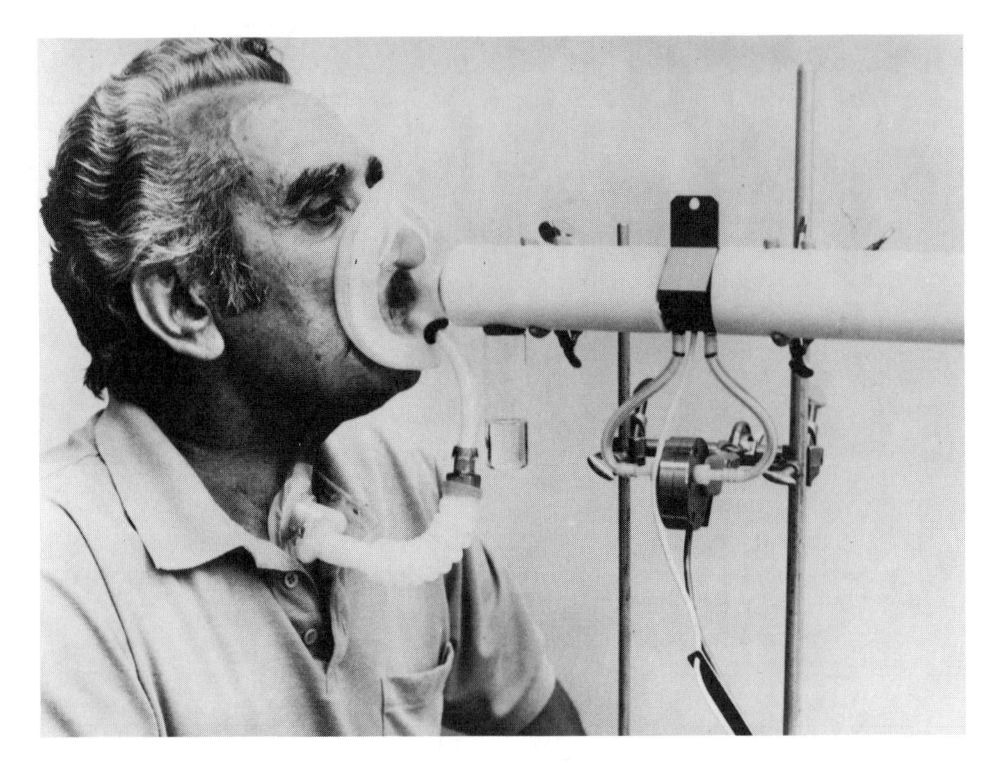

**FIGURE 2** The pneumotachometer used to trace patient sniff air flow profiles. See text for description. (Schwartz et al., 1987; reproduced by permission of The Laryngoscope Journal.)

as shown in Fig. 2 (Mozell et al., 1983, 1986; Schwartz et al., 1987). The pneumotachometer was sandwiched between two Teflon flow straighteners, the more proximal of which was drilled with a port to accept an odorant bottle. Provision was made with the mask to monitor the patient's ability to generate air flow both with and without the larynx bypass; Fig. 3 shows a sample of this ability. The lower trace was the sniff profile of a patient trying to detect an odorant without the bypass. As exemplified by this patient, many laryngectomees can draw at least some low volume at some low flow rate into the nose by using a number of maneuvers such as dropping the lower jaw to produce a negative pressure in the oral cavity. These maneuvers produce "sniffs" of much less magnitude than those produced using the larynx bypass as shown for the same patient in the upper trace. For comparison the inset shows the sniff profile of a nonlaryngectomized subject. It should be noted that even our nonlaryngectomized subjects varied considerably in the exact air flow patterns they generated. Especially considering

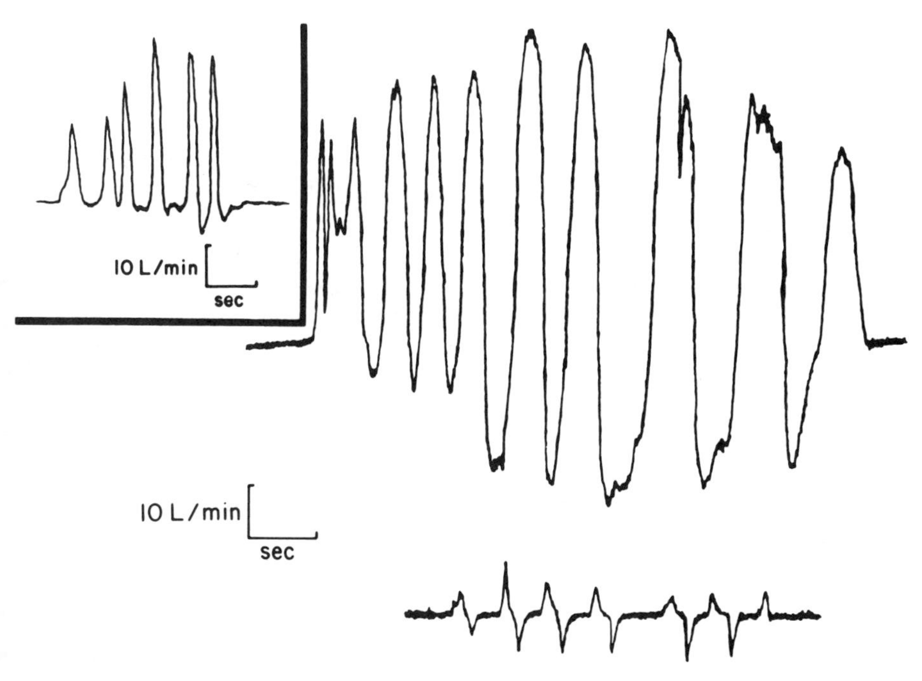


**FIGURE 3** Pneumotachographs of the air flow profiles generated by a patient using (upper trace) and not using (lower trace) the larynx bypass. Inset: airflows of a "normal," nonlaryngectomized patient. (Mozell et al., 1983; reproduced by permission of Oxford University Press.)

this wide variation, it appeared to us that laryngectomized patients with the bypass could approximate, if not actually duplicate, normal sniffing behavior.

## III. DETECTION THRESHOLD STUDY I

We determined the detection thresholds of 17 laryngectomized patients to ammonia and vanillin (Mozell et al., 1986). They included three females, average 63 years of age, and ranged from 2 months to almost 17 years postsurgery. In addition to the laryngectomized patients, the detection thresholds of 10 nonlaryngectomized subjects representing similar age and gender distributions were also measured. To determine the detection threshold for each odorant, we used a two-interval forced-choice tracking procedure with descending concentration. Beginning with standard solutions (household ammonia and 1% vanillin), each of the successive dilution steps was threefold more dilute than its predecessor. With each laryngectomized patient we determined for each odorant two detection thresholds: one without the bypass and the other with the bypass.

Our choice of these two odorants, ammonia and vanillin, stemmed from Henkin and Larson's (1972) admonition that without the proper precautions odorants can be detected by inputs other than the olfactory nerve. Since we judged that to reach the primary olfactory region odorant molecules would have to travel over some accessory olfactory region (at least that supplied by the trigeminal nerve) we decided to use vanillin as one of the odorants. We made this choice because Doty et al. (1978) reported evidence that strongly suggested that responsiveness to vanillin in humans is mediated predominantly, and maybe even exclusively, by the olfactory nerve input. Having made the vanillin choice, we then also chose ammonia as an example of a strong, though not exclusive, trigeminal irritant. Thus, we could begin to explore whether laryngectomy and the restored air flow with the larynx bypass affects the detection thresholds of odorants differently, depending on what their neural inputs might be.

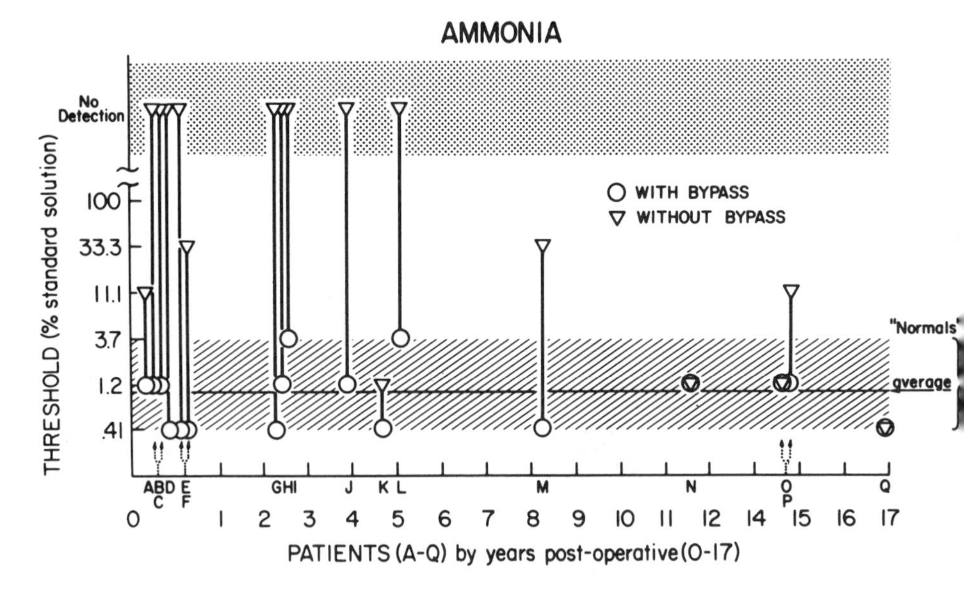


**FIGURE 4** Patient-by-patient detection thresholds for ammonia with and without the larynx bypass. See text for description. Each fork along the abscissa depicts a pair of patients who were postoperative for the same length of time. (Mozell et al., 1986; reproduced by permission of Oxford University Press.)

Figure 4 gives an overall, patient-by-patient look at the results for ammonia. Along the abscissa each patient (A-Q) is plotted in accordance with his or her years postoperative, and the threshold (in percent standard solution) is given on the ordinate. The triangles represent the thresholds for each patient without the bypass and the circles represent them with the bypass. The slant line area shows the range of thresholds and the average for the nonlaryngectomized, "normal" subjects.

This figure shows that the majority of laryngectomized patients could not detect ammonia without the bypass. It was not simply that they had higher thresholds but rather that with the two-interval forced-choice paradigm they could not even detect the highest concentration. This is depicted by the stippled area in Fig. 4. On the other hand, some patients could detect ammonia without the bypass and some of these could even detect ammonia in the normal

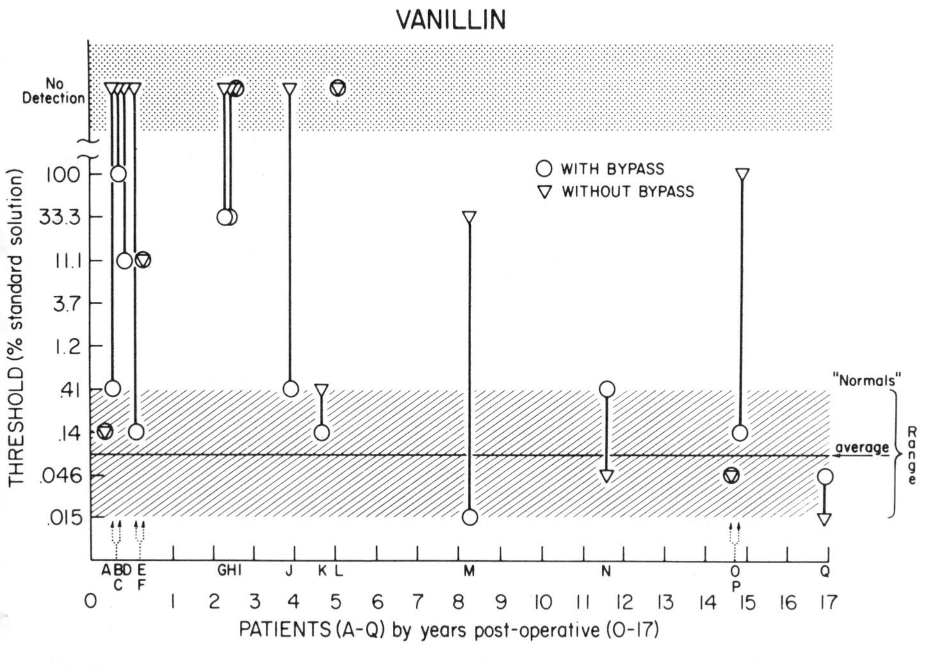


**FIGURE 5** Patient-by-patient detection thresholds for vanillin with and without the larynx bypass. See text for description. Each fork along the abscissa depicts a pair of patients who were postoperative for the same length of time. (Mozell et al., 1986; reproduced by permission of Oxford University Press.)

**TABLE 1** A Comparison (for Both Ammonia and Vanillin) of How the 17 Patients Fall Into Three Threshold Levels When Tested Without and With the Larynx Bypass

| Threshold level | Ammonia | | Vanillin | |
|---|---|---|---|---|
| | Without bypass | With bypass | Without bypass | With bypass |
| No detection | 9 | 0 | 9 | 2 |
| Between no detection and 'normal' range | 4 | 0 | 3 | 5 |
| 'Normal' range | 4 | 17 | 5 | 10 |

(Mozell et al., 1986; reproduced by permission of Oxford University Press.)

range. Most importantly, however, as shown by the lines connecting the triangles and circles, the larynx bypass with its augmented nasal flow markedly improved the sensitivity of laryngectomized patients to ammonia.

Figure 5 gives the patient by patient results for vanillin. Most patients (the same ones as with ammonia) could not detect vanillin without the bypass, but some could do so, and some could even detect it as well as the nonlaryngectomized subjects. As shown by the lines connecting the triangles and the circles, the use of the larynx bypass greatly increased sensitivity to vanillin, but this improvement was somewhat less marked than it was for ammonia.

Table 1 numerically compares how the 17 patients fell into each of three threshold levels (no detection, normal range detection, and detection at levels between no detection and normal detection) when tested with and without the bypass for both ammonia and vanillin. Without the bypass the same nine patients could detect neither ammonia nor vanillin. On the other hand, eight patients could detect both ammonia and vanillin without the bypass with five in the normal range for ammonia and five in the normal range for vanillin. With the bypass, however, not only did all 17 patients detect ammonia, they all did so in the normal range.

For vanillin with the bypass, 15 patients (compared to 17 for ammonia) could detect it with only 10 (rather than 17) detecting it in the normal range. In addition, two patients remained unable to detect vanillin at all. These same two patients could, however, detect ammonia in the normal range. Thus the improvement with the bypass for vanillin, though substantial, was not as impressive as it was for ammonia. A sign test of the paired difference in improvement for ammonia vs. vanillin in the 12 patients initially falling into the two non-normal threshold levels showed a nominally significant difference for ammonia over vanillin ($P = 0.016$).

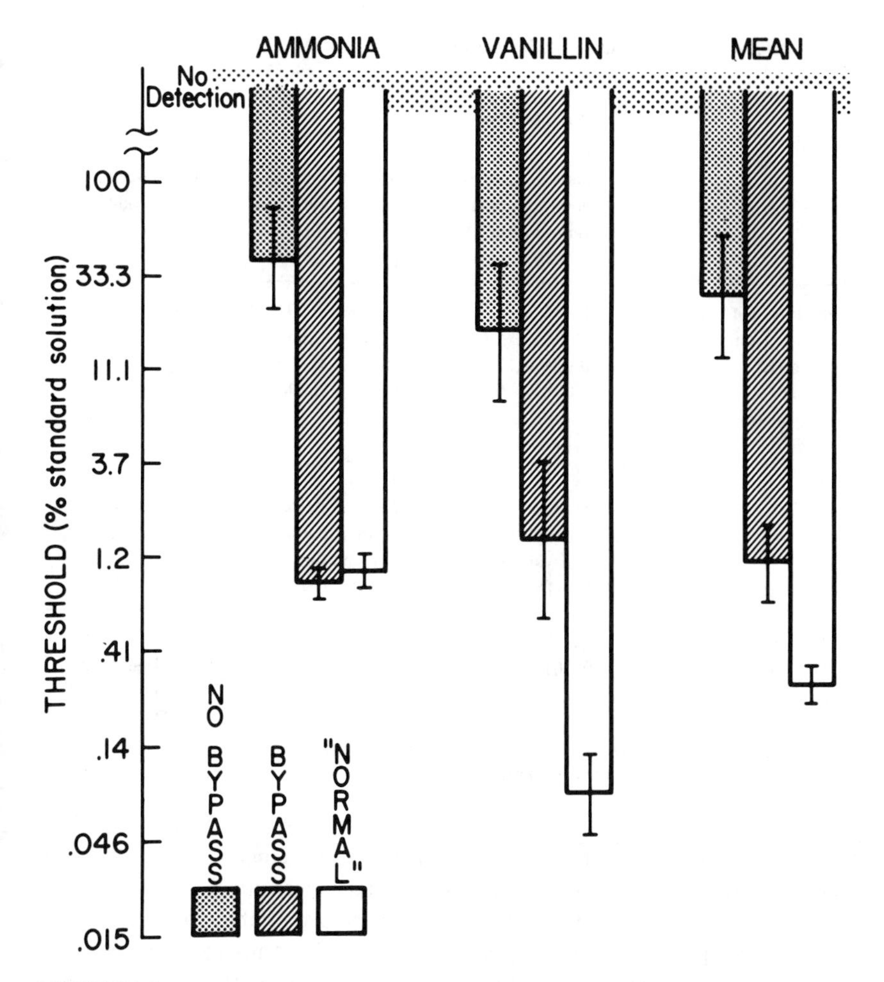


**FIGURE 6** Average detection thresholds for ammonia, vanillin, and their combined means with and without the larynx bypass. The average ammonia and vanillin thresholds are also given for the nonlaryngectomized, normal subjects. The ordinate is oriented from top to bottom to show decreasing thresholds. (Mozell et al., 1986; reproduced by permission of Oxford University Press.)

By focusing on the threshold values themselves rather than the number of patients achieving various threshold levels, Fig. 6 further emphasizes this difference between the two odorants. Note that although the average bypass thresholds for both ammonia and vanillin were lower than the average no-bypass thresholds, the improvement from no-bypass to bypass was greater for the ammonia. Furthermore, with the bypass the laryngectomees' average ammonia detection threshold essentially equaled that of normal subjects

**TABLE 2** Threshold Data in Relation to Air Flow

| | Without bypass | | |
|---|---|---|---|
| | I<br>Patients detecting both ammonia and vanillin<br>($n$=8) | II<br>Patients detecting neither ammonia nor vanillin<br>($n$=9) | III<br>Patients detecting only one odorant<br>($n$=0) |
| Peak flow-rate (l/min) | 12.34 ± 2.78 | 4.90 ± 1.50 | – |
| Mean flow-rate (l/min) | 6.31 ± 1.61 | 2.41 ± 0.96 | – |
| Volume (L) | 0.025 ± 0.007 | 0.012 ± 0.003 | – |
| Duration (s) | 0.26 ± 0.054 | 0.36 ± 0.076 | – |
| | With bypass | | |
| | IV<br>Patients detecting both ammonia and vanillin<br>($n$=15) | V<br>Patients detecting neither ammonia nor vanillin<br>($n$=0) | VI<br>Patients detecting only ammonia<br>($n$=2) |
| Peak flow-rate (l/min) | 32.77 ± 3.38 | – | 33.15 ± 5.49 |
| Mean flow-rate (l/min) | 22.61 ± 2.32 | – | 21.54 ± 5.13 |
| Volume (L) | 0.436 ± 0.064 | – | 0.371 ± 0.032 |
| Duration (s) | 1.31 ± 0.16 | – | 1.04 ± 0.28 |

(Mozell et al., 1986; reproduced by permission of Oxford University Press.)

whereas for vanillin, in spite of a substantial improvement, the normal subjects were still considerably better than the laryngectomees.

Table 2 addresses how these threshold data are related to the air flows that the laryngectomees could generate. This table compares some sniffing characteristics (average peak flow rate, average mean flow rate, average sniff volume, average sniff duration) between patients who could detect and patients who could not detect the odorants with and without the bypass. Without the bypass there was a two- to threefold difference in most of the sniff characteristics between those eight patients detecting both ammonia and vanillin and those nine patients detecting neither. Thus even without the bypass the detecting patients produced somewhat more vigorous sniffs than the nondetecting patients, but this increase in vigor was small compared to the increase achieved with the larynx bypass. Certainly as far as this experiment is concerned these large changes in sniffing characteristics brought about by the larynx bypass converted nondetecting patients into detectors, albeit more so for ammonia than for vanillin. However, one wonders whether the same improvement could have been achieved by bringing the sniffs of the nondetectors up to the more modest level of those patients who without the bypass still detected the odorants. It is possible that a certain base level of sniff is required for detection and that little is to be gained with more vigorous levels.

Up to this point several conclusions can be drawn. First, these data specifically argue against the disruption of larynx-dependent neural networks as the basis for laryngectomy-induced hyposmia. This cannot be correct since all 17 patients underwent this type of neural disruption and all but two could detect both odorants. Second, these data confirm that the compromised nasal transport of odorants is a major contributor to the hyposmia of laryngectomy and may be the only contributor for trigeminal irritants like ammonia. That is, once the nasal air flow was reestablished with the larynx bypass, all the patients detected ammonia at the level of the nonlaryngectomized subjects with no residual deficiency. On the other hand, for vanillin, the nontrigeminal odorant, the compromised nasal air flow may be the whole story for some patients but not for others. That is, with the reestablished nasal air flow 41% of the patients did not reach the nonlaryngectomy, normal threshold and two patients could not detect vanillin at all.

This shortfall for vanillin may be the result of the unmasked effects of the other possible contributors to the hyposmia of laryngectomy suggested earlier, such as disuse atrophy of the receptors and changes in airway patency. However, before we can seriously entertain the existence of these mechanisms or otherwise try to interpret these vanillin results, we must rule out another, perhaps less interesting, possibility. First note that in this cross-sectional study with all the patients already laryngectomized there was no way to know their presurgical olfactory abilities. It is conceivable that some of the poor detectors of vanillin in this study were also poor detectors of vanillin before their laryngectomies. Perhaps candidates for laryngectomy have some predisposition for hyposmia, maybe even specific hyposmias for certain types of odorants. They tend, for instance, to be extremely heavy smokers and drinkers. If, for whatever reason, these patients have been poor detectors of vanillin all along, the reestablished air flow given by the larynx bypass could not be expected to return the vanillin thresholds to the nonlaryngectomy, normal levels but rather only to the presurgical levels. If this is what was occurring in our study, it would mean that for vanillin, like ammonia, the loss of nasal air flow was the sole contributor to the laryngectomy-induced hyposmia.

## IV. DETECTION THRESHOLD STUDY II

To pursue this possibility of presurgical detection problems, we did a longitudinal study (Schwartz et al., 1987) in which the ammonia and vanillin de-

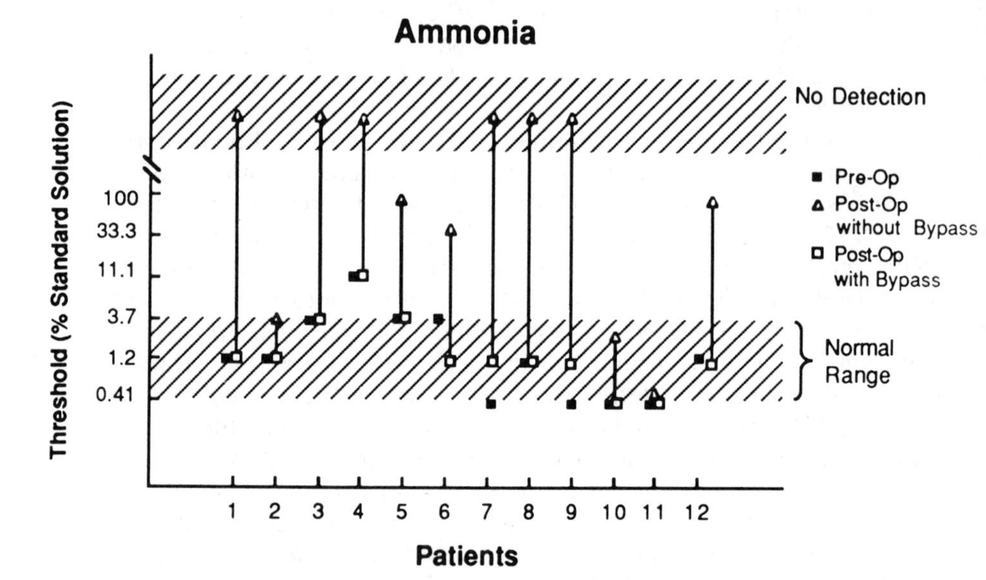


**FIGURE 7** Patient-by-patient detection thresholds for ammonia. The detection thresholds were determined three times for each patient: prior to surgery, postsurgery without the larynx bypass, and postsurgery with the larynx bypass. (Schwartz et al., 1987; reproduced by permission of The Laryngoscope Journal.)

tection thresholds of 12 patients were determined both before laryngectomy (during the week prior to surgery) and after laryngectomy (during the second month postsurgery). Postsurgically the thresholds were determined both with and without the bypass.

The results for ammonia are shown in Fig. 7, which resembles the previous Figs. 4 and 5 except that there are three symbols instead of two. The filled-in squares represent the preoperative thresholds; the triangles represent the postoperative thresholds without the larynx bypass; and the open squares represent the postoperative thresholds with the larynx bypass. Note that with the reestablished air flow afforded by the larynx bypass, only two of the 12 patients (17%) did not reach their preoperative threshold levels and only one (8%) was not in the normal range. This contrasts with the vanillin data shown in Fig. 8. Note that eight of the 12 patients (75%) did not reach their preoperative threshold levels and five (42%) were not in the normal range. A sign test showed a nominally significant difference ($P = 0.032$) between ammonia and vanillin in how many patients were brought back to their presurgical detection thresholds by the bypass.

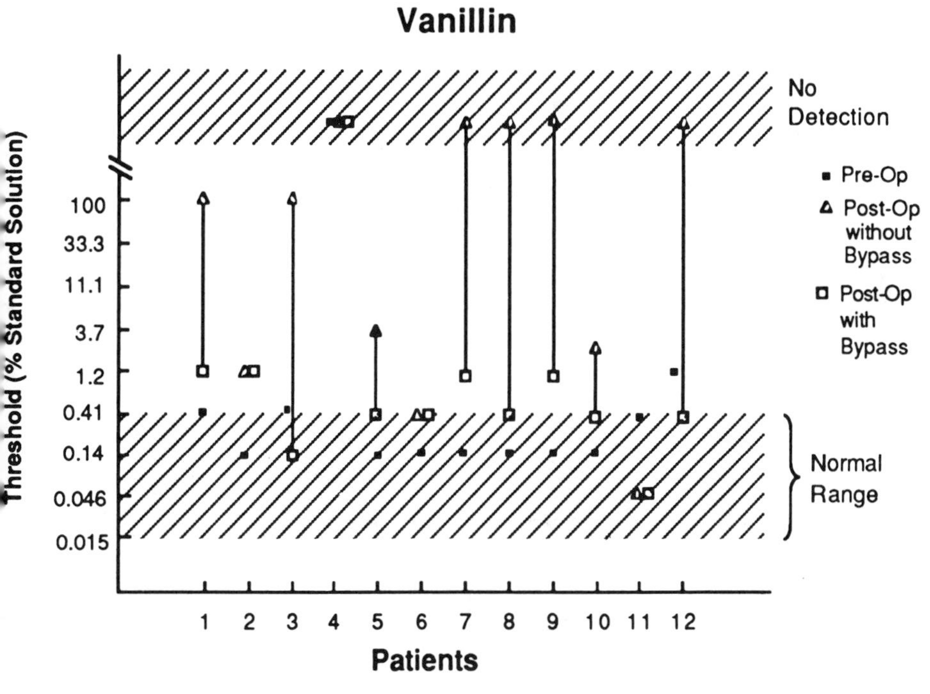


**FIGURE 8** Patient-by-patient detection thresholds for vanillin. The detection thresholds were determined three times for each patient: prior to surgery, postsurgery without the larynx bypass, and postsurgery with the larynx bypass. (Schwartz et al., 1987; reproduced by permission of The Laryngoscope Journal.)

Thus this study reenforces the previous one, i.e., the thresholds for ammonia and vanillin were both improved by the reestablished air flow but the improvement was more complete for ammonia than for vanillin. This continues to leave open the possibility that although the loss of nasal air flow may by itself explain the laryngectomy-induced hyposmia for trigeminal nerve odorants, additional mechanisms may contribute to the hyposmia for olfactory nerve odorants. [In an independently done study based on our larynx bypass and the strategy it affords, Tatchell et al. (1985) also observed that the enhanced air flow dramatically improved the olfactory ability of their laryngectomized subjects, but they did not specifically analyze their data in regard to any trigeminal nerve vs. olfactory nerve differences.)

One of these mechanisms might rest with the patients' sniffing behavior. Laing (1982, 1983), referring to normal subjects, reported that each person

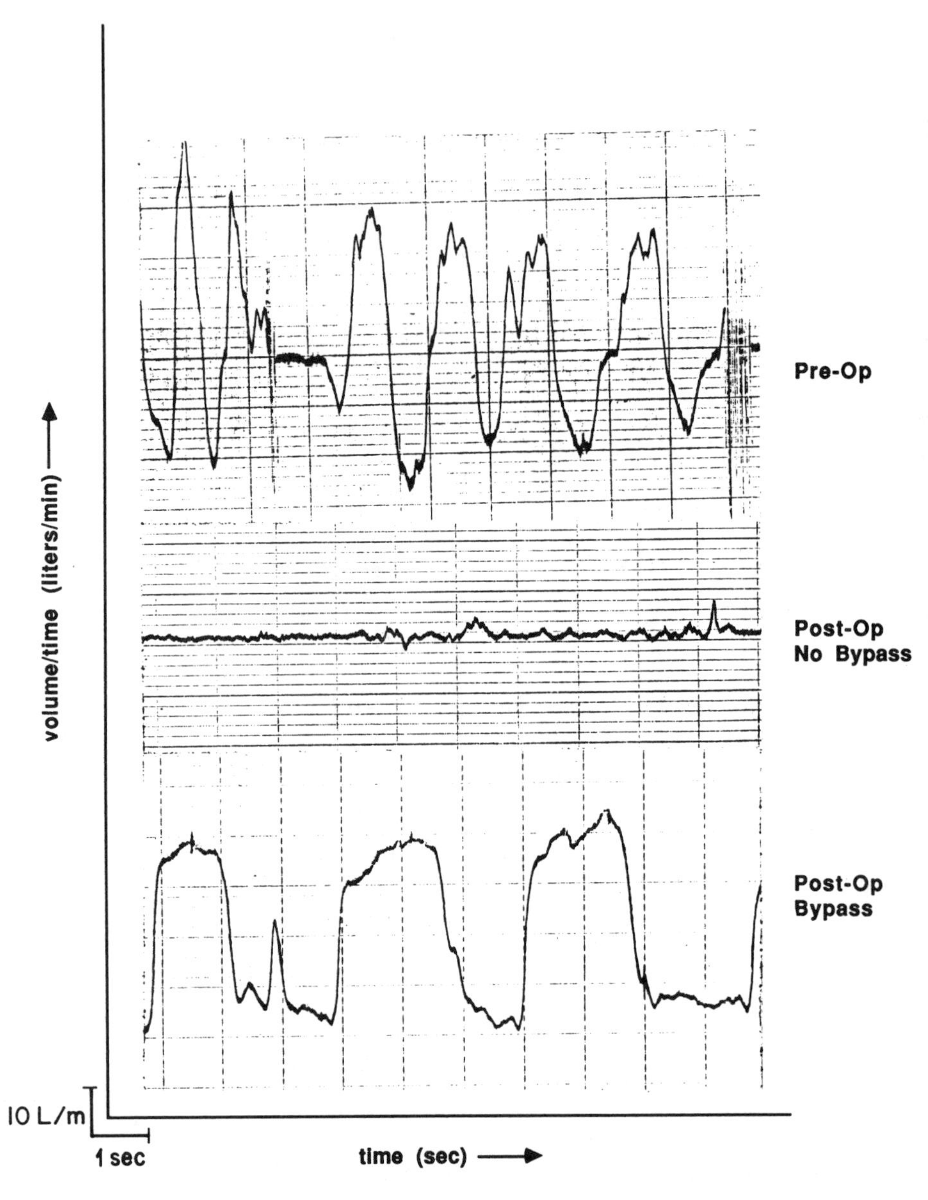


**FIGURE 9** Pneumotachographs of the nasal air flows generated by a representative laryngectomee entered in the study. The upper trace is this patient's preoperative sniffing airflow profile. The middle trace is this patient's postoperative sniffing air flow profile without the larynx bypass. The lower trace is this patient's postoperative sniffing air flow profile with the larynx bypass. Note that although this patient can generate significant air flow with the larynx bypass, the sniff profile generated is quite different from the preoperative profile. (Schwartz et al., 1987; reproduced by permission of The Laryngoscope Journal.)

produces a rather individualized sniff which for him or her seems to maximize olfactory ability. It is possible that the larynx bypass did not allow the laryngectomees to restore their individual optimal sniffing behaviors even though against the entire spectrum of human sniffing profiles they appeared normal. Indeed, Fig. 9 shows that even though postsurgically a patient can generate very respectable nasal air flows with the larynx bypass, the sniff profiles can still differ markedly from those generated presurgically. An argument can be made that this difference in sniffing might affect the olfactory nerve input more than the trigeminal nerve input,* i.e., perhaps any reasonable approximation of normal nasal air flow would assure odorant access to the rather widely distributed trigeminal receptors whereas the olfactory receptors, being located in the uppermost reaches of the nasal vault, require a closer approximation to optimal sniffing behavior.

## V. FURTHER IMPLICATIONS OF THE THRESHOLD STUDIES

The above explanation, though appealing in its apparent simplicity, leads to a number of conundrums. To begin with, for this explanation to be correct, the ammonia detection threshold would have to depend rather exclusively on the trigeminal nerve input. That is, if, instead, it depended on the olfactory nerve input, there should have been, as for vanillin, a deficient return toward the normal threshold when the bypass was used. However, can we accept the possibility of an exclusive trigeminal input for ammonia at its detection threshold if at this low-intensity level the stinglike quality of ammonia does not appear to be present? Some might argue that the sting for ammonia is the hallmark of its trigeminal nerve involvement.

Of course, we cannot be absolutely certain in the experiments reported here that there was no sting since, with other major goals in mind, we did not pursue the sensory qualities experienced by our laryngectomees at detection threshold. However, judging from common experience and anecdotal observation, the presence of a sting seems unlikely. At any rate, by just considering these possibilities we raise again an often raised controversial question: Does the trigeminal input for airborne chemical stimuli give rise to discriminable sensation qualities other than those which, like "sting," "burn," and "cool," are traditionally considered somatosensory? This would be one of the suggestions if, as we have been speculating from our data, the ammonia threshold is exclusively based on the trigeminal input.

An opposing view for these data, also giving rise to conundrums, would be that the ammonia detection threshold, like that proposed for vanillin,

*We use the term "trigeminal nerve input" generically to include, if necessary, any one or any combination of the accessory olfactory regions.

depends predominantly if not exclusively on the input of the olfactory nerve. This would mean that the threshold for the olfactory nerve's discharge to ammonia would be lower than the trigeminal nerve's discharge threshold to ammonia. In this case, the differences seen between ammonia and vanillin in how closely they regain their normal thresholds with the bypass would be independent of their being olfactory nerve vs. trigeminal nerve stimuli. Instead, these would depend on some other differentiating properties referable strictly to the primary olfactory region.

One way to determine whether the ammonia threshold for the olfactory nerve input is indeed below that for the trigeminal nerve input would be to compare the ammonia detection thresholds of normal subjects to the ammonia detection threshold of subjects who, like Kallman syndrome patients, lack the olfactory nerve input. If the ammonia thresholds for the two groups are similar, it would mean that the trigeminal input can account for the ammonia detection threshold even in normal subjects. This in turn would keep open the possibility of an olfactory nerve vs. trigeminal nerve explanation for the threshold differences reported above. If, on the other hand, the patients lacking the olfactory input had higher ammonia thresholds, it would mean that in normal subjects the ammonia detection threshold depends on the olfactory nerve input, and this would indicate a need for another explanation of the threshold differences reported above.

## VI. AN ODORANT IDENTIFICATION STUDY

It is of course rather presumptuous in the above interpretations to invoke an olfactory nerve vs. trigeminal nerve effect from just one representative odorant for each. Furthermore, we have used only one measure of olfactory ability, namely, detection thresholds. To address both these shortcomings, we did another cross-sectional study (Schwartz et al., 1987) which, using five different odorants in a confusion matrix (Wright, 1987), focused on the ability of previously laryngectomized patients to make odorant identifications both with and without the bypass. That is, the patients had to choose from a list of the five odorants which one in a randomized sequence was being presented. The analysis of the data was based on five presentations of each odorant under each of the two bypass conditions (with and without) to each of the 30 patients entered in the study. The patients were randomized for the bypass condition under which they would first be tested. The same confusion matrix was administered to a group of 25 nonlaryngectomized subjects. The five odorants were ammonia and vinegar, both of which are strong trigeminal irritants; vanillin and rose (phenylethyl alcohol), both of which are predominantly, if not exclusively, olfactory nerve stimulants (Doty et al., 1978); and licorice, which, although it combines both trigeminal and olfactory input

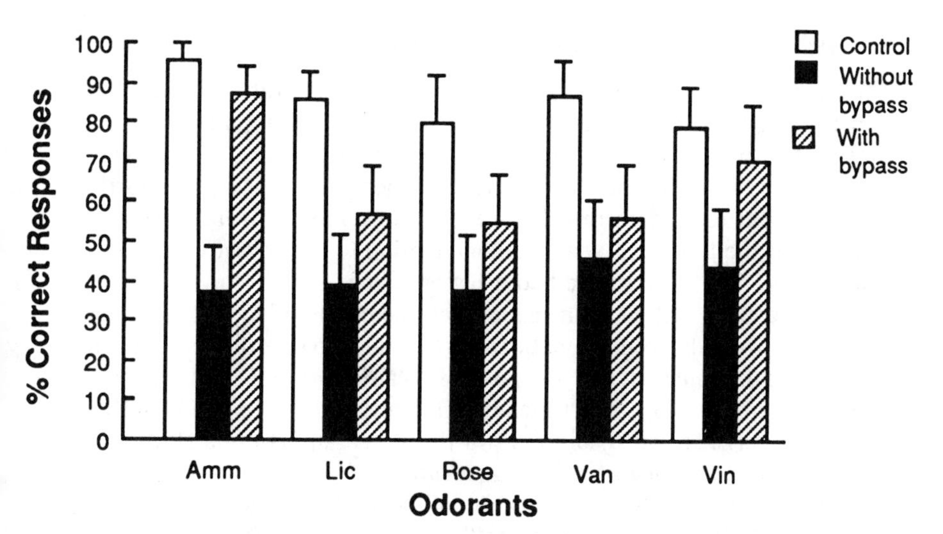


**FIGURE 10** The average percent correct responses (i.e., identifications) given by the laryngectomees for each of the five odorants presented in the confusion matrix. The average percent correct responses for each of the five odorants is also given for the group of nonlaryngectomized subjects. (Schwartz et al., 1987; reproduced by permission of The Laryngoscope Journal.)

(Doty et al., 1978), does not appear to have the strongly irritating sting of ammonia and vinegar.

The results of this study are given in Fig. 10. The ordinate gives the percentage correct and the open bars represent the scores of the nonlaryngectomized group of subjects for each odorant. The black bars and the slanted line bars represent respectively the laryngectomized patients without and with the reestablished air flow afforded by the larynx bypass. Note that in all cases the laryngectomized patients correctly identified the odorants more often with the bypass than without it. However, the improvement was much greater with the bypass for the strong trigeminal irritants (ammonia and vinegar) than for the predominantly olfactory stimuli and licorice. Furthermore, when the laryngectomees used the bypass with the strong trigeminal odorants, there was no statistical evidence for a difference between their percentage of correct scores and the scores of the nonlaryngectomized subjects, whereas for the other odorants there was evidence that the bypass-aided laryngectomees achieved lower scores than did the nonlaryngectomized subjects.

## VII. INTERPRETING THE ODORANT IDENTIFICATION STUDY

Thus using another type of measure and additional odorants, we again appear to have uncovered evidence that the lack of nasal air flow is the sole contributor to the laryngectomy-induced olfactory dysfunction for the trigeminal nerve input but not for the olfactory nerve input. However, further thought about these data raised some questions about this interpretation. Note that if these patients were paying attention to the presence or absence of the sting, they would have had a better chance of choosing the correct strong trigeminal irritant than the correct one of the other odorants since there were two of the former and three of the latter. To take this into account, we replotted the data such that a response was called correct if for the presentation of one of the two strong trigeminal irritants, either one was identified. A response was also called correct if for the presentation of one of the other three odorants, any one of them was identified.

This replotting of the data is shown in Fig. 11. This figure gives the percent correct for the strong "trigeminal" and the predominantly "olfactory" odorants with and without the bypass. Again it appears that the bypass improvement is greater for the trigeminal odorants, but this graph points out

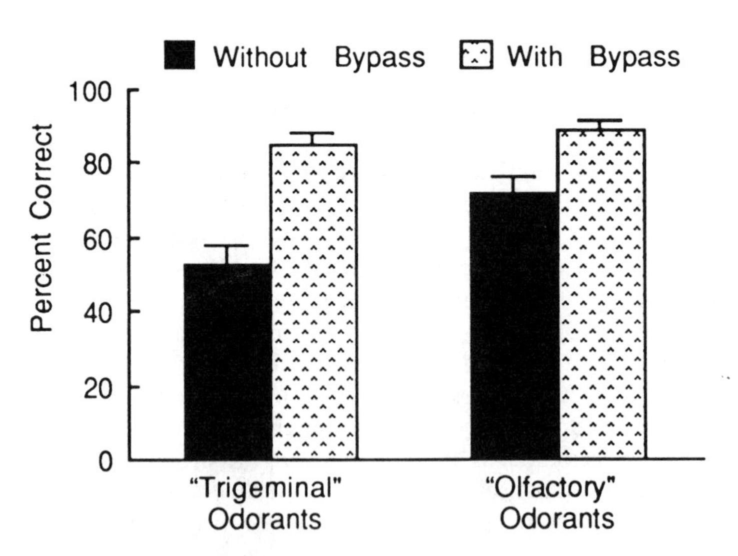


**FIGURE 11** The laryngectomees' average percent responses in the correct category (i.e., "trigeminal" or "olfactory") when using and not using the larynx bypass. See text for further explanation.

that this greater improvement is mostly based on the poorer scores without the bypass for the trigeminal odorants than for the olfactory odorants. Thus, the trigeminal odorants had greater room for improvement. One interpretation of these results is that people expect ammonia and vinegar to be intense and have a stinging quality. If, without the augmented air flow from the bypass, this quality is markedly attenuated for laryngectomized patients, more incorrect identifications of these odorants would ensue. It is well to keep in mind in this regard that many of these patients had been laryngectomees for up to 20 years. Therefore, as many of them complained, their recall of what odorants should smell like may have been somewhat hazy. [Recognizing that licorice is not a strictly "olfactory" odorant (Doty et al., 1978), we generated histograms like Figs. 11 and 12 excluding the licorice responses. The histograms including and excluding licorice showed essentially the same trends.]

Some such interpretation as the above seems consistent with yet another way of looking at the data as shown in Fig. 12. Figure 12, like Fig. 11, displays

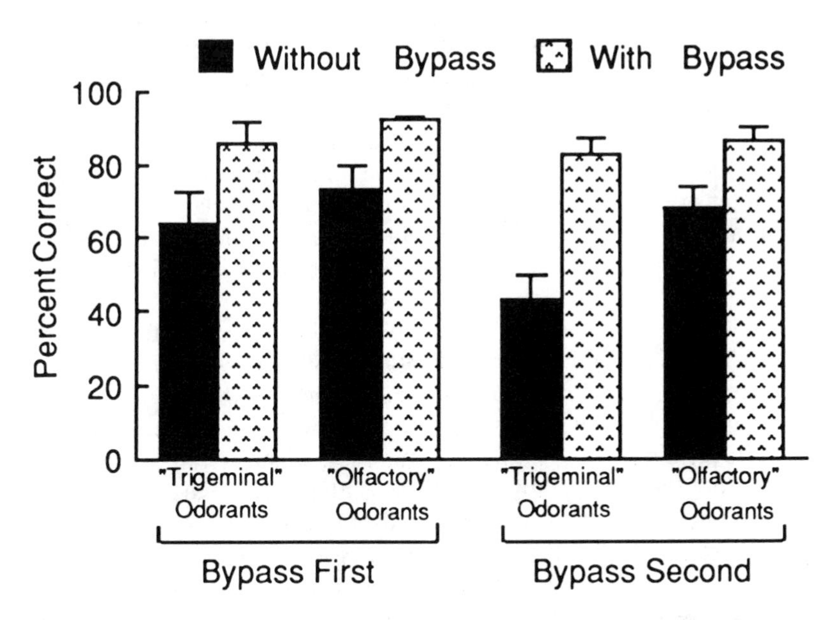


**FIGURE 12** Same as Fig. 11 but keeping separate the patients who were tested first using the bypass (bypass first) from those who were tested first not using the bypass (bypass second). See text for further explanation.

for the trigeminal and olfactory odorants how well the laryngectomees identified the odorants in each category without and with the bypass. This allows a comparison between the two categories as to how much the bypass helped. The difference between Figs. 11 and 12, however, is that the latter separates the scores of those laryngectomees who, for the two bypass conditions, were tested first with the bypass from the scores of those who were tested first without the bypass. (These latter patients are called "bypass second" in Fig. 12.) It can be argued from Fig. 12 that the greater improvement for the trigeminal odorants as compared to the olfactory odorants seen in Fig. 11 was mainly due to the patients who were tested first without the bypass. Indeed, for those who were first tested with the bypass the improvement for the trigeminal odorants was not much different from that for the olfactory odorants. That is, the laryngectomees' ability to identify odorants in the trigeminal category without the bypass was not much different from their ability to identify odorants in the olfactory category when the bypass was used first. (Note that for the olfactory odorants the improvement was about the same no matter which bypass conditions came first.) Perhaps without the augmented air flow of the bypass the intense stinging quality of ammonia and vinegar were attenuated enough as to not match the patient's expectations, thus making their identification as trigeminal odorants less probable. This would be particularly true for the patients first experiencing the listed odorants in the test battery without the bypass since they would be approaching the test from a history of very limited olfactory experience. On the other hand, those patients who first experienced the battery of odorants with the bypass had a chance to become more familiar with them. Perhaps they familiarized themselves with the other qualities of ammonia and vinegar besides the high intensity and irritation sting, so that when they later confronted them without the bypass, identification was more probable. That is, experience with odorants may allow them to be more readily identified with reduced cues.

Obviously, the interpretation we have been giving above for the identification of odorants by laryngectomized patients with and without the bypass needs a number of disclaimers. For instance, we have invoked a number of concepts, such as odorant expectancy, odorant memory, reduced odorant cues, and "trigeminal" vs. "olfactory" qualities, which certainly need clarification before they can rightly be used as explanations. Furthermore, we have left a number of experimental loose ends. For instance, we did not know whether laryngectomized patients with the bypass sense the sting of ammonia whereas without the bypass they do not. When this experiment was done this whole topic was not a major issue, so that such determinations were not made. Thus, for a variety of reasons, we think of this work as preliminary. However, we do suggest that the laryngectomized patient can offer new dimensions to olfactory research. Among the intriguing possibilities is a nasal air flow which can be turned on and off with the larynx bypass and a long-term history of compromised odorant stimulation which, again by using this larynx bypass, can be reversed in seconds.

## ACKNOWLEDGMENT

The work reported in this chapter from the SUNY Upstate Clinical Olfactory Research Center was supported by NIH grant NS 19658.

## REFERENCES

DeBeule, G., and Damste, P. H. (1972). Rehabilitation following laryngectomy; the results of a questionnaire study. *Br. J. Disorders Commun.* 7: 141-147.

Diedrich, W. M., and Youngstrom, K. A. (1966). *Alaryngeal Speech.* Charles C Thomas, Springfield, IL.

Dixon, F. D., Hoerr, N. L., and McCall, J. W. (1949). The nasal mucosa in laryngectomized patients. *Ann. Otol. Rhinol. Laryngol.* 58: 535-547.

Doty, R. L., Brugger, W. E., Jurs, P. C., Orndorff, M. A., Snyder, P. J., and Lowry, L. D. (1978). Intranasal trigeminal stimulation from odorous volatiles: Psychometric responses from anosmic and normal humans. *Physiol. Behav.* 20: 175-185.

Gilchrist, A. G. (1973). Rehabilitation after largynectomy. *Acta Otolaryngol.* 75: 511-581.

Henkin, R. I., Hoye, R. C., Ketcham, A. S., and Gould, W. J. (1968). Hyposmia following laryngectomy. *Lancet* 2: 479-481.

Henkin, R. I., and Larson, A. L. (1972). On mechanisms of hyposmia following laryngectomy in man. *Laryngoscope* 82: 836-843.

Hoye, R. C., Ketchan, A. S., and Henkin, R. I. (1970). Hyposmia following paranasal sinus exenteration or laryngectomy. *Am. J. Surg.* 120: 485-491.

Laing, D. G. (1982). Characterization of human behavior during odor perception. *Perception* 11: 221-230.

Laing, D. G. (1983). Natural sniffing gives optimal odor perception for humans. *Perception* 12: 99-117.

Mozell, M. M., Hornung, D. E., Leopold, D. A., and Youngentob, S. L. (1983). Initial mechanisms basic to olfactory perception. *Am. J. Otolaryngol.* 4: 238-245.

Mozell, M., Schwartz, D., Leopold, D., Youngentob, S., and Hornung, D. (1985). Reversal of hyposmia in laryngectomized patients. *Chem. Senses* 10: 4. (Abstr.)

Mozell, M., Schwartz, D. N., Youngentob, S. L., Leopold, D. A., Hornung, D. E., and Sheehe, P. R. (1986). Reversal of hyposmia in laryngectomized patients. *Chem. Senses* 11: 397-410.

Read, G. F. (1961). The long term follow-up care of laryngectomized patients. *J. Am. Med. Assoc.* 175: 980-985.

Ritter, F. N. (1964). Fate of olfaction after laryngectomy. *Arch. Otolaryngol.* 79: 169-171.

Schwartz, D. N., Mozell, M. M., Youngentob, S. L., Leopold, D. L., and Sheehe, P. R. (1987). Improvement of olfaction in laryngectomized patients with the larynx bypass. *Laryngoscope* 97: 1280-1286.

Tatchell, R. H., Lerman, J. W., and Watt, J. (1985). Olfactory ability as a function of nasal airflow volume in laryngectomees. *Am. J. Otolaryngol.* 6: 426-432.

Toppozada, H. H., and Gaarfar, H. A. (1974). Human nasal ser-mucinous glands after permanent tracheostomy. Electron-microscope study. *ORL* 38: 299-305.

Wright, H. N. (1987). Characterization of olfactory dysfunction. *Arch. Otolaryngol.* 113: 163-168.

# Chapter 5 Discussion

**Dr. Cain:** We see practice effects in thresholds. When we test people over 3-4 days, they get better. In one study we published a couple of years ago, they went down by an average of 13-fold. So I wondered if you thought of running these people repeatedly on the vanillin to see if they would get better.

**Dr. Mozell:** You run into a problem with laryngectomized people—they are uncooperative. It is difficult to ask them to do that. However, you raise an important question: which one of the things we did first, the bypass or the no-bypass. I did not go into the whole other set of data that will come out in a published report, where practice plays a role. But it showed up rather heavily in the part of the study which had to do with the identification of odors, and it depended very strongly on what condition they had first, as to how well they could identify the odor. So practice does play an important role as you will see in the published report.

**Dr. Tepper:** A question about some of your data which looked very interesting in that there were a number of laryngectomized subjects that seemed to be more sensitive than the normal subjects. Can you comment on that please?

**Dr. Mozell:** I wish I could. Those happen to be the people who could get at least some moderate amount of air into the nose, and we made the point that it might be just enough. I think they were getting flow rates of 12 liters per minute compared to the normal of 27. But maybe 12 is enough. We made the point that if we just brought it back to 12 instead of the normal amount it would be sufficient. I don't think anyone was better than the normal range, but they fell in the normal range. There were four that were better than normal average but not normal range. It may very well be that they were among those who reached the flow level needed. It's amazing how much air some of these people can get into their noses by a number of maneuvers. They can also swallow air into their stomachs keeping their mouths closed. I didn't mention the fact there was a positive correlation between how well they did and how long they were laryngectomized. It wasn't an important point to this talk, but it is true, the longer they were laryngectomized, the better they got. It may be that they learn a lot of these techniques and part of the thing we are doing now is teaching new laryngectomized patients how to get air into their nose.

# 6
# Responses of Normal and Anosmic Subjects to Odorants

**James C. Walker and John H. Reynolds IV**
R. J. Reynolds Tobacco Co.
Winston-Salem, North Carolina

**Donald W. Warren and James D. Sidman**
University of North Carolina
Chapel Hill, North Carolina

## I. INTRODUCTION AND HISTORICAL OVERVIEW

Any attempt to elucidate the biological bases of the responses of animals or humans to odorants must take into account at least one afferent pathway in addition to the olfactory. The evidence for the importance of nonolfactory pathways in the responses to odorants was reviewed by Tucker (1963b, 1971), Silver and Maruniak (1981), and Keverne et al. (1986). In all vertebrates some branches of the ophthalmic and maxillary divisions of the trigeminal nerve respond to odorants. In some animals, the vomeronasal (accessory olfactory) system must also be considered. There has been little investigation of the possible role of this system in the behavioral responses to odorants. It is not present in adult humans or birds, the primary subjects of this chapter, although it is found in amphibians, reptiles, and several mammalian orders (see Wysocki, 1979, for a review of the phylogenetic distribution of the vomeronasal system). As discussed by Silver and Maruniak (1981), there is much speculation concerning the role of the septal organ of Masera and the terminal nerve in the response to odorants, but there is little direct evidence in support of this idea. In practice, then, almost all of the information concerning the nonolfactory mediation of responses to vapor phase odorants has dealt with the trigeminal nerve and that will be the major focus of this chapter.

Following a brief historical overview, odor psychophysical experiments designed to elucidate odorant responses of the trigeminal nerve by comparing the responses of normal and anosmic subjects will be discussed. Recent work with humans, which illustrates the value of recording respiratory responses to odorants and shows the effects of odorant stimulation of the eyes, will then be described. Some suggestions as to how some long-standing questions in this area may be answered unequivocally are included at several points in the chapter.

Most of the research efforts in this area are concerned with the following questions: (a) To what compounds is the trigeminal nerve sensitive? (b) What biophysical or biochemical mechanisms underlie trigeminal stimulation? (c) Can the trigeminal nerve support the qualitative discrimination between odorants matched for apparent intensity? (d) What perceptual qualities and/or reflexive responses to odorants may be linked exclusively to trigeminal stimulation? (e) What role does the stimulation of ocular receptors play in response to odorants? and (f) What are the neural bases of olfactory-trigeminal interaction (both peripheral and central)?

As is clear from Tucker's (1971) review, significant progress was made on most of these questions in the 1920s and 1930s. W. F. Allen performed a variety of experiments in which he assessed the relative importance of the olfactory and trigeminal nerves in the control of reflexive responses to inhaled odorants. He compared the responses of dogs in which the olfactory nerve had been transected with those in which nasal trigeminal input was removed. In some cases the vagus and other nerves in the respiratory airway were severed along with the nasociliary and maxillary nerves of the trigeminal system. These other nerves appeared to play a less important role in respiratory reflexes.

Although Allen provided little information on the intensity of the odorants presented to his subjects, he was able to draw some qualitative conclusions about the relative stimulatory effectiveness of different odorants for the olfactory and trigeminal systems. Many compounds were at least as stimulatory for the trigeminal as for the olfactory nerve, although several odorants (e.g., extracts of orange, oil of cloves, lavender, anise, asafetida) appeared to stimulate only the olfactory nerve. Odorants that stimulated only the olfactory nerve caused much smaller changes in respiration and the responses that were observed were readily abolished by anesthesia (Allen, 1936). Almost all compounds that were stimulatory for the trigeminal nerve also stimulated the olfactory nerve. Allen (1929a,b) also conducted a limited amount of work with humans. One such subject, rendered anosmic by skull fracture, could still recognize some odorants and also exhibited respiratory responses to others which he did not detect consciously.

The other pioneer of this time was Elsberg, who worked exclusively with human subjects. He first used a blast injection technique (Elsberg et al., 1935a) to present odorants to the nose but later (Elsberg et al., 1935b) developed a technique whereby the subjects breathed through the mouth and were presented with a continuous stream of odorant into the nose. He had normal subjects, through introspection, rate the degree of "trigeminal" sensations (e.g., stinging, pain, cooling) from different odorants. Based on the results of these tests, he concluded that the following odorants stimulated only the olfactory nerve: coffee, musk ketone, phenylethyl alcohol. Like Allen, Elsberg did not provide information about the intensity of the odor stimuli which he used.

Elsberg used this stream injection technique to measure the responses of both normal and anosmic human subjects. Apparently, he did not verify that these anosmics were unable to detect the three odorants which he had concluded, from his work with normal subjects, were pure olfactory stimuli. Interestingly, anosmic subjects could discriminate, apparently qualitatively, between different chemicals. For example, ether could be discriminated from ammonia and benzaldehyde could be discriminated from xylol.

Until the 1970s, intermittent research on the role of the trigeminal nerve in odor perception, following the efforts of Allen and Elsberg, was largely limited to electrophysiological investigations. Following an early report by Beidler and Tucker (1955), Tucker (1963a,b) compared simultaneously the sensitivity of the olfactory and trigeminal nerves to a large number of odorants. Although he reported electrophysiological responses of the olfactory and/or trigeminal nerve in many animals, the comparisons of olfactory and trigeminal sensitivity were based on the gopher tortoise. He reported that for most compounds the sensitivity of the olfactory exceeded that of the trigeminal. The greatest difference in sensitivity was seen with amyl acetate, which stimulated the olfactory nerve at concentrations at least two log units below the trigeminal threshold. For a few odorants, trigeminal responses appeared at lower concentrations than did olfactory; the clearest example of this was benzylamine. Tucker's approach of simultaneous olfactory-trigeminal recording allowed him to observe directly the role of odorant stimulation of the trigeminal nerve on autonomic effectors in the nasal cavity. Trigeminal stimulation caused, through the cervical sympathetic nerve, changes in the access of odorants to the olfactory receptors and, through the greater superficial petrosal nerve, changes in the flow of nasal secretions. These results clearly show several mechanisms, in addition to those illustrated by Allen and Elsberg, through which the trigeminal system could play a role in determining the responses to odorants.

Stone and colleagues provided yet another way in which the trigeminal nerve could play a role in responses to odorants. They measured several physiological responses in rabbits while manipulating the excitatory state of the trigeminal system. In general, these experiments suggest that blockade of the trigeminal input increases the response of the olfactory bulb (Stone et al., 1968) while trigeminal stimulation decreases olfactory bulb excitability (Stone, 1969). Stone (1969) suggested that the trigeminal system may, through pathways involving the reticular formation or prepyriform cortex, either inhibit or facilitate the response of the olfactory system to odorants. Given the findings of Allen, Elsberg, and Tucker, the magnitude of such an influence would be expected to differ widely across odorants.

Modern odor psychophysical studies in this area have been motivated by two major goals. For one, information from psychophysical studies of trigeminal odorant sensitivity is important in interpreting odor psychophysical data from experiments on normal subjects. For each odorant, it *may* be possible to determine that in normal subjects responses to concentrations below the trigeminal psychophysical thresholds are not influenced by trigeminal stimulation. For this purpose it is necessary only to determine the absolute sensitivity to odorants in animals in which olfactory (and, with some species, the vomeronasal) input has been removed. Clearly such information is a necessary component of research programs in which the goal is to relate odor psychophysical performance to the structure and/or function of the olfactory system alone.

A second goal of odor psychophysical studies of the trigeminal system is to fully characterize the responses of the trigeminal system to odorant stimulation and to determine how elaborate its potential role is in odor perception under normal, "real-world" conditions. Experimental approaches motivated by this goal include both the measurement of physiological responses and a thorough characterization of the kinds of sensations resulting from trigeminal stimulation. It will be apparent from the following summary of human and animal psychophysical studies that experiments in this latter category, though likely to provide extremely valuable information, are seldom conducted.

## II. ANIMAL STUDIES

As pointed out by Graziadei and Okano (1979) and Walker et al. (1979), the pigeon is an ideal subject for use in psychophysical studies of the trigeminal responses to odorants. Adult forms have no vomeronasal system and the olfactory nerves may be easily transected or resected with no damage to the nasal cavity or olfactory bulb and with no vascular damage. Experiments by

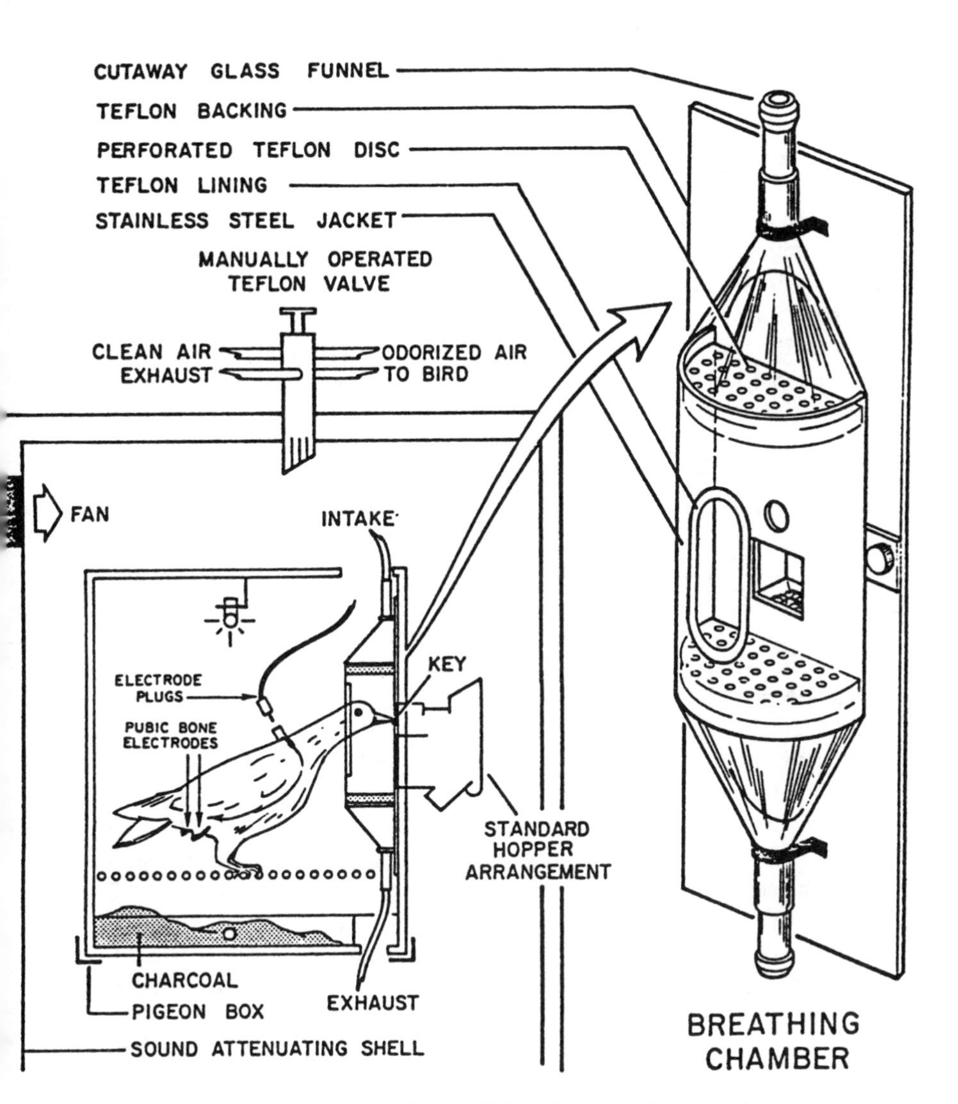


**FIGURE 1** Apparatus for using the conditioned suppression technique to measure psychophysical responses to odorants in pigeons. After training, the introduction of odorized air into the breathing chamber causes the pigeon to suppress its food-reinforced key-pecking behavior. (From Recovery of structure and function following transection of the primary olfactory nerves in pigeons, by N. Oley, R. S. DeHan, D. Tucker, J. C. Smith, and P. P. C. Graziadei, 1975, Journal of Comparative and Physiological Psychology, Vol. 88, p. 477. Copyright 1975 by American Psychological Association. Reprinted by permission.)

Henton et al. (1969) suggested that the trigeminal nerve in this animal became increasingly sensitive to amyl acetate with continued postoperative (olfactory nerve transection) testing. These authors also reported that anosmic pigeons could discriminate between amyl acetate and butyl acetate, when each odorant was presented at the same concentration ($10^{-1}$ vapor saturation).

Both of these key findings prompted Walker et al. (1979) to replicate this work with one key exception: Instead of a simple transection, a radical resection, in which most of the 1 cm length of the olfactory nerve was removed, was performed. This was necessary since, after Henton's work was completed, it became clear that the olfactory nerve in the pigeon and every other animal that has been examined replaces itself following surgery or other damage (see Graziadei and Graziadei, 1978 for review). Pigeons were first trained using the same technique and apparatus (see Fig. 1) as was used in previous work (Henton et al., 1969; Oley et al., 1975) to suppress key-pecking behavior in the presence of odor stimuli. A suppression of food-reinforced key-pecking behavior in response to odor stimuli was produced by the repeated presentation of odorant (usually amyl acetate) followed by mild, unavoidable electric shock. With this procedure, the bird eventually begins to slow or stop his pecking when odorant is presented.

Once each pigeon exhibited reliable suppression to odorant, and not to clean air presentations, successively lower concentrations were presented and the degree of suppression at each concentration was quantified. Following this preoperative threshold determination, olfactory nerve resections were performed and the sensitivity to odorant was again determined. Table 1 summarizes the pre- and postoperative sensitivity of three pigeons to amyl acetate. The odorant concentrations are presented in terms of both parts per million (ppm; v/v) and molarity (M). Postoperative thresholds were elevated by at least 2.6 log units and remained so with repeated testing. The finding that the trigeminal nerve is much less sensitive to amyl acetate than the olfactory nerve is consistent with electrophysiological work by both Tucker (1963a,b) and Silver and Moulton (1982).

**TABLE 1** Psychophysical Sensitivity of Pigeons to Amyl Acetate Before and After Olfactory Nerve Resection

| Subject | Preoperative threshold (ppm/M) | Postoperative threshold (ppm/M) |
|---|---|---|
| 1 | $0.34/10^{-7.82}$ | $219/10^{-5.01}$ |
| 2 | $0.36/10^{-7.79}$ | $200/10^{-5.05}$ |
| 3 | $0.59/10^{-7.58}$ | $479/10^{-4.67}$ |

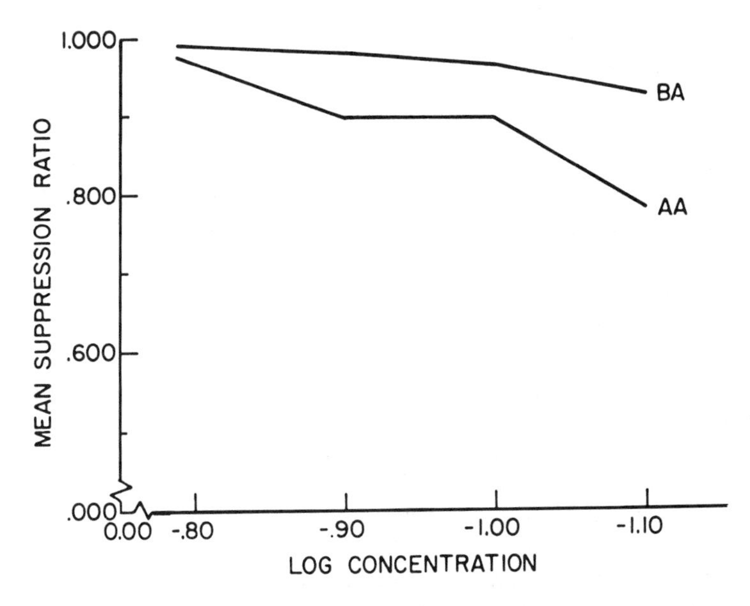


**FIGURE 2** Representative results of odor discrimination testing in anosmic pigeons. The degree of suppression to each odorant at the same four fractions of vapor saturation is shown. (From Walker et al., 1979.)

The results of unsuccessful attempts to establish a true odor quality discrimination between amyl acetate and butyl acetate following olfactory nerve resection are shown in Fig. 2. Data for only one pigeon are shown but similar results were obtained for all three subjects. Note that the two odorants were matched in terms of physical intensity. Although the bird was shocked on only the amyl acetate trials, it showed greater suppression to the butyl acetate. These results are opposite to what would be expected if the two odorants had been discriminable by the anosmic pigeons. Psychophysical work prior to and after this would indicate that the concentration ranges of these two odorants should have overlapped in terms of apparent intensity but that at each fraction of vapor saturation the intensity of the butyl acetate should be greater. Thus the most reasonable interpretation of these results is that the pigeons were not able to discriminate between the two odorants but simply exhibited greater suppression to those stimuli which were higher in apparent intensity. According to this view, the other factor causing lower suppression on amyl acetate trials would be the fact that shock was presented on only 50% of the trials. Work by Mason et al. (1981) first indicated that in the tiger salamander the trigeminal nerve could mediate odor quality discrimination, but this view was not supported by later work (Silver et al., 1986) in which odorants were matched for apparent intensity.

Although the pigeon offers clear morphological advantages over other animals in this regard, psychophysical work in this area has been hindered by the fact that the technique used in the experiment described above (conditioned suppression; see Smith, 1970, for review) is rather tricky to use. Many hours of training are required before a pigeon can reliably discriminate between air and odor trials. Therefore, it would be extremely difficult to use this procedure to, for example, screen a large number of odorants in terms of their stimulatory effectiveness for the trigeminal nerve. For this reason, a procedure for using classical conditioning of cardiac acceleration as a tool to measure odor sensitivity in this model animal was developed (Walker et al., 1986). The animal was simply placed in a chamber similar to that used for the conditioned suppression work and given odorant presentations paired with mild, unavoidable electric shock. Figure 3 shows the type of data that can be obtained. Absolute thresholds to the four odorants shown in this figure were obtained before and after radical olfactory nerve resection. For those odorants that have been tested using this approach and the conditioned suppression technique, both the normal and postoperative thresholds are quite similar. However, the training and testing time is greatly reduced with the cardiac acceleration technique. These data also support the idea that the trigeminal nerve is far less sensitive to odorants than is the olfactory.

The approach of comparing odor psychophysical results from normal and anosmic animal subjects would seem to offer a straightforward and valid means of characterizing nonolfactory responses to odorants. At the very least, one might expect that this information could be used to set concentration limits for each odorant below which responding could be attributed entirely to the olfactory system. However, there are at least three reasons for caution in this regard.

For one, these studies employed only psychophysical measures. The experiments by Allen on animals and humans suggest that odorant stimulation of the trigeminal nerve causes reflexive responses and that the subject may not be aware of at least some effects of this stimulation. This idea is consistent with the finding by Cain (1974) that the trigeminal nerve contributes

**FIGURE 3** Comparison of odor sensitivity of pigeons before and after olfactory nerve resection. The magnitude of the increase in heart rate is plotted as a function of odorant concentration in units of log vapor saturation. In each panel, the pair of functions on the left represent the preoperative performance of two subjects and those on the right depict the postoperative sensitivity of these same two pigeons. (From Walker et al., 1986.)

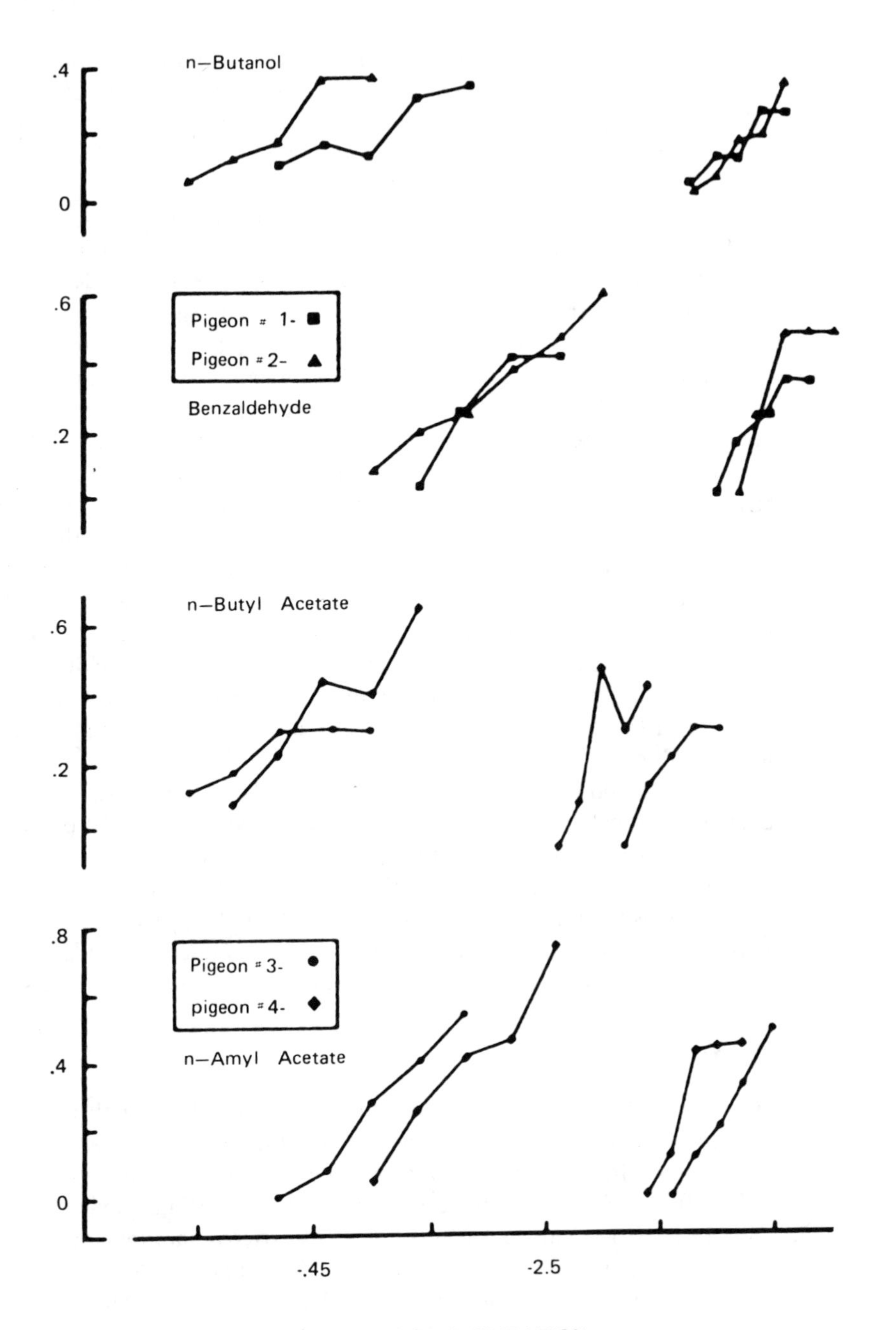
n–Butanol
.4
0
.6
Pigeon # 1-
Pigeon # 2-
Benzaldehyde
.2
n–Butyl Acetate
.6
.2
.8
Pigeon # 3-
pigeon # 4-
n–Amyl Acetate
.4
0
MEAN CARDIAC ACCELERATION
-.45
-2.5
LOG VAPOR SATURATION

to perceived odor magnitude, even at quite low, "nonirritating" odorant concentrations. Based on these limited results, it is probably unwise to conclude that this nerve plays no role (through effects on breathing and/or through neural modulation of the olfactory bulb) in the perception of odor stimuli below the psychophysical threshold of anosmic subjects.

Second, these studies provide little information as to the relative importance of different nonolfactory pathways in mediating behavioral responses to odorants. Walker et al. (1979, 1986), Mason et al. (1981), and Silver et al. (1986) assumed that the response which they observed was due to nasal trigeminal afferents, but this assumption was not directly tested. This could have been done by examining the effect of transection of the ethmoid and nasopalatine branches of the ophthalmic and maxillary divisions (of the trigeminal system), respectively, on psychophysical performance. In all of these studies the odorants could have stimulated ocular trigeminal afferents as well as nerve endings of the facial, glossopharyngeal, and/or vagus (cranial nerves VII, IX, and X) nerves (Widdicombe, 1986). Knowledge of the relative contribution of these different "candidate" pathways is an important component of any attempt to explain psychophysical data from anosmic subjects based on electrophysiological studies of the sensitivity of different nonolfactory pathways. Additionally, this information is important in elucidating the neural mechanisms of both reflexive responses to odorants and possible modulation of the processing of odor information by the olfactory system of normal subjects.

A third shortcoming of the manner in which these animal studies were conducted is that no study of the role of the trigeminal system in the processing of odor information by the olfactory system was included. For example, none of these studies compared suprathreshold scaling of odorant intensity in intact animals with that observed in uni- or bilaterally trigeminectomized animals. Such experiments could be designed to complement and extend Cain's (1974) work with unilaterally trigeminectomized human subjects.

## III. HUMAN STUDIES

All of the drawbacks discussed above can be removed by straightforward changes in the design and conduct of animal odor psychophysical studies. However, regardless of the improvements that are made in the conduct of animal studies, such experiments can only reveal changes in behavioral abilities that result from removal of the olfactory or trigeminal input. One cannot, for example, use animal experiments to determine which odor stimuli cause nasal irritation in humans. Of course, this would not prevent one from deriving empirical relationships between animal responses and human perception, as has been proposed by Alarie and Luo (1986). In order to answer

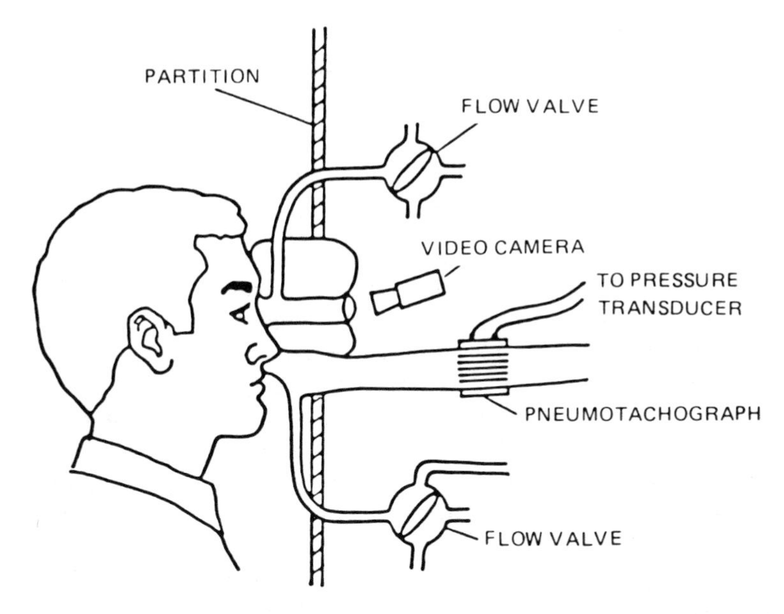


**FIGURE 4** Schematic diagrams of subject in place at custom-fitted odor mask. Separate olfactometers and flow valves in combination with this mask are used to effect separate and independent stimulation of the nose and eyes. A pneumotachograph and pressure transducer are used to record respiratory behavior and a video camera records the responses of the eyes.

fundamental questions about human odor perception and irritation, it will be necessary to carefully integrate psychophysical and physiological data (including, ideally, odorant-evoked potentials) collected on normal human subjects with data that are similarly collected from both anosmic and trigeminectomized human subjects.

A device for the automated measurement of the psychophysical and physiological responses of normal, anosmic, and trigeminectomized subjects was recently developed (Walker et al., 1988). A custom-fitted mask is first made for each subject. In test sessions the subject presses his face into this mask, as shown in Fig. 4. The pneumotachograph placed downstream of the nasal port of the mask is used to measure the subject's breathing. Signals from this device are processed so that a voltage that is linearly related to the instantaneous volume flow rate of the subject's inhalations and exhalations is sent to the computer. The output of one olfactometer is sent through a flow valve to the eye ports of this mask. Similarly, a second, completely independent olfactometer is used to supply the nasal port of the mask. The combined use of a custom-fitted mask and two independent olfactometers makes it possible to present the following four types of trials:

a. Air control—air to both nose and eyes
b. "Eye only"—odor to eyes, clean air to nose
c. "Nose only"—odor to nose, clean air to eyes
d. "Eye + nose"—odor to both nose and eyes

On each trial the subject is in place at the mask for at least 20 sec, throughout which his eyes are videotaped and his respiratory behavior recorded. This 20 sec interval is divided into a "pre" period, when only clean air is presented to the nose and eyes, and a "during" period, in which one of the four conditions above (trial types a-d) is in effect. Data for the "pre" period can be used to determine the average baseline of responding for both eye blinks and respiratory behavior. At the end of each trial the subject uses a "mouse" connected to a microcomputer to enter his responses on the following sensory attribute scales: odor strength, nasal irritation, eye irritation, eye feel, and overall acceptability.

We have observed almost no reports of eye irritation or eye feel on any type of trial and no significant changes in eye blink rates under any condition. As explained below, the failure of these measures to reveal an effect of ocular stimulation is contrasted with clear respiratory effects of ocular stimulation. A group of 16 male nonsmokers (18-30 years of age and with no history of any condition associated with taste or smell dysfunction) served as the normal control group. Selected aspects of the psychophysical and physiological responses of this group to ranges of concentrations of four odorants are presented below.

Figure 5 is a summary of the odor strength ratings of these subjects to each of the four odorants. For trial types b-d, each subject was given four trials at each concentration of each odorant. This figure shows only the trial type c ("nose only") results but the data for the "eye + nose" trials are indistinguishable. Odor strength ratings of the two acids, but not of amyl acetate and nicotine, appear to be simple linear functions of log odorant concentration.

The relationship between the overall acceptability of the air and the other, more specific, attributes was examined. For both "nose-only" and "eye + nose" trial types, the magnitudes of the declines in overall acceptability were predicted much better by increases in odor strength than by increases in nasal irritation. This is illustrated in Fig. 6, in which the responses to the four concentrations of propionic acid for the "eye + nose" trials are shown. With decreasing concentrations, odor strength declines smoothly and nasal irritation shows only minor decreases as overall acceptability increases. This finding, which was seen with all four odorants, may suggest that the major

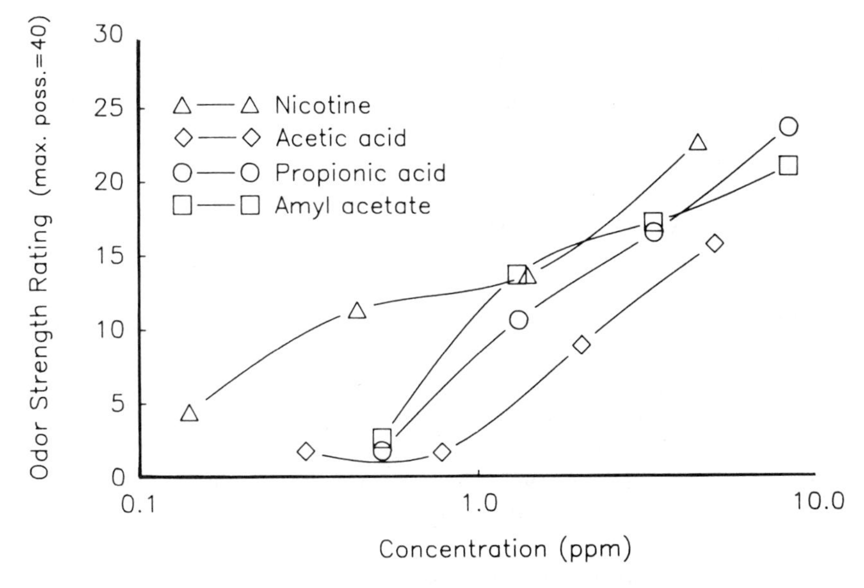


**FIGURE 5** Summary of the odor strength ratings of 16 normal subjects to a range of concentrations of four odorants presented to the nose only.

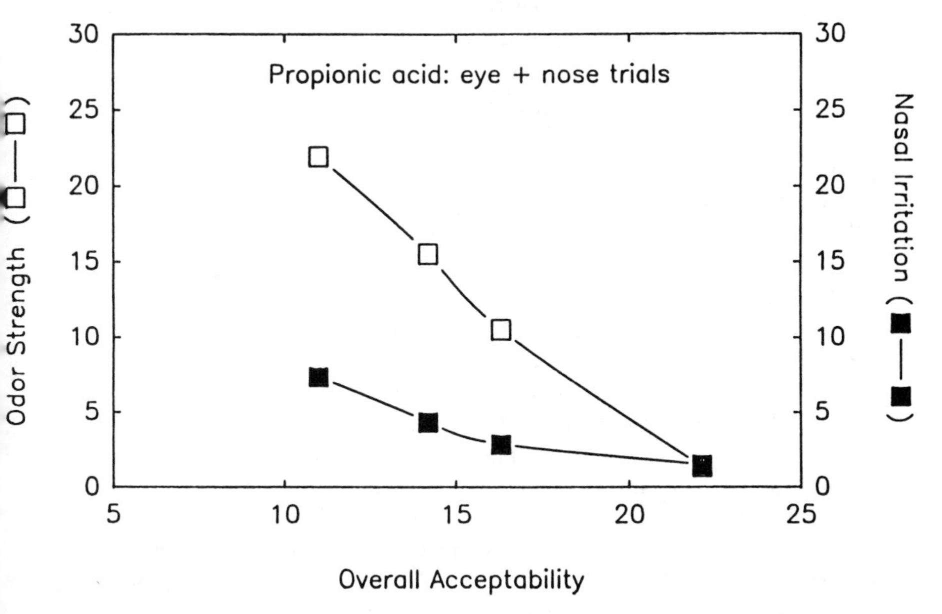


**FIGURE 6** Relationship of both odor strength (□) and nasal irritation (■) to overall acceptability. These three attribute ratings are shown for each of four concentrations of propionic acid presented to both the nose and eyes of normal human subjects.

biological basis of complaints about indoor air environments is the stimulation of the olfactory nerve. Furthermore there may be no level of odor strength that people will tolerate without reporting some decreased overall acceptability.

Although the levels of nasal irritation were always much lower than the odor strength ratings, each concentration which was given a nonzero odor strength rating was also given a nonzero nasal irritation rating. The ratio between nasal irritation and odor strength was lowest for propionic acid and amyl acetate. It should be noted that we used 0% relative humidity (RH) in this work, i.e., clean, dry air was used as the carrier gas. Based on the small and rather inconsistent set of studies of the role of this variable in the perception of odor and/or irritation, one might expect that the lack of humidity could have caused subjects to "overreport" nasal irritation (e.g., McIntyre, 1978, Eng, 1979). There is less reason to suspect that the 0% RH altered odor strength ratings; Tucker's (1963a) work demonstrated that the functioning of the peripheral olfactory system is not altered by variations in the RH from 0 to 80%.

Assuming that this finding (that some nasal irritation is seen even at low concentrations that humans can barely smell) is not due entirely to our use of 0% RH, then how should this result be interpreted? There are at least two ways. One could argue that the perception of nasal irritation does not require stimulation of the trigeminal nerve. This would be consistent with the small amount of neurophysiological and psychophysical data that suggest that the trigeminal is much less sensitive than the olfactory. Alternatively, one could take the position that, given the rather commonly held view that trigeminal stimulation is necessary and/or sufficient for perceptual irritation, the trigeminal nerve is as sensitive as the olfactory to these odorants. It is clear that this issue cannot be resolved simply by more complex psychophysical experiments or by introspective "dissecting out" of the trigeminal components of sensation. Instead, it will likely be necessary to test the responses of normal subjects with those of both anosmic and trigeminectomized subjects tested in the same manner. It would be especially interesting to test uni- or bilaterally trigeminectomized subjects with the same four odorants that we have tested with normals and are testing with anosmics.

An odor-testing apparatus that allows completely separate and independent stimulation of the nose and eyes makes it possible to study, in a well-controlled fashion, the role of ocular stimulation in response to odorants. Additionally, the combined measurement of physiological and psychophysical responses to odorants makes it possible to investigate the biological bases of psychophysical responses and to evaluate the relative merits of each. The data shown in Figs. 7 and 8 illustrate why both of these features are important and they provide some conceptual framework for the simultaneous

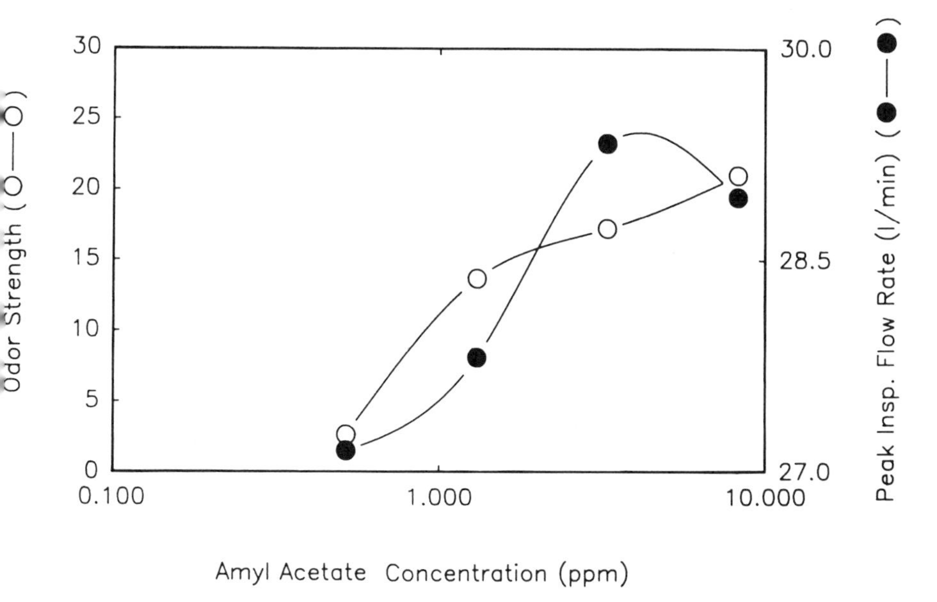


**FIGURE 7** Odor strength and peak inspiratory flow rate as a function of amyl acetate concentrations presented in "nose-only" trials to normal human subjects.

measurement of psychophysical and physiological responses to odorants. Figure 7 shows, for the nose-only trials, both the odor strength ratings and the peak inspiratory flow rate (the maximum inhalation rate during 10 sec odorant presentation) throughout the concentration range of amyl acetate. These results indicate that respiratory responses may be equal in sensitivity to psychophysical ones as a measure of the responses to odorants. They also suggest that the general idea of being able to use nonverbal response measures (e.g., respiratory behavior, eye blink rate) in place of, or in conjunction with, traditional verbal psychophysical responses may have merit. This would require that the relationship between one or more respiratory parameters and psychophysical response first be established for a given odorant or, preferably, a class of odorants. A second, extremely appealing possibility here is that this approach could ultimately make it possible to integrate human and animal psychophysical results in a straightforward manner, i.e., identical physiological measures, collected in humans and animals, could be used to bridge between two currently separate sets of data.

Comparison of the data in Fig. 7 with those in Fig. 8, in which the "eye + nose" data for amyl acetate are shown, provides strong evidence that

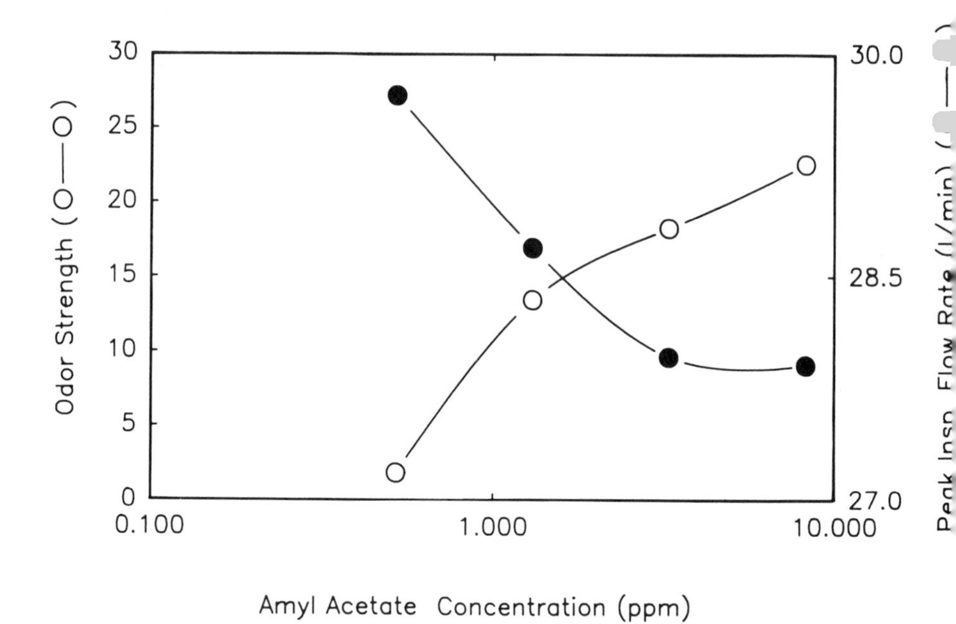


**FIGURE 8** Same as Fig. 7 except that data are from "eye + nose" trials.

corneal chemoreceptors must be taken into account in any attempt to understand human responses to odorants or to relate these responses to animal data. These corneal afferents project to the brain through the ophthalmic branch of the trigeminal nerve. Although the inclusion of ocular stimulation does not affect psychophysical responses to nasal stimulation, differences between the respiratory results in Figs. 7 and 8 can be attributed only to odorant stimulation of the eyes. When ocular stimulation is included, the relationships between peak inspiratory flow rate and both odorant concentration and odor strength ratings are essentially the opposite of those seen on nose-only trials. These data, as well as those collected with other odorants, also indicate that the effects of ocular stimulation are best "unmasked" by testing its effect against the background of nasal stimulation. In many cases where no evidence of eye-only stimulation is seen, resiratory data on nose-only and eye + nose trials are different. This kind of finding is a further indication that caution should be exercised when interpreting (especially in the case of trigeminal chemoreceptors) a lack of psychophysical response to a particular odorant concentration as proof that a particular set of receptors does not normally play a role in the response to that stimulus. These results also illustrate our general finding that, for each odorant, one or more respiratory parameters can usually be used to predict odor strength ratings on nose-only and/or eye + nose trials.

The data presented above illustrate the value of studying both the separate and interactive effects of nasal and ocular stimulation with odorants. The inclusion of physiological responses adds an important dimension to the investigation of the responses of humans to odorants. Some important questions are raised by these experiments with normal subjects. For example:

1. To what degree is the perception of nasal irritation at low concentrations due to stimulation of the trigeminal nerve?
2. What is the relative importance of the olfactory and trigeminal nerves in mediating respiratory responses to odorants?
3. To what degree do odorant-induced changes in respiratory behavior alter the perception of odor and/or irritation?
4. In view of the role of ocular chemoreceptors in the respiratory responses to odorants even at low concentrations, how does the sensitivity of these receptors compare to those of the nasal trigeminal and olfactory receptors?
5. Through what pathways of the central nervous system do ocular chemoreceptors alter respiratory behavior?

There are at least two general strategies that could be used to resolve some of these questions. One is the use of olfactory and trigeminal neurophysiological data when reaching conclusions about the neural mediation of psychophysical and/or physiological responses of humans to odorants. Ideally, for example, one could turn to a (validated) animal model and test its olfactory and trigeminal sensitivity to different odorants. These data could then be used to estimate the ranges of each compound at which olfactory and/or trigeminal stimulation serves as the basis for the responses of humans. A second approach to elucidating the roles of different neural pathways is to interpret the results from normals in light of data, collected in exactly the same way, from both anosmic and trigeminectomized subjects. We recently began to test anosmics with the same four odorants as were used in the work that I have just described. Since only one anosmic was tested for nicotine, data for this odorant will not be included.

The sensitivity of the two anosmics that we tested, as compared to that of the normal controls, is shown in Table 2. These subjects showed virtually no psychophysical evidence of detecting any of these compounds. However, examination of the respiratory responses by these subjects showed that with every odorant this type of response showed much better sensitivity than did the psychophysical measures. That is, nonolfactory receptors in the nose and eyes respond to each of these chemicals and caused respiratory changes at

**TABLE 2** Comparison of Psychophysical Sensitivity of Normal and Anosmic Humans to Three Odorants

| Odorant | Normal (n = 16) thresholds (ppm/M) | Anosmic (n = 2) thresholds (ppm/M) |
|---|---|---|
| Acetic acid | $2.0/10^{-7.05}$ | $527/10^{-4.63}$ |
| Amyl acetate | $1.3/10^{-7.23}$ | $>201/>10^{-5.81}$ |
| Propionic acid | $4.7/10^{-7.2}$ | $>49/>10^{-6.18}$ |

concentrations that were not perceived. This finding is illustrated below through presentation of some of the data for amyl acetate and propionic acid.

Figure 9 shows the overall acceptability results when amyl acetate was the odorant. There is only slight evidence for a change in the overall acceptability measure throughout the concentration range for two of the trial types; the other four psychophysical attributes provided even less evidence that the

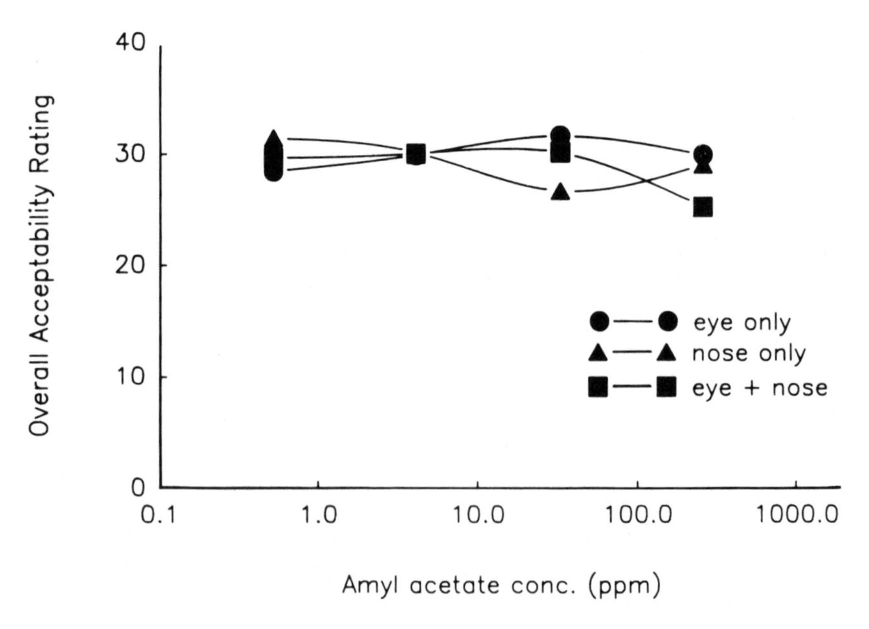


**FIGURE 9** Ratings by two anosmics of the overall acceptability of a range of concentrations of amyl acetate.

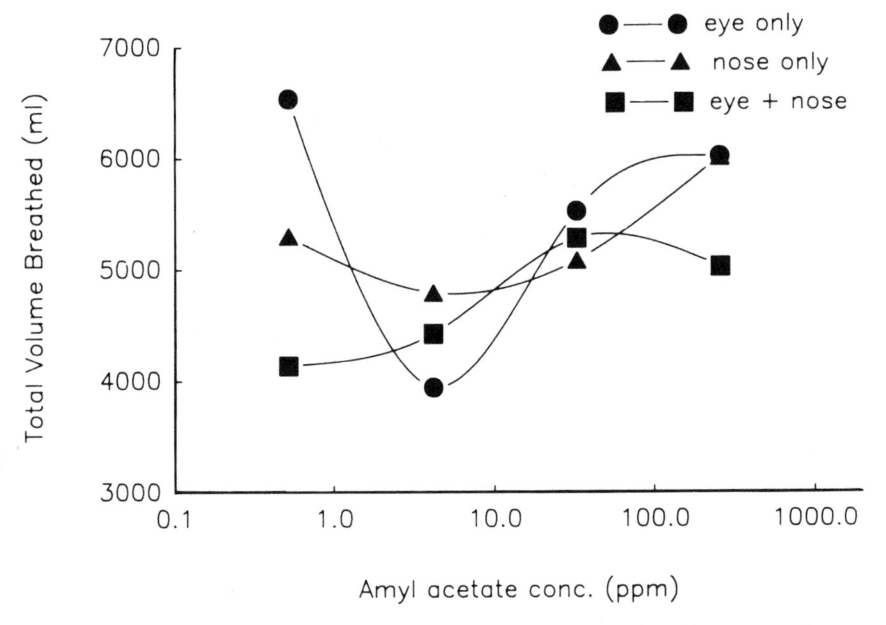


**FIGURE 10** Total volume of air breathed by two anosmics when exposed to a range of concentrations of amyl acetate.

anosmics detected this compound at any concentration. Note that the highest concentration of amyl acetate presented was over 200 times higher than the lowest concentrations that normals can detect. In contrast to the lack of psychophysical responding, clear changes in respiratory behavior are seen throughout the concentration range of this odorant. The total volume of air breathed (volume inspired + volume expired during 10 sec odorant presentation) for different trial types and concentrations of amyl acetate is shown in Fig. 10. Both the eye-only and the nose-only curves are concave upward with a minimum at 4.1 ppm while the eye + nose curve is concave downward with a maximum at 32.8 ppm.

Across odorants and trial types, we found that the changes in different respiratory measures were nonmonotonically related to odorant concentration. For example, Fig. 11 shows the net volume of air breathed (volume inspired − volume expired during 10 sec odorant presentation) when propionic acid was presented. No psychophysical evidence of detection of any concentration of this odorant was obtained with either anosmic. Again there are respiratory responses to odor stimuli at concentrations comparable to the psychophysical thresholds of normal subjects.

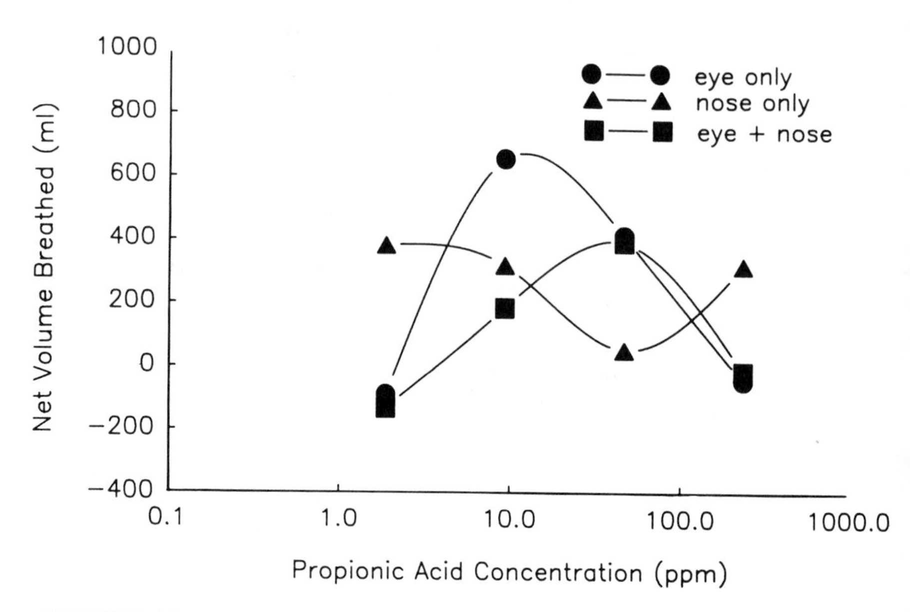


**FIGURE 11** Net volume of air breathed by two anosmics when exposed to a range of (undetected) concentrations of propionic acid.

These data, though limited, do provide strong evidence that in anosmic humans one or more nonolfactory system(s) are stimulated by odorants at concentrations at least as low as the psychophysical thresholds measured in normal subjects. This stimulation results in changes in several aspects of respiratory behavior. Based on the reports of Allen and on more recent work on the neural pathways involved in reflexes originating from the upper respiratory airway (see Widdicombe, 1986 for an excellent review), the most reasonable interpretation of these results is that they were due primarily to stimulation of trigeminal receptors on the surface of the cornea and in the nasal cavity. Measurement of the psychophysical and physiological responses of human or animal subjects lacking trigeminal input from one or both of these locations would provide valuable information for the interpretation of the present data from anosmics. An unequivocal answer in this area may require the testing of subjects that are both anosmic and trigeminectomized.

In future work it is essential to record psychophysical and physiological responses of additional anosmics to these odorants and to relate data from normal and anosmic human subjects to those from subjects that have been uni- or bilaterally trigeminectomized. Such a comparison could be used, for example, to determine the role of the trigeminal system in modulating the processing of odor information by the olfactory system. The recording of odorant-evoked potentials from the scalp, as pioneered by Kobal (see his contribution in this volume), is likely to be especially useful in this effort.

## ACKNOWLEDGMENTS

The authors thank D. B. Kurtz, V. Clayborn, and F. M. Shore for assistance in the construction of the automated human odor testing apparatus. The comments of W. L. Silver on an earlier version of this chapter are greatly appreciated, as is the assistance of R. A. Jennings, K. L. Shore, and M. W. Stancill in the preparation of the manuscript. Portions of the data from normal human subjects were presented at the Tenth Annual meeting of the Association for Chemoreception Sciences, Sarasota, FL.

## REFERENCES

Alarie, Y., and Luo, J. E. (1986). Sensory irritation by airborne chemicals: A basis to establish acceptable levels of exposure. In: *Toxicology of the Nasal Passages.* C. Barrow (Ed.). Hemisphere, New York, pp. 91-100.

Allen, W. F. (1929a). Effect on respiration, blood pressure, and carotid pulse of various inhaled and insufflated vapors when stimulating one cranial nerve and various combinations of cranial nerves. *Am. J. Physiol.* 88: 117-129.

Allen, W. F. (1929b). Effect of various inhaled vapors on respiration and blood pressure in anesthetized, unanesthetized, sleeping and anosmic subjects. *Am. J. Physiol.* 88: 620-632.

Allen, W. F. (1936). Studies on the level of anesthesia for the olfactory and trigeminal respiratory reflexes in dogs and rabbits. *Am. J. Physiol.* 115: 579-587.

Allen, W. F. (1937). Olfactory and trigeminal conditioned reflexes in dogs. *Am. J. Physiol.* 118: 532-540.

Beidler, L. M., and Tucker, D. (1955). Response of nasal epithelium to odor stimulation. *Science* 122: 76.

Cain, W. S. (1974). Contribution of the trigeminal nerve to perceived odor magnitude. *Ann. N. Y. Acad. Sci.* 237: 28-34.

Elsberg, C. A., Brewer, E. D., and Levy, I. (1935a). The sense of smell V. The relative importance of volume and pressure of the impulse for the sensation of smell and the nature of the olfactory stimulation process. *Bull. Neurol. Inst. N.Y.* 4: 264-269.

Elsberg, C. A., Levy, I., and Brewer, E. D. (1935b). The sense of smell VI. The trigeminal effects of odorous substances. *Bull. Neurol. Inst. N.Y.* 4: 270-285.

Eng, W. G. (1979). Survey on eye comfort in aircraft: 1. Flight attendants. *Aviat. Space Environ. Med.* 50: 401-404.

Graziadei, P. P. C., and Graziadei, G. A. Monti (1978). Continuous nerve cell renewal in the olfactory system. In: *Handbook of Sensory Physiology, Vol. IX.* M. Jacobson (Ed.) Springer-Verlag, Berlin, New York, pp. 55-83.

Graziadei, P. P. C., and Okano, M. (1979). Neuronal degeneration and regeneration in the olfactory epithelium of the pigeon following transection of the first cranial nerve. *Acta Anat.* 104: 220-236.

Henton, W. W., Smith, J. C., and Tucker, D. (1969). Odor discrimination in pigeons following section of the olfactory nerves. *J. Comp. Physiol. Psychol.* 69: 317-323.

Keverne, E. B., Murphy, C. L., Silver, W. L., Wysocki, C. J., and Meredith, M. (1986). Non-olfactory chemoreceptors of the nose: Recent advances in understanding the vomeronasal and trigeminal systems. *Chem. Senses* 11: 119-133.

Mason, J. R., Meredith, M., and Stevens, D. A. (1981). Odorant discrimination by tiger salamanders after combined olfactory and vomeronasal nerve cuts. *Physiol. Behav.* 27: 125-132.

McIntyre, D. A. (1978). Response to atmospheric humidity at comfortable air temperature: A comparison of three experiments. *Ann. Occup. Hyg.* 21: 177-190.

Oley, N., DeHan, R. S., Tucker, D., Smith, J. C., and Graziadei, P. P. C. (1975). Recovery of structure and function following transection of the primary olfactory nerves in pigeons. *J. Comp. Physiol. Psychol.* 88: 477-495.

Silver, W. L., Arzt, A. H., and Mason, J. R. (1987). Trigeminal chemoreceptors cannot discriminate between equally intense odorants. In: *Olfaction and Taste IX*. S. D. Roper and J. Atema (Eds.). Annals of the New York Academy of Sciences, Vol. 510, New York, pp. 616-618.

Silver, W. L., and Maruniak, J. A. (1981). Trigeminal chemoreception in the nasal and oral cavities. *Chem. Senses* 6: 295-305.

Silver, W. L., and Moulton, D. G. (1982). Chemosensitivity of rat nasal trigeminal receptors. *Physiol. Behav.* 28: 927-931.

Smith, J. C. (1970). Conditioned suppression as an animal psychophysical technique. In: *Animal Psychophysics*. W. C. Stebbins (Ed.). Academic Press, New York, pp. 125-159.

Stone, H. (1969). Effect of ethmoidal nerve stimulation on olfactory bulbar electrical activity. In: *Olfaction and Taste III*. C. Pfaffman (Ed.). Rockefeller University Press, New York, pp. 216-220.

Stone, H., Williams, B., and Carregal, E. J. A. (1968). The role of the trigeminal nerve in olfaction. *Exp. Neurol.* 21: 11-19.

Tucker, D. (1963a). Physical variables in the olfactory stimulation process. *J. Gen. Physiol.* 46: 453-489.

Tucker, D. (1963b). Olfactory, vomeronasal and trigeminal receptor responses to odorants. In: *Olfaction and Taste. I*. Y. Zotterman (Ed.). Pergamon Press, New York, pp. 45-69.

Tucker, D. (1971). Nonolfactory responses from the nasal cavity: Jacobson's organ and the trigeminal system. In: *Handbook of Sensory Physiology, Vol. IV, Chemical Senses, Part 1, Olfaction*. L. M. Beidler (Ed.). Springer-Verlag, Berlin, New York, pp. 151-181.

Walker, J. C., and Kurtz, D. B. (1988). Psychophysical and physiological responses of humans to odorant stimulation of the nose and eyes. Paper presented at the Tenth Annual Meeting of the Association for Chemoreception Sciences.

Walker, J. C., Tucker, D., and Smith, J. C. (1979). Odor sensitivity mediated by the trigeminal nerve in the pigeon. *Chem. Senses Flav.* 4: 107-116.

Walker, J. C., Walker, D. B., Tambiah, C. R., and Gilmore, K. S. (1986). Olfactory and nonolfactory odor detection in pigeons: Elucidation by a cardiac acceleration paradigm. 38: 575-580.

Walker, J. C., Kurtz, D. B., Shore, F. M., and Clayborn, V. (1988). Automated human odor testing. Paper presented at the Tenth Annual Meeting of the Association for Chemoreception Sciences.

Widdicombe, J. G. (1986). Reflexes from the upper respiratory tract. In: *Handbook of Physiology. Sect. 3: The Respiratory System, Vol. II Control of Breathing, Part 1.* N. S. Cherniack and J. G. Widdicombe (Eds.). Waverly Press, Baltimore, pp. 363-394.

Wysocki, C. J. (1979). Neurobehavioral evidence for the involvement of the vomeronasal system in mammalian reproduction. *Neurosci. Biobehav. Rev.* 3: 301-341.

# Chapter 6 Discussion

**Dr. Wysocki:** Jim, you showed us some very interesting data for the human psychophysical responses. I have four points I want to address and perhaps you can clarify them. One, you said if they can smell anything, there is a decrease in acceptability, and I know it was for proprionic acid. Do you mean to generalize that to all substances?

**Dr. Walker:** Only to those four.

**Dr. Wysocki:** Two, you seem to equate strength with detection.

**Dr. Walker:** If it was zero odor intensity. We ask odor strength or intensity, nasal irritation, eye irritation, we even ask something called eye feel—just in case there is some sensation in their eye that they are not calling irritation.

**Dr. Wysocki:** When they say there is no strength you are equating that with nondetection?

**Dr. Walker:** Yes.

**Dr. Wysocki:** Do you think that's appropriate? Do you think you could drive those thresholds lower using a forced-choice method, for example?

**Dr. Walker:** It is possible. I would agree. You can always say it is possible to drive them down with some procedure.

**Dr. Wysocki:** My third point is (did I get it right?) you have two patients you looked at that were anosmic?

**Dr. Walker:** That's all so far.

**Dr. Wysocki:** How many observations per point on those patients? Is there some measure of variation around those points so that we could assess the variation within the curves you have shown us?

**Dr. Walker:** We have computed standard errors—they do not overlap in any case where I was making a point. There are eight measurements per concentration for each trial type.

**Dr. Wysocki:** And a point of clarification, on one of the figures you showed a net volume flow for proprionic acid. That would suggest that the patients are exhaling?

**Dr. Walker:** That's simply during the 10 sec stimulation when they are getting the odorant; it is simply quantity of air inhaled minus quantity exhaled, so that's what net is.

**Dr. Kurtz:** You made the point that odor strength relates to overall acceptance. What would you predict for an odorant whose predominant characteristic was not one of irritation as opposed to odor, like ammonia?

**Dr. Walker:** Then I would start to talk like Bill Cain and say whatever is the biggest is the most important. If the nasal irritation predominates I would say that it will certainly do the job, that it's going to start driving down the acceptability.

**Dr. Eccles:** I found the paper very stimulating and I have a whole host of questions, but I'm not going to ask all of them. One thing I am interested in is that you found the total volume breathed changed presumably before the persons could detect an odorant. Did you notice any changes in the frequency of respiration or the tidal volume?

**Dr. Walker:** We don't even write frequency to the file. We should probably change that, but frequency has never shown us anything. You have to remember it is short-term stimulation.

**Dr. Eccles:** Are you recording over a minute?

**Dr. Walker:** No. Only 10 sec.

**Dr. Eccles:** Only 10 sec? The other point was about the net volume when you were looking at inspired and expired. We would expect if you are not correcting for changes in temperature and saturation, that you would get a 10% reduction in volume anyway.

**Dr. Walker:** That would be true on every trial.

**Dr. Eccles:** You would expire 10% less than you inspired.

**Dr. Walker:** Right, I see what you're saying. The point I'm making is that it would not be dependent on the concentration of stimulus that you gave.

**Dr. Alarie:** I am tremendously worried about your data in terms of the change that you claim occurs with respiration. You show that the psychophysical response ranges from 1 to 15, and that's a big change. Then the changes in respiration shown on the slide look very impressive, but you have changed the axis on us. The change you have is actually only 10%. You just plot from 30 to 36, not 0 to 36. The second thing is that the presentation of the substance is for such a short period of time, 10 sec. Humans can modify respiration so easily from outside stimuli without having to imply that this is stimulating the trigeminal nerve or olfactory nerve or anything else. You just present something to me and I change my pattern of breathing just because you present something to me. I would like to see the presentation of the stimulus be more continuous so that you can achieve a steady state against which to measure the change of respiration.

**Dr. Walker:** Well, trial type was randomized and concentration was randomized, so I'm not sure how simply giving it to them longer would necessarily allow you to eliminate external variation.

# 7

# Brain Responses to Chemical Stimulation of the Trigeminal Nerve in Man

**Gerd Kobal and Thomas Hummel**
University of Erlangen-Nürnberg
Erlangen, Federal Republic of Germany

## I. INTRODUCTION

Based on the experiments of Finkenzeller (1966) and Allison and Goff (1967), who first described cerebral evoked potentials in man to chemical stimuli of the nasal mucosa, we developed a technique (Kobal and Plattig, 1978) by which odorants were presented in a constantly flowing moist and warm air stream, thus avoiding undesired stimulation of thermo- and mechanoreceptors, while steep stimulus onsets were maintained ($<20$ msec). Although the avoidance of excitation of other than chemosensitive receptor cells was guaranteed, it remained at first difficult to decide whether chemoreceptors of the olfactory or the trigeminal nerve had been excited. The fact that no potentials could be elicited when anosmics were stimulated with vanillin and phenylethyl alcohol made it possible to establish the origin of these cortical chemosensory evoked potentials (Kobal, 1982). These findings were concordant with the results of Doty et al. (1978), who reported that these substances were not perceived by anosmics. A further approach to the problem of how to differentiate between olfactory and somatosensory responses was to choose odorless carbon dioxide as the stimulus in order to obtain a monomodal stimulation of the nerve endings of the trigeminal nerve. Carbon dioxide applied to the nasal mucosa produces a sharp localized pain without any olfactory sensations. Of 60 anosmics who were investigated, all yielded responses to carbon dioxide, which we have since defined as chemosematosensory evoked potentials (CSSEP) (Kobal and C. Hummel, 1988) in order to distinguish them from olfactory evoked potentials (OEP).

Recently we succeeded in obtaining a further clue in order to differentiate between olfactory and somatosensory stimuli. That is, we were able to prove that after randomized dichotomic stimulation of both nostrils, it is possible to distinguish between two kinds of chemical stimulants: (a) those after application of which subjects are able to localize the stimulated side, and (b) those after application of which the subjects are unable to localize the stimulated nostril. Belonging to the first group are carbon dioxide, menthol (Kobal et al., 1987), ammonia, and sulfur dioxide (Kobal and Hummel, unpublished). Belonging to the second group are vanillin, acetaldehyde (in low concentrations), and hydrogen sulfide, i.e., substances which cannot be perceived by anosmics. From these results we concluded that for single momentary odorous sensations, directional olfactory orientation can only be assumed when the olfactory stimulants simultaneously excite the trigeminal somatosensory system. In other words, directional smelling exclusively mediated by the olfactory nerve does not exist. On the other hand, it was shown without a doubt that the trigeminal nerve conveys information about the location of a source of smell (Kobal et al., 1989).

As early as 1979 there were indications that the topography of chemosensory evoked potentials reveals a difference between olfactory and trigeminal responses (Plattig and Kobal, 1979). Indeed, we are now in a position to distinguish between olfactory and chemosomatosensory responses by utilizing evoked potential mappings. The differences were concordant with the grouping of the stimulants in relation to their property of being locatable. Nonlocatable stimuli yield a typical olfactory distribution, whereas locatable stimuli yield a typical somatosensory distribution (Kobal et al., 1987).

Prior to the employment of a novel biosignal, it is expedient to study all its characteristic features. To meet this prerequisite we endeavored in recent years to (a) establish the cortical generator of the CSSEPs by using magnetoencephalography (Huttunen et al., 1986), (b) characterize their dependence on the stimulus intensity (Müller, 1988), (c) describe their adaptation and habituation (Hummel and Kobal, unpublished), and (d) investigate new ranges in which the CSSEPs might be used. The following shall give a survey of our main endeavors in this field and at the same time point out the as yet unsolved problems.

## II. MATERIAL AND METHODS

For nasal stimulation an olfactometer was employed which delivered the chemical stimulants without altering the mechanical or thermal conditions at the stimulated mucosa (Kobal, 1981, 1985). This monomodal chemical stimulation was achieved by mixing pulses of the stimulants in a constantly flowing air stream with controlled temperature (36.5 °C) and humidity (80% relative humidity). The carrier stream was led to the nasal cavities by way of Teflon tubing (6 cm length, 4 mm outer diameter). The total flow rate was 140 ml/sec. Stimulus duration was 200 msec with a rise time below 20 msec. The subjects had no additional cues, such as tactile, thermal, or acoustic sensations, which would have provided extraneous information about the timing of the stimulus presentation. This was demonstrated by applying chemical stimuli that were below threshold, while other experimental conditions remained constant.

Stimuli were presented nonsynchronously to breathing. To avoid respiratory flow of air in the nose, subjects were trained to practice velopharyngeal closure, breathing through the mouth (Kobal, 1981, 1985). They were comfortably seated in an accoustically and odor-shielded chamber. White noise of approximately 50 dB SPL was used to mask any switching clicks of the stimulator which might have been audible. The interstimulus intervals varied in the different experiments and will be specified when reporting the results.

The chamber was equipped with a video monitor and a joystick. Using these, subjects were instructed to indicate after the application of each stimulus the intensity of the perceived stimulus in relation to a standard stimulus given at the start of the experimental session. Estimates of intensity were made by using a visual analog scale (Kobal, 1985). During interstimulus intervals two squares of different sizes appeared on the video screen. Subjects were instructed to keep the smaller square, which could be controlled by the joystick, inside the larger one, which moved around unpredictably. Tracking performance was checked by counting how often and by measuring how long the subjects lost track of the independently moving square.

The EEGs (bandpass 0.2-70 Hz, impedance 1-2 kOhm) were recorded from eight positions of the international 10/20 system (Fz, Cz, Pz, F3, F4, C3, C4, and Fp2, all referenced to A1). EEG records of 2048 msec duration were digitized (sampling frequency 250 Hz) and averaged. Data were evaluated by OFFLAB (Kobal, 1981) and DATAN programs (Brandl, unpublished). EEG records contaminated by eye blinks (Fp2/A1) or motor artifacts were discarded. The remaining data were subjected to analysis of variance looking for main effects of the factors under investigation. Since the variance-covariance matrices were not equal (Box-test), data were submitted to the conservative F test (Geisser and Greenhouse, 1958). The measurements of magnetic responses were made in a magnetically shielded room (Kelhä et al., 1982). The magnetic field perpendicular to the skull was measured with a first-order SQUID gradiometer (Ilmoniemi et al., 1984) simultaneously from three channels. In four subjects, responses were measured (bandpass 0.5-100

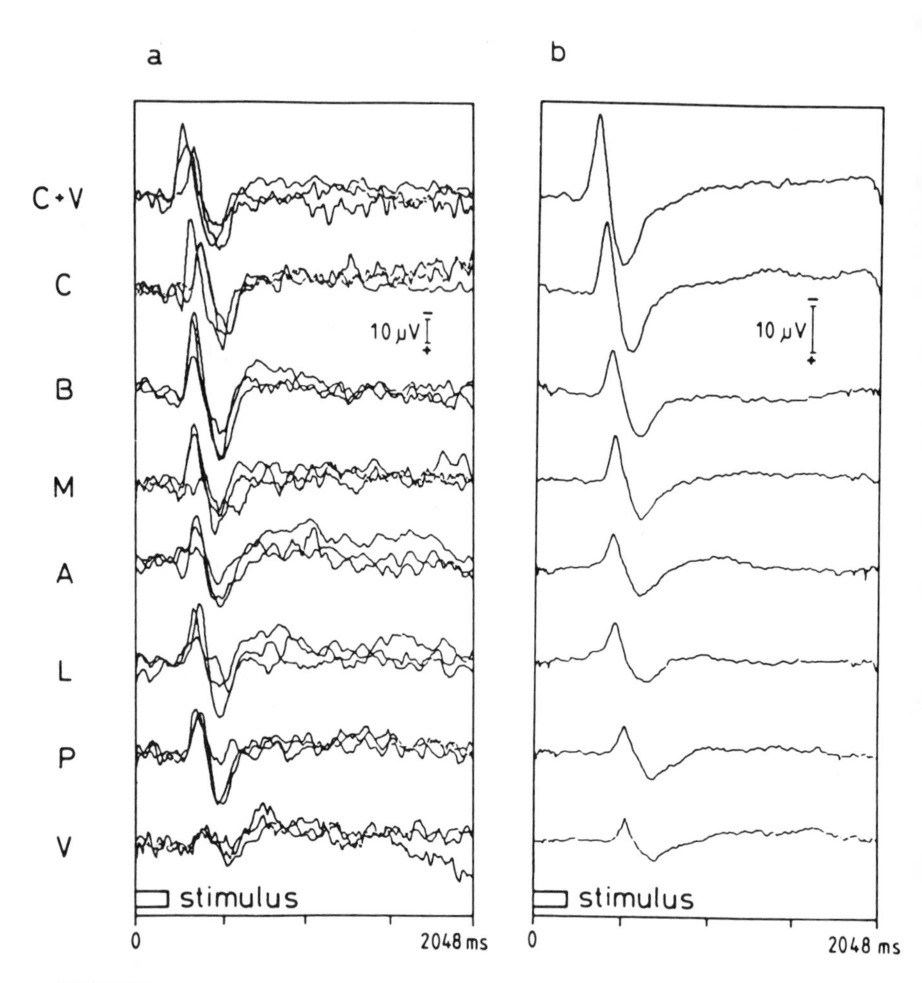


**FIGURE 1** Chemosensory evoked potentials. (a) Example from one subject. Three evoked potentials to each stimulant obtained on different days are superimposed. N = 10, Cz/A1. (b) Grand means to each stimulant over all subjects. N = 380. The main upward negative deflection is termed N1 and the main downward positive deflection is termed P2. V = vanillin, P = phenylethyl alcohol, L = limonene, A = anethol, M = menthol, B = benzaldehyde, C = carbon dioxide, C + V = carbon dioxide mixed with vanillin. (Modified from Kobal and C. Hummel, 1988.)

Hz) in 15-87 locations to cover an area of 60-200 $cm^2$ over the right hemisphere. Isofield contour maps were calculated using a weighted least squares approximation and bicubic spline interpolation. Equivalent current dipoles were founded by a least squares fit (Huttunen et al., 1986).

Fully informed volunteer subjects participated in the experiments, usually with some slight recompense. The experiments were conducted in accordance with the Helsinki-Tokyo-Venice declaration and the subjects were not exposed to any risks.

## III. RESULTS

Figure 1 represents the result of an experiment in which eight different stimuli were presented to 13 subjects (Kobal and C. Hummel, 1988). Among these stimuli was carbon dioxide (C), which, as stated above, causes a pricking pain when applied to the nostril. The averaged potentials evoked by chemical stimulation of the nasal mucosa are of a shape that is typical for late event-related potentials. They commence with a positive deflection followed by a series of negative and positive deflections of different amplitudes. The result of this study was that an increased excitation of the somatosensory system effected reduced latencies and enhanced amplitudes. Responses to the mixture of carbon dioxide and vanillin (C + V) appeared significantly earlier (50-150 msec) than responses to either substance alone. However, it is evident that there are no fundamental differences in the potentials' shapes due to the activation of different sensory systems (i.e., olfactory and somatosensory) after stimulation with vanillin and carbon dioxide. Most differences in amplitude and latency must be attributed to the different subjective perception of the stimulus intensities. The recording from only one position—in this case vertex (Cz) referenced to the left ear lobe (A1)—obviously is not sufficient for a differentiation between the sensory systems involved.

Initially we wanted to establish the site of the cortical generators of the CSSEPs. For this purpose magnetoencephalographic measures were made in the laboratory of R. Hari, Helsinki (Huttunen et al., 1986). The magnetoencephalography (MEG) is superior to EEG in locating cortical current sources. It detects the component of the cortical current flow that is tangential to the surface of the skull (for details, see Williamson and Kaufman, 1981). The maximal deflection of the magnetic evoked responses coincided with the negative vertex potentials at 350-400 msec after onset of the carbon dioxide stimuli (Fig. 2). Isofield maps were dipolar at the peak latencies of the electrical potentials in all subjects, with maximal amplitudes between 0.3 and 0.65 pT. The equivalent dipoles causing a negative potential deflection at the vertex were situated within a 2.5 × 2.5 $cm^2$ area in a region where the

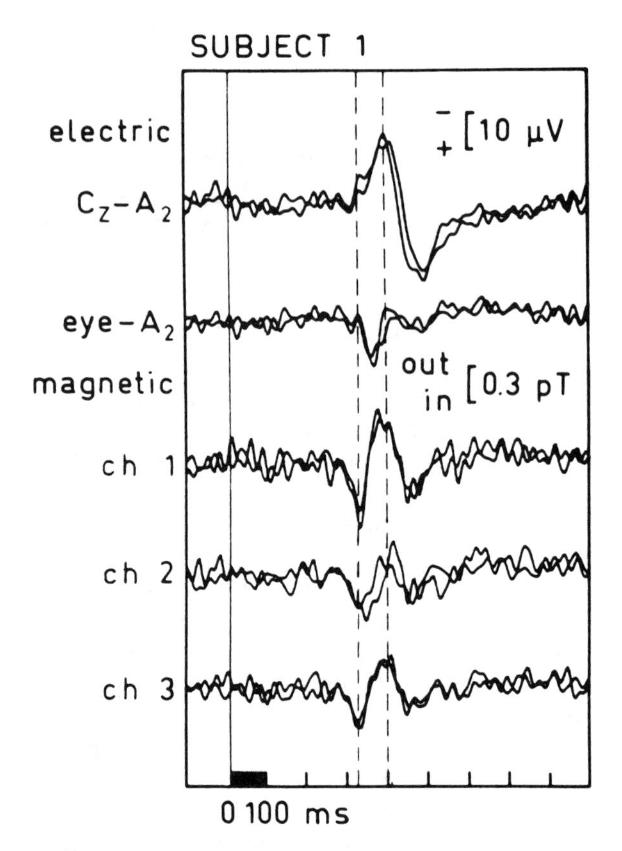


**FIGURE 2** Simultaneously measured electric and magnetic responses evoked by stimulation of the nasal mucosa with painful $CO_2$ pulses (N = 32). The magnetometer was placed 5 cm upward and 3 cm forward from the right-ear canal. The three magnetometer channels were separated by 16 mm. The black bar at the bottom shows the stimulus duration.

central sulcus and the sulvian fissure are contiguous with each other. This corresponds to an activation of SII. According to Chudler et al. (1985), this area is closely connected to the stimulation of nociceptors and the sensation of pain. As yet no other chemical stimulants have been investigated utilizing this method, so that the olfactory areas still have to be established by this method.

Another technique that aids in differentiating between cortical responses is multichannel recording, which is used in "brain mapping." For this purpose evoked potentials are recorded from 8, 16, or 32 positions and the topo-

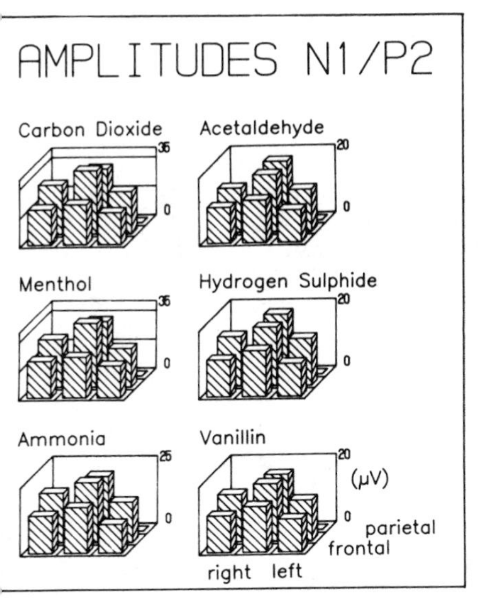


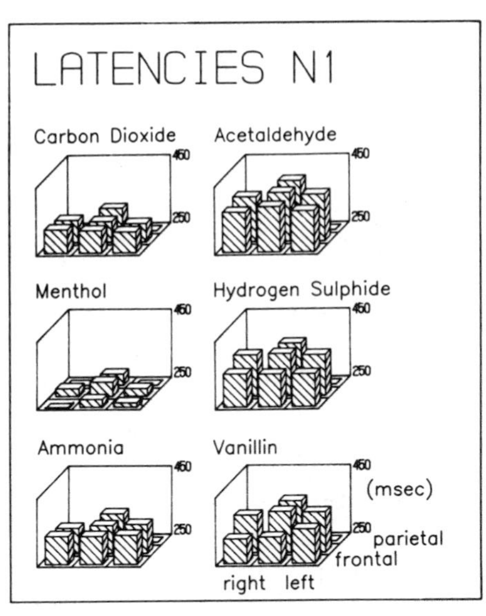


**FIGURE 3** Topographical distribution of the peak-to-peak amplitude N1/P2 and the latency of N1 on the skull. The recording positions from left to right are frontal: F3, Fz, F4; central (between frontal and parietal): C3, Cz, C4; parietal: Pz. The term vertex used in the text is synonymous to Cz. Note different scalings in the amplitudes plot.

graphical distribution of the amplitudes and latencies is investigated. Figure 3 surveys data obtained in different studies. The figure illustrates the distribution of the peak-to-peak amplitude N1/P2 between the main negativity (N1) and the main positivity (P2), as well as the latency of N1. The plots are arranged so that the left sides represent results pertinent to the stimulants which the subjects were able to locate (number of subjects between 12 and 22) after both nostrils had been stimulated. The right sides show the responses to stimulants which the subjects were unable to localize. Beside the interesting differences found between the right and left hemisphere—the demonstrated results in Fig. 3 were obtained by stimulating the left nostril—a topographical pattern was found by which both groups of stimulants could be differentiated. Specific to responses to carbon dioxide, menthol, and ammonia is the fact that N1/P2 is significantly ($p < 0.01$) largest at the vertex (Cz), represented by the columns in the middle part of the boxes. For the stimulants acetaldehyde (in lower concentrations), hydrogen sulfide, and vanillin these maxima at the vertex disappear and other significant maxima can be observed at the

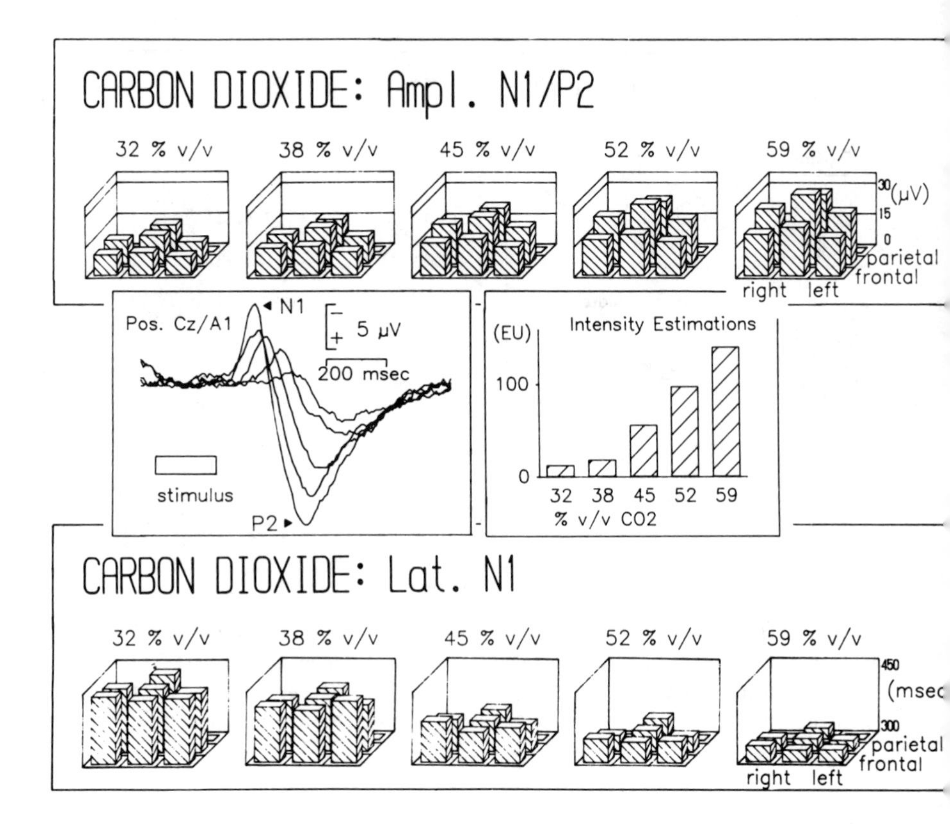


**FIGURE 4** Influence of different concentrations of carbon dioxide on the CSSEP. Superimposed grand means (N = 600) and mean values (N = 600) of intensity estimates in the middle part of the figure. Topographical distributions of the peak-to-peak amplitude N1/P2 and of the latency of N1. Same electrode positions as in Fig. 3. Latencies decrease with increasing amplitudes due to higher concentrations of the stimulus. Stimulus duration = 200 msec. Different concentrations were randomly presented within one session 10 times. EU = estimation units according to the length of the visual analog scale. 100 EU = standard stimulus (45% v/v) that was presented as the first stimulus in the experiments.

parietal position Pz. Also, the latencies of this group of stimulants are shortest at the frontal sites and increase in the parietal positions. Except for menthol, this cannot be observed for stimulants of the first group, i.e., for locatable stimulants, where a tendency to shortened latencies can be found in the central sites of recording (between frontal and parietal). These different topographical patterns indicate different underlying generators. So far, in order to determine the ability of a chemical stimulant to activate the somatosensory system, three indicators are thus available, which all exceed a mere verbal description of the subjective sensation and which are methodologically independent of each other: (a) perceptibility by anosmics, (b) localization after dichotomic stimulation, and (c) the typical somatosensory pattern of the evoked potential mappings.

One of the main characteristic features of event-related potentials is their dependence on stimulus intensity (Fig. 4) (Keidel and Spreng, 1965). Therefore it was necessary to examine whether the chemosomatosensory evoked potentials (CSSEP) likewise possess these properties. Twenty volunteers participated in these experiments. Indeed, it was observed that with increasing stimulus intensities, CSSEPs appeared earlier and were of larger amplitudes ($p < 0.001$). For component N1/P2 an interaction between concentration and position was evident ($p < 0.05$). This indicated that the relationship between concentrations and amplitudes was dependent on the site of recording. The increase in potentials obtained from central sites and sites along the central line was distinctly steeper than in those obtained from the frontal positions. Finally, a significant relationship ($p < 0.01$) was found between the potential components (excepting the latency of N2) and their corresponding estimates of intensity (Müller, 1988).

In a study of 22 subjects, we investigated the influence of repeated stimulation on the CSSEPs as well as the subjective perception of intensity (Hummel and Kobal, unpublished). Series of six stimuli were presented, using three different interstimulus intervals (8, 4, and 2 sec). The series was repeated 16 times with an interseries interval of 50 sec. Figure 5 illustrates the results of the experiments utilizing the 4 sec interstimulus interval. A decrease in amplitude of the potentials induced by repeated stimulation is evident. The latencies, on the other hand, changed only slightly. The superimposed grand means of the evoked potentials show that the amplitudes most conspicuously decreased between presentation of the first and the second stimulus at the position Cz but not at the frontal leads. Hari et al. (1982), who obtained similar results after acoustic stimulation, were of the opinion that different underlying cortical generators occasioned the decrease in amplitudes at the vertex relative to the decrease at the frontal or temporal positions. Future magnetoencephalographic experiments will reveal whether this is also true for noxious stimulation with carbon dioxide.

Since carbon dioxide elicits a painful sensation, we made use of this effect and have employed CSSEPs in algesimetry. In contrast to painful electrical stimuli, painful chemical stimuli excite nociceptors. Thus it is possible to determine the action of analgesics, which act at the level of nociceptors. In a single-blind, controlled, threefold crossover study of 13 volunteers, we were able to demonstrate the analgesic action of acetylsalicylic acid (Kobal

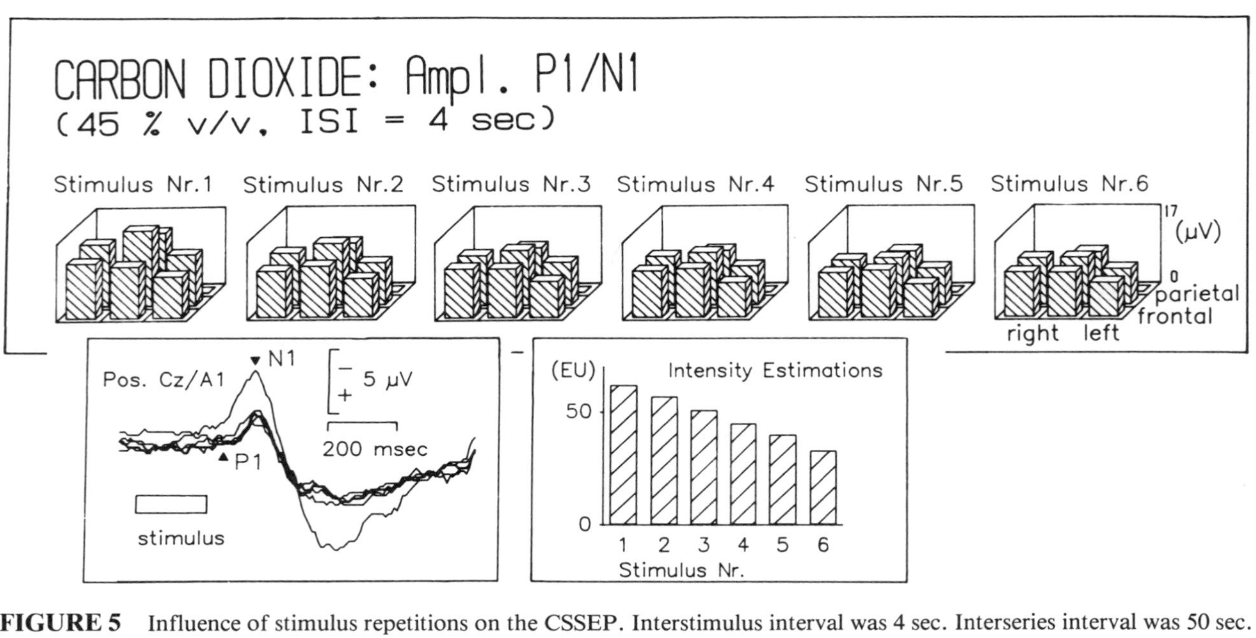


**FIGURE 5** Influence of stimulus repetitions on the CSSEP. Interstimulus interval was 4 sec. Interseries interval was 50 sec. The series was presented 16 times. Superimposed grand means (N = 704) and mean values of intensity estimates (N = 704) in the lower part of the figure. Topographical distribution of the peak-to-peak amplitude in the upper part. Same electrode positions as in Figs. 3 and 4. While amplitudes decrease with repeated stimulation, latencies remain unchanged. Decrease of amplitudes is more pronounced in the central than in the frontal leads.

et al., 1986). In comparison to placebo, the amplitudes were significantly smaller in the frontal leads. On the other hand, the centrally acting partial morphine agonist pentazocine occasioned significant decreases in the frontal, central, and parietal electrode positions. Moreover, latencies were prolonged. Currently we are investigating several different analgesics in order to determine typical patterns that might indicate the peripheral or central site of action of the substances (Kobal and Hummel, 1988).

## IV. DISCUSSION

This chapter presents an overview of our work with chemosomatosensory evoked potentials. To date we are able to clearly distinguish them from olfactory evoked potentials. However, additional investigations of the utilized stimulant as well as a mapping of the evoked potentials by multichannel recordings are prerequisite for a successful discrimination. Chemosomatosensory evoked potentials have a topographical distribution that is different from that of OEPs. The larger amplitudes were found contralaterally to the stimulated nostril, but maximal amplitudes were recorded at the vertex. At this site generally the shortest latencies were observed. When trigeminal stimulants as well as stimulants with trigeminal and olfactory components were birhinally presented, subjects had no difficulties in identifying the stimulated nostril. Thus it is evident that humans area able to obtain spatial orientation information from the somatosensory system. Topographical patterns of the CSSEPs and the property of being locatable help to establish the character of a stimulant in regard to which chemical sensory system it will activate. This has taken us a step further in solving the old problem in olfactometry, i.e., how to discover a pure odorant.

Magnetoencephalographic data indicate that the CSSEPs are generated at the site where nociceptive information reaches the cortical level. Animal studies have shown that there are neurons specifically responding to noxious stimuli in both the primary (SI) and the secondary (SII) somatosensory cortices (Kenshalo and Isensee, 1983; Lamour et al., 1983). Partial removal of the postcentral gyrus has relieved chronic pain in patients, although often the relief has been transient (Lewin and Phillips, 1952). Thus it seems that the CSSEP is generated in parts of the somatosensory cortex which is involved in the perception of pain in man. Recent findings in experimental animals after desensitization with capsaicin (Silver et al., 1986) also indicate that chemoreceptors of the trigeminal nerve are mainly of a nociceptive nature. The responses of this system, which include the peripheral mucosal potentials (Kobal, 1985), can be very effectively employed in algesimetry and analgesimetry.

The CSSEPs might also help in solving problems concerning the chemical senses. It was very impressive to observe that the mixture of vanillin and carbon dioxide drastically reduced the latencies and slightly increased the amplitudes compared to the responses elicited by either substance alone. At the same time intensity estimates to pain increased and intensity estimates to odor decreased (Kobal and C. Hummel, 1988). It seems as if the olfactory system supports the somatosensory system, whereas we could demonstrate that in contrast the someatosensory system inhibits the olfactory system (see also Cain and Murphy, 1980). It is evident that the technique of CSSEP recording helps to further extensive studies investigating the interaction between the two systems.

In the area of clinical olfactometry, CSSEPs are routinely recorded in our laboratory in order to obtain an internal standard from the patient. As yet we have not investigated a person with an intact trigeminal nerve who has not yielded potentials after stimulation with carbon dioxide. However, it was observed that anosmic patients tended to yield smaller amplitudes of CSSEPs than normosmic subjects (Kobal, 1982). The future role of CSSEPs in neurological and rhinological routine measurement will depend on the efficacy of the available stimulators. To date the requisite EEG recording facilities are part of every clinical neurophysiological standard equipment.

## ACKNOWLEDGMENTS

This research was supported by the Deutsche Forschungsgemeinschaft grant 812/1-1,2 and Ko 812/2-1.

## REFERENCES

Allison, T., and Goff, W. R. (1967). Human cerebral evoked potentials to odorous stimuli. *Electroenceph. Clin. Neurophysiol.* 14: 331-343.

Chudler, E. H., Dong, W. K., and Kawakami, Y. (1985). Tooth pulp evoked potentials in the monkey: Cortical surface and intacortical distribution. *Pain* 22: 221-233.

Doty, R. L., Brugger, W. P. E., Jurs, P. C., Orndorff, M. A., Snyder, P. J., and Lowry, L. D. (1978). Intranasal trigeminal stimulation from odorous volatiles: Psychometric responses from anosmic and normal humans. *Physiol. Behav.* 20: 175-185.

Finkenzeller, P. (1966). Gemittelte EEG-Potentiale bei olfaktorischer Reizung. *Pfluegers Arch.* 292: 76-80.

Geisser, S., and Greenhouse, S. W. (1958). An extension of Box's results on the use of the F distribution in multivariate analysis. *Ann. Math. Stat.* 29: 885-891.

Hari, R., Katila, K., Tuomisto, T., and Varpula, T. (1982). Interstimulus interval dependence of the auditory vertex response and its magnetic counterpart: implications for their neural generation. *Electroenceph. Clin. Neurophysiol.* 54: 561-569.

Huttunen, J., Kobal, G., Kaukoranta, E., and Hari, R. (1986). Neuromagnetic responses to painful $CO_2$ stimulation of nasal mucosa. *Electroenceph. Clin. Neurophysiol.* 64: 347-349.

Ilmoniemi, R., Hari, R., and Reinikainen, K. (1984). A four-channel SQUID magnetometer for brain research. *Electroenceph. Clin. Neurophysiol.* 58: 467-473.

Keidel, W. D., and Spreng, M. (1965). Neurophysiological evidence for the Stevens power function in man. *J. Acoust. Soc. Am.* 38: 191-195.

Kelhä, V. O., Pukki, J. M., Peltonen, R. S., Penttinen, A. A., Ilmomiemi, R. J., and Heino, J. J. (1982). Design, construction and performance of a large volume magnetic shield. *IEEE Trans. Magn.* 18: 260-270.

Kenshalo, D. R., and Isensee, O. (1983). Responses of primate SI cortical neurons to noxious stimuli. *J. Neurophysiol.* 50: 1479-1496.

Kobal, G. (1981). *Elektrophysiologische Untersuchungen des menschlichen Geruchssinns.* Thieme Verlag, Stuttgart, p. 171.

Kobal, G. (1982). A new method for determination of the olfactory and the trigeminal nerve's dysfunction: Olfactory (OEP) and chemical somatosensory (CSEP) evoked potentials. In: *Event-Related Potentials in Children.* A. Rothenberger (Ed.). Elsevier, Amsterdam, pp. 455-461.

Kobal, G. (1985). Pain-related electrical potentials of the human respiratory nasal mucosa elicited by chemical stimulation. *Pain* 22: 151-163.

Kobal, G., Hummel, C., and Nürnberg, B. (1986). The effect of pentazocine and ASA on the human pain-related chemo-somatosensory evoked potentials (CSEP). *Naunyn-Schmiedeberg's Arch. Pharmacol.* 332: R87.

Kobal, G., Hummel, Th., and Van Toller, C. (1987). Olfactory and chemosomatosensory evoked potentials from stimuli presented to the left and right nostrils. *Chem. Senses* 12(1): 183.

Kobal, G., and Hummel, Th. (1988). Effect of flupirtine on the pain-related evoked potential and the spontaneous EEG. *Agents Actions* 23(1/2): 117-119.

Kobal, G., and Hummel, C. (1988). Cerebral chemosensory evoked potentials elicited by chemical stimulation of the human olfactory and respiratory nasal mucosa. *Electroenceph. Clin. Neurophysiol.* 71:241-250.

Kobal, G., and Plattig, K.-H. (1978). Methodische Anmerkungen zur Gewinnung olfaktorischer EEG-Antworten des wachen Menschen (objektive Olfaktometrie). *Z. EEG-EMG* 9: 135-145.

Lamour, Y., Willer, J. C., and Guilbaud, G. (1983). Rat somatosensory (SmI) cortex. I. Characteristics of neuronal responses to noxious stimulation and comparison with responses to non-noxious stimulation. *Exp. Brain Res.* 49: 35-45.

Lewin, W., and Phillips, C. G. (1952). Observations on partial removal of the postcentral gyrus for pain. *J. Neurol. Neurosurg. Psychiatr.* 15: 143-147.

Müller, G. (1988). Schmerzkorrelierte evozierte Potentiale nach Stimulation der menschlichen Nasenschleimhaut mit verschiedenen Kohlendioxid-Konzentrationen. Medical thesis, University of Erlangen, FRG.

Plattig, K.-H., and Kobal, G. (1979). Spatial and temporal distribution of olfactory evoked potentials and techniques involved in their measurement. In: *Human*

*Evoked Potentials*. D. Lehmann and E. Callaway (Eds.). Plenum Press, New York, pp. 285-301.

Silver, W. L., Mason, J. R., Marshall, D. A., and Maruniak, J. A. (1986). Rat trigeminal, olfactory, and taste responses after capsaicin desensitization. *Brain Res.* 333: 45-54.

Williamson, S. J., and Kaufman, L. (1981). Biomagnetism. *J. Magn. Magn. Mat.* 22: 129-202.

# Chapter 7 Discussion

**Dr. Cain:** It is true our psychophysical latencies are much longer than the evoked potential latencies you have obtained. But I think the question of latency really is a concentration-dependent thing. I'm not saying it is always one second; in fact one of the graphs I showed this morning of $CO_2$ and amyl butyrate showed that if you extrapolate it, it would come out to be maybe 50-100 msec or so. It is concentration-dependent and maybe to some degree substance-dependent. Your latencies in general are very short though, and I'm wondering if it has something to do with the means of stimulus presentation. Can you say a little more about what flow rates you use?

**Dr. Kobal:** I think the short latencies are a result of the way in which stimuli are presented, i.e. steep onset of the stimulus and a rather high flow rate of 140 ml/sec.

**Dr. Cain:** Do you inject the stimulus into the flowing stream so that the stream is flowing ahead of time, or does the stimulus begin in the wave front of the stream?

**Dr. Kobal:** For stimulus presentation a technique was used, which was developed in my laboratory and was patented several years ago. Virtually, two gaseous streams are being switched, comparable to the functioning of a railway switch. The switching is induced by the applying of vacuums so that, on the one hand, odorants of preestablished concentrations and on the other hand, during interstimulus intervals, non-odorous air reach the nose. By carefully tuning the switching of the vacuums, the stimulus characteristics can be optimized in such a way that the rise time does not exceed 20 msec.

Thus, it is guaranteed that the subjects have no additional cues, such as tactile, thermal or acoustic sensations, which would provide extraneous information about the timing of the stimulus presentation.

**Dr. Cain:** I'm sorry but the data went by fast and its hard to remember it all. With respect to vanillin and $CO_2$ evoked responses, you had some evoked responses to vanillin, too? I'm not saying you should not get any enhancement. As you know, my psychophysical data show that when you add the odorant to the irritant the sensation does go up; it just doesn't go up in an additive fashion. So when you look, so to speak, inside the sensation you find that one or another thing has given way. So I wouldn't predict that you'd see no change. If you have a change in response to A and a change in response to B, I'm not saying you wouldn't see some kind of additivity; it just wouldn't be perfect, simple additivity.

**Dr. Kobal:** I agree. The actual findings somewhat surprised us. Responses to the mixture of carbon dioxide and vanillin appeared significantly earlier (50-150 msec) than responses to either substance alone. The reason for this cannot be a simple addition, since the psychophysical responses represented themselves in such a way that a decrease was observed in the olfactory sensations and an increase was apparent in the pain (trigeminal) sensation.

**Dr. Walker:** Let's say you're taking about 5% of one and 5% of another. If you are doing half volumes, you would add 10% and 10%. Do you see what I'm getting at?

**Dr. Kobal:** No, I am not quite sure what your point is. In those experiments I just described we used a mixture of 66% v/v carbon dioxide and 0.034 mg/l vanillin. In case of stimulation with the single substances the same stimulus strengths were utilized.

**Dr. Walker:** So it makes intuitive sense that you had a more intense stimulus.

**Dr. Kobal:** Altogether it was a more intense stimulus.

**Dr. Walker:** You have done evoked potentials in anosmics, is that right?

**Dr. Kobal:** Yes.

**Dr. Walker:** Do you see the same foci or focus of activity? Do you still see somatosensory II?

**Dr. Kobal:** We always see responses to $CO_2$.

**Dr. Walker:** What about vanillin?

**Dr. Kobal:** We never found responses to vanillin in anosmics.

**Dr. Walker:** You never see a response of any type?

**Dr. Kobal:** Not of any type. The averaged EEG only revealed base line activity, just as if a stimulus had not been presented. The same is true for stimulation with phenylethyl alcohol. But in these cases we find responses using menthol and carbon dioxide as stimuli.

**Dr. Walker:** You assume that for all of these potentials no olfactory component was present?

**Dr. Kobal:** No, I did not say that. I would favour the statement that vanillin and phenylethyl alcohol responses were mediated by the olfactory nerve. This is why they are absent in anosmic patients. The above described results obtained after presentation of the mixed stimuli were found in healthy subjects.

**Dr. Walker:** Can you use the onset of inspiration as a synchronizer?

**Dr. Kobal:** Yes that is the way I proceeded previously. But then I gave up this technique, because it resulted in unwanted synchronization with respiratory potentials in the EEG, which superimposed the olfactory evoked potentials.

**Dr. Szolscanyi:** Just a short question. You have a really elegant method of measuring evoked potentials and you have already shown the correlation between the intensity of sensation and the evoked potential. Using this very sensitive method, could you detect evoked potentials at subliminal levels of perception?

**Dr. Kobal:** That is a very interesting question. In case vanillin or the other above mentioned odorants in concentrations below the subjective threshold are presented we do not get any responses. But, we are planning an experiment, in which we try to find out whether there are substances which evoke responses, although they have not been perceived by the subjects.

# 8
# Capsaicin, Irritation, and Desensitization
## Neurophysiological Basis and Future Perspectives

**J. Szolcsànyi**
University Medical School of Pécs
Pécs, Hungary

## I. INTRODUCTION

Capsaicin is the pungent principle in red pepper. Therefore the burning hot sensation induced by this agent is commonly experienced. Nevertheless, this chapter does not focus on this oral, gustatory aspect of capsaicin. Instead it deals with a new scope which has emerged during the last decades related to the usage of capsaicin under experimental conditions as a pharmacobiological tool to re-
the functions of a major group of somatosensory primary afferent neurons.

The usefulness of a pharmacological means resides in its selectivity. This chapter will discuss selectivity in detail, and the interesting new horizons which have appeared in this regard will be outlined briefly. The major aim is to support further research, and therefore a frame of conception of the author, who has been working in this field for more than a quarter of a century, is summarized. For more information the reader can rely on my earlier reviews (Szolcsànyi, 1982, 1984a,b, 1985) as well as on very useful major review articles written by others in recent years (Nagy, 1982; Monseerenusorn et al., 1982; Fitzgerald, 1983; Russel and Burchiel, 1984; Hori, 1984; Buck and Burks, 1986; Maggi and Meli, 1988; Holzer, 1988). Two volumes comprising the full papers of two symposia (*Antidromic Vasodilatation and Neurogenic Inflammation*. L. A. Chahl, J. Szolcsànyi, and F. Lembeck, Akadémiai kiadò, Budapest, 1984; *Acta Physiologica Hungarica* Vol. 6 No. 3-4, 1987) complete the list of recent major references.

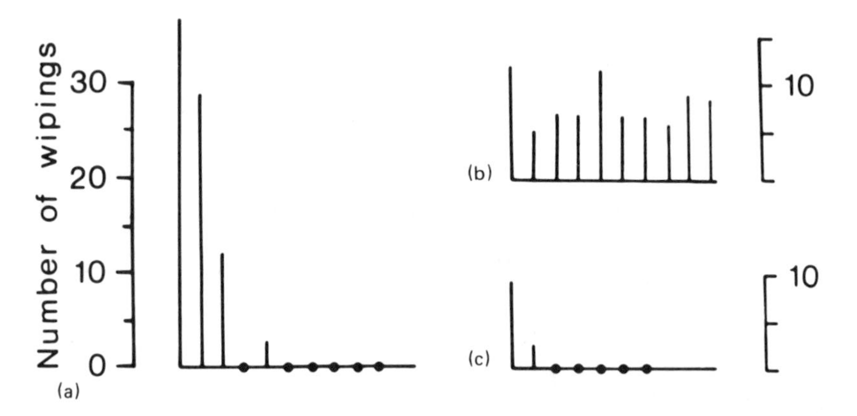


**FIGURE 1** Number of wiping movements elicited by instillation of capsaicin (100 μg/ml 10 times at 3 min intervals) into the right eye of a conscious rat (a). Responses 2 hr later to a nondesensitizing chemical stimulus (zingerone 10 mg/ml at 1 min intervals) (b): left eye (control); (c) right eye (capsaicin-treated).

## II. CHARACTERISTICS OF THE NOCICEPTIVE AND ANTINOCICEPTIVE EFFECTS OF CAPSAICIN

### A. Topical Application

Solutions of capsaicin or other pungent agents elicit protective nocifensor reactions when applied to exteroceptive mucosal areas (eye, nose, oral cavity, penis, anus) of the rat, guinea pig, and other mammalian species (see later). The evoked behavioral response is reproducible if the concentration of capsaicin is low. For example, 10 instillations of 33 μM capsaicin into the eye of the rat at 3 min intervals provoke reproducible wiping movements with the foreleg (Szolcsànyi et al., 1975; Szolcsànyi and Jancsò-Gàbor, 1976). Instillation of capsaicin in 10 times higher concentration, however, results in a desensitizing effect (Fig. 1). Desensitization is the accepted term to denote the decreased responsiveness of the sensory receptors (Paintal, 1973) to stimuli, e.g., to chemonociceptive ones. As a descriptive term, capsaicin desensitization is also used to denote a decreased responsiveness to the irritant and other effects of capsaicin-type pungent agents irrespective of the mode of action (Jancsò, 1955, 1968; Szolcsànyi, 1982). To be precise the former definition of desensitization is adopted in this chapter.

The evidence that after topical capsaicin application the loss of behavioral response is due to a reduced responsiveness of the sensory receptors and not to blockade of axonal conduction or some central nervous system interactions is as follows:

1. Instillation of local anaesthetics into the eye abolished the corneal reflex evoked by mechanical probes. Instillation of capsaicin into the eye (Szol-

csànyi et al., 1975), however, impairs chemonociception without any apparent inhibition of the effectiveness of mechanical stimuli.

2. The desensitizing effect lasts for hours (Fig. 1) or days, depending on the concentration of the drug, contact time, and penetration barriers. The development of this local desensitization is not altered by pretreating the area with local anesthetics (Jancsò, 1955).

3. One sign of excitation of these endings (e.g., neurogenic inflammation or bronchoconstriction) cannot be inhibited (Jancsò et al., 1968; Szolcsànyi, 1982, 1983b, 1984a,b) or abolished (Lundberg et al., 1984) by local anesthetics or tetrodotoxin. It is noteworthy that neither the generative potential of sensory receptors nor chemically evoked excitation-secretion coupling in presynaptic nerve terminals is inhibited by local anesthetics or tetrodotoxin (cf. Szolcsànyi, 1982, 1984b).

4. After local desensitization of the rat's eye, ultrastructural changes were observed in the naked portion of some intraepithelial and subepithelial fibers. Axons ensheathed by Schwann cells were more resistant to the effect of capsaicin and the fine structure of nonneural elements remained unchanged (Szolcsànyi et al., 1975).

5. A characteristic sign of desensitization with capsaicin is the fatigue-like phenomenon of the nocifensor responses when a nondesensitizing chemical stimulus (e.g., zingerone) is applied repeatedly (Jancsò, 1968; Szolcsànyi et al., 1975). It can be seen in Fig. 1 that 2 hr after pretreatment of the right eye with capsaicin, the first chemical challenge was as effective as on the control untreated side in evoking protective wiping reactions. The loss of response manifests itself when stimulations are repeated at short intervals.

It is concluded that capsaicin excites and desensitizes the sensory receptors when it is applied to mucosal areas.

Selective or nonselective blockade of axonal conduction ensues when the drug is applied directly around the nerve trunk (for reference, see Petsche et al., 1983; Handwerker et al., 1984; Lynn et al., 1984; Jancsò et al., 1987). This mechanism has been invoked to explain the antinociceptive effect of capsaicin when other types of drug administration were used (Baranowsky and Lynn, 1986). In light of the above data and other considerations, it has been suggested that after topical application the primary site of action of capsaicin is the sensory receptor. Conduction blockade of preterminal axons protected by Schwann cells manifests itself only when high concentrations of the drug are applied for a prolonged period of time. This conclusion is supported also by single-unit studies (Szolcsànyi, 1987a), to be discussed later.

## B. Systemic Application in the Adult

Subcutaneous injections of capsaicin to rats, mice, and guinea pigs induce a long-lasting or irreversible reduction of responsiveness of the animals to various chemonociceptive stimuli. This effect can be attributed only partly to a

**TABLE 1** Effect of Capsaicin on Noxious Heat Threshold Under Different Conditions

| Pretreatment | Increase in threshold on day 1-3 (ΔT °C) | Recovery time (days) | Ref.[a] |
|---|---|---|---|
| Rat systemic | | | |
| 150 mg/kg s.c. | 2.5 | 4 | 1 |
| 400 mg/kg s.c. | 3.3 | 21 | 1 |
| Rat perineural | | | |
| 1 mM | 1.2 | >120 | 2 |
| 3.3 mM | 2.6 | >120 | 2 |
| 33 mM | 4.1 | >360 | 1 |
| Guinea pig perineural | | | |
| 1 mM | 2.2 | >60 | 2 |
| 3.3 mM | 4.2 | >60 | 2 |
| 33 mM | 5.6 | >60 | 2 |
| Human skin topical | | | |
| 33 mM | 9.0 | 10 | 3 |

[a]Ref. 1, Szolcsànyi, 1985, 1987b; Ref. 2, Szolcsànyi, unpublished; Ref. 3, Szolcsànyi et al., 1985.

desensitizing effect of the drug at the level of the sensory receptors. There is an involvement of other parts of the neuron, particularly its central terminals (Yaksh et al., 1979; Jhamandas et al., 1984). The loss of function cannot be entirely due to a functional impairment of the affected primary sensory neurons (*sensory neuron blocking effect*). *A neurotoxic effect*, particularly after higher doses in the guinea pig or rat, is also involved in the impaired responsiveness to nociceptive stimuli.

Assessment of the antinociceptive effect of capsaicin pretreatment against noxious heat and mechanical stimuli resulted in conflicting findings when conventional hot-plate, tail-flick, and Randall Sellito tests were used (for reference, Nagy, 1982; Szolcsànyi, 1985, 1987b; Buck and Burks, 1986).

The former two methods, however, do not provide information about the noxious heat threshold; instead the reflex latencies to suprathreshold stimuli are measured. These techniques are certainly useful for testing the analgesic effect of opiates, but seem to be inadequate when the site of action is on the primary afferent neurons.

Therefore a new technique was introduced with which the noxious heat threshold could be measured on the paws of rats or guinea pigs as well as on the tails of mice or rats (Szolcsànyi, 1985, 1987b). There was a definite, dose-dependent, significant increase in noxious heat threshold after systemic capsaicin treatment of the rat. However, the antinociceptive effect was never complete and there was a recovery of the original responsiveness within a few days (Table 1). These features provide an explanation for many of the earlier conflicting data.

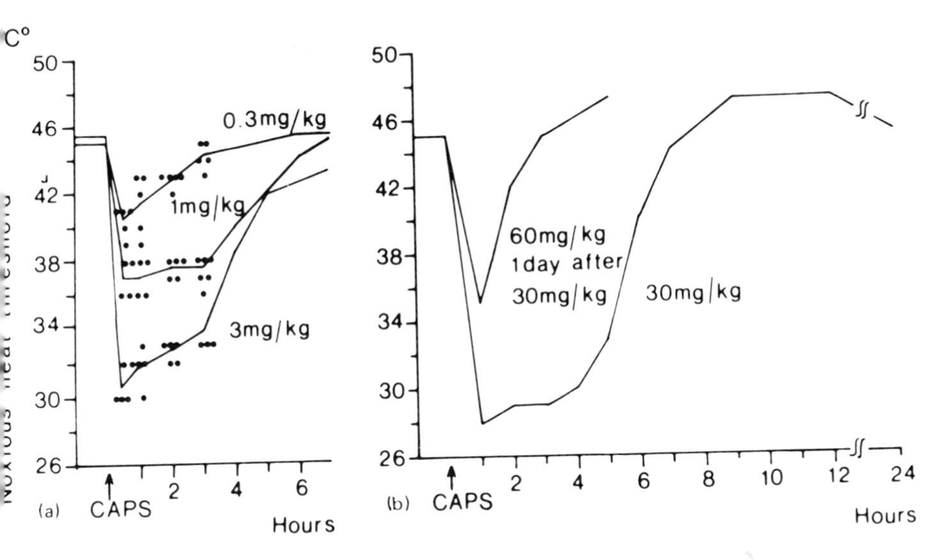


**FIGURE 2** Effect of subcutaneous injection of capsaicin in different doses on the noxious heat threshold as measured on the hindpaw (Szolcsànyi, 1985, 1987). Note the dose-dependent hyperalgesic response as the average of 5-5 experiments (a) and the desensitizing effect (b).

Noxious heat threshold measurements have also revealed a dose-dependent acute thermal hyperalgesia lasting for several hours (Fig. 2). How this marked thermal hyperalgesia is related to the acute fall in body temperature induced by the drug (Szolcsànyi, 1982) needs further investigation. Nevertheless, any analgesic assessment during this severe hyperalgesic state has rather questionable value.

Further important characteristics of capsiacin-induced antinociception are illustrated in Fig. 2b. The rat which received 30 mg/kg s.c. capsaicin after recovering from the profound thermal hyperalgesic state responded with an increased threshold to noxious heat stimuli for a few hours. However, 24 hr later the noxious heat threshold was again at the control value, indicating that nociception to noxious heat remained unchanged after this dose of capsaicin. In this situation a subsequent injection of capsaicin in a higher dose induced a much less pronounced thermohyperalgesic effect. The threshold dose of capsaicin for antinociception, as tested on the wiping test and on the noxious heat threshold test 1-2 days after the pretreatment, corresponded to 10-20 mg/kg and 150 mg/kg, respectively. Furthermore, the recovery of nociception evoked by painting the skin of the paw with xylene is much slower than the recovery of the noxious heat threshold (Szolcsànyi, 1985, 1987b). Together these findings imply that chemonociception is more effectively inhibited by systemic capsaicin pretreatment than is thermonociception.

Mechanonociception has been assessed in different laboratories by methods based on threshold measurements, with variable results. It should be considered, however, that capsaicin does not affect the cutaneous mechanonociceptors while the polymodal nociceptors, which can be desensitized by capsaicin, respond to mechanonociceptive stimuli (see later). Therefore, it is conceivable that after capsaicin pretreatment the effectiveness of noxious mechanical stimulation depends on the role of the affected cutaneous polymodal nociceptors and other capsaicin-sensitive receptors in mechanonociception. This might differ considerably in different skin areas and in different species, as well as when different stimulation techniques (e.g., pinching, pressure with a blunt probe) are used. The dose-dependent recovery of sensitivity to nociceptive stimulation should also be considered.

### C. Perineural Application

Capsaicin in aqueous or oily solutions applied around the sciatic and saphenous nerves of the rat or guinea pig induces a practically irreversible antinociception to chemonociceptive and noxious heat stimuli (Table 1).

In contrast to earlier assumptions, there is no complete thermoanalgesia and the effect is dose-dependent in the range of 1-33 mM in both the rat and guinea pig (Table 1). The development of the antinociceptive effect is not due to a capsaicin-induced acute conduction block of C fibers. Capsaicin applied around the sciatic nerve is much more effective in blocking the conduction of C fibers in the rat than in the guinea pig. A concentration of 0.1 mM elicited a 70% reduction in the compound action potential in the rat, while the maximum reduction in guinea pig was only 30% when 33 mM was applied (Baranowsksy and Lynn, 1986). On the contrary, the increase in noxious heat threshold was higher in the guinea pig than in the rat (Table 1).

Perineural capsaicin application around the sciatic nerve did not induce autotomy of the leg in rats or guinea pigs, nor was there any impairment of tactile or noxious cold sensitivity in the rat (Coderre et al., 1984).

### D. Intrathecal, Intracisternal, and Epidural Application

In 1975 on the basis of ultrastructural and functional evidence it was suggested that the antinociceptive effect of systemic capsaicin treatment was not due solely to an action on peripheral axons and nerve terminals. The hypothesis was put forward that capsaicin affects the "whole primary sensory neuron" (Szolcsànyi et al., 1975). Subsequent studies have supported this conclusion and revealed that capsaicin applied to central portions of the primary afferent neurons by means of intrathecal, intracisternal, or epidural drug administration results in a long-lasting chemo- and thermoanalgesia (Yaksh et al., 1979; Gamse et al., 1984; Jhamandas et al., 1984; Eimerl and Papir-Kricheli, 1987).

After the treatments there was no detectable alteration either in the response to mechanical stimuli or in motor coordination.

The immediate response to these modes of capsaicin injection resembled the effect of topical application of the drug in the sense that it manifested itself in a nocifensive "scratching, biting" behavior. This reaction can be attributed to excitation of the dorsal roots and/or their terminal spinal processes, since it is directed to dermatomes corresponding to those levels affected by the injection and because various morphological, biochemical, and electrophysiological data do not support a site of action on spinal neurons.

It is worthy of mention that intrathecally vs. peripherally administered congeners of capsaicin do not differ significantly in their ability to evoke the respective nocifensive behaviors (scratching/biting, eye wiping) or to produce subsequent desensitization. Therefore, a similar pharmacological receptor site and mode of action has been postulated for both the peripheral and central terminals of the pseudounipolar capsaicin-sensitive sensory neurons (Szolcsànyi, 1982; Jhamandas et al., 1984).

### E. Systemic Treatment of Neonatal Animals

Systemic treatment of newborn rats and mice induces more profound antinociceptive effects than occur in adult rats (Jancsò et al., 1977, 1987). In spite of the complete destruction of a considerable portion of the primary afferent neurons in these immature animals, antinociception as detected in the adult animals showed a contradictory picture (for reference, see Buck and Burks, 1986; Szolcsànyi, 1985, 1987). The conflicting results observed after neonatal administration of capsaicin even within the same group of animals was described as a paradox (Cervero and McRitchie, 1981).

A detailed description of the differences between the consequences of neonatal and adult pretreatments will be discussed later in order to facilitate the appropriate usage of this technique in further experiments.

### F. Species Differences

Apparently capsaicin acts as an irritant in all mammalian species. Nocifensive reactions to this agent have been described not only in the rat, mouse, guinea pig, rabbit, cat, dog, goat, and golden hamster (Szolcsànyi, 1982), but also in a range of wild animals where it has been used as a spray repellant (Rogers, 1984).

In the two most commonly used species (rat and guinea pig) a similar thresh-dose of capsaicin (20-30 ng) was required for evoking nociception by close arterial injection (Donnerer and Lembeck, 1982; Szolcsànyi et al., 1986). Nevertheless, respiratory impairment, which often develops in both species in response to high systemic doses, is more severe in the guinea pig (Jancsò-Gàbor et al., 1970; Gamse et al., 1981; Buck et al., 1983; Papka et al., 1984).

The rabbit is about 20 times less sensitive to capsaicin than the rat (Szolcsànyi, 1982, 1987); therefore this species is less suitable for systemic desensitization.

Topical or local desensitization has been achieved in all mammalian species tested so far, including the rabbit. The limiting factor of effective systemic desensitization in various species is related to the severity of acute cardiovascular and respiratory reflexes and responses during the treatment.

Phylogenetically lower species, such as birds, are extremely insensitive to the stimulatory and desensitizing effects of the drug. The threshold dose of capsaicin for nocifensive reaction was 10,000-fold higher in pigeons than in guinea pigs (Mason et al., 1983; Szolcsànyi et al., 1986; Pierau et al., 1988).

## G. Effects on Human Subjects

A natural source of interest in oral chemical irritation induced by capsaicin is related to the fact that the agent is the hot principle of red peppers, which are consumed by many hundred of millions of humans (Rozin et al., 1981; Sizer and Harris, 1985; Green, 1986). Psychophysical and other studies concerning the "spice" aspect of the topic is not addressed in this chapter.

The detection threshold of capsaicin on the tongue (Szolcsànyi, 1977) or in the oral cavity (Rozin et al., 1981; Sizer and Harris, 1985) is around 0.2 $\mu$g/ml (range ≅ 0.09-0.35 $\mu$g/ml). A small but reliable and significant increase in the detection threshold (desensitization) was observed in subjects who like to eat chili peppers (Rozin et al., 1981). The burning sensation induced by capsaicin can be inhibited or even abolished by cooling not only in the mouth but also on the skin. Warming had an opposite effect. Complete abolition of the burning sensation was observed at 24 °C in the mouth (Green, 1986) and at 28 °C on the skin. At and above 30-31 °C the appearance of tactile hyperalgesia with definite burning pain sensation was also remarkable (Szolcsànyi, 1977, 1982).

It is important to note that the phenomenon can be reproduced by recording action potentials from the sensory nerves of rats (Szolcsànyi, 1977). Thus, the thermodependence of the capsaicin action ensues at the level of the sensory receptors.

The quality of sensation capsaicin produces is characteristically hot or burning; but a sensation of warmth on the tongue at threshold concentrations, a pricking-like feeling in the throat, or a sharp pain without a definite thermal component in response to intracutaneous capsaicin injection or application to the blister bases all indicate that a variety of sensation qualities can be evoked by the agent. Furthermore, capsaicin activates receptors mediating itch sensation from the skin, sneezing from the nasal mucosa, and coughing from the respiratory airways (Szolcsànyi, 1982, unpublished observations; Collier and Fuller et al., 1985; Geppetti et al., 1988; Barnes et al., 1988). Intravenous injection of capsaicin in man (0.5-4 $\mu$g/kg) evoked "hot-flushing" sensations in the chest, face, rectum, and extremities (Winning et al., 1986).

Topical application of capsaicin in high concentration (1%) desensitized the receptors of the affected area (tongue, skin, blister base) to subsequent stimuli applied to the area for several hours or days (Szolcsànyi, 1977, 1988; Carpenter and Lynn, 1981; Szolcsànyi et al., 1985; Tòth-Kàsa et al., 1986). Testing after a longer time course had the advantage of avoiding the masking effect of the acute irritation produced by capsaicin.

Chemically induced burning or pricking pain on the tongue (capsaicin, piperine, zingerone, mustard oil), sharp pain on the blister base (bradykinin, acetylcholine, capsaicin), itching (histamine) or burning pain (capsaicin) on the skin were strongly inhibited by capsaicin desensitization. Pain induced by KCl on the blister base or by means of pin prick on the tongue or skin was not inhibited. The recognition thresholds for quinine sulfate, glucose, sodium chloride, ascorbic acid, and menthol remained unchanged, indicating the unimpaired function of specific taste chemoreceptors and that of the chemical excitability of cold receptors.

On the tongue (Szolcsànyi, 1977) the difference limen of temperature discrimination became broader in the warm range (36-45 °C) and remained unaltered in the cold range (23-24 °C). Tactile discrimination measurements indicated no impairment in function.

On the volar skin of the forearm desensitization was achieved by five applications of a 1% alcoholic solution of capsaicin after dimethyl sulfoxide (DMSO) treatment for 1 min (Szolcsànyi et al., 1985). Figure 3 shows the marked increase and recovery of the heat pain threshold after the treatment.

In this series of experiments a thermocontrolled brass thermode with a contact area of 2.5 $cm^2$ touched to the skin for exactly 2 sec in order to test the sensitivity of the presumably more desensitized superficial layers. By this technique reproducible assessments were obtained, and the analgesic effect was more pronounced than was reported earlier (Carpenter and Lynn, 1981).

Measurements of the sensitivity to temperature were made by asking the subject to compare the temperature of two thermodes attached to the solvent-treated arm and capsaicin-treated arm simultaneously. The task of the subject was to decide which thermode felt warmer or cooler. The $p = 0.75$ probability of correct answers (20-20 testings at each temperature difference on five subjects) before the treatment was $0.9 \pm 0.5$ °C ($x \pm SD$) in the warm range (35-40 °C). One day after topical capsaicin desensitization the warmth sensitivity of the skin decreased and a higher temperature ($4.6 \pm 0.6$ °C) was needed on the desensitized area to reach a $p = 0.75$ probability of correct answers. The $p = 0.75$ value in the cold range ($1.8 \pm 1.1$ °C at 19-25 °C) did not differ significantly from the value estimated before the treatment ($2.4 \pm 0.45$ °C).

There was no difference in tactile or mechanonociceptive thresholds between the solvent-treated and capsaicin-treated areas of the forearms. Mech-

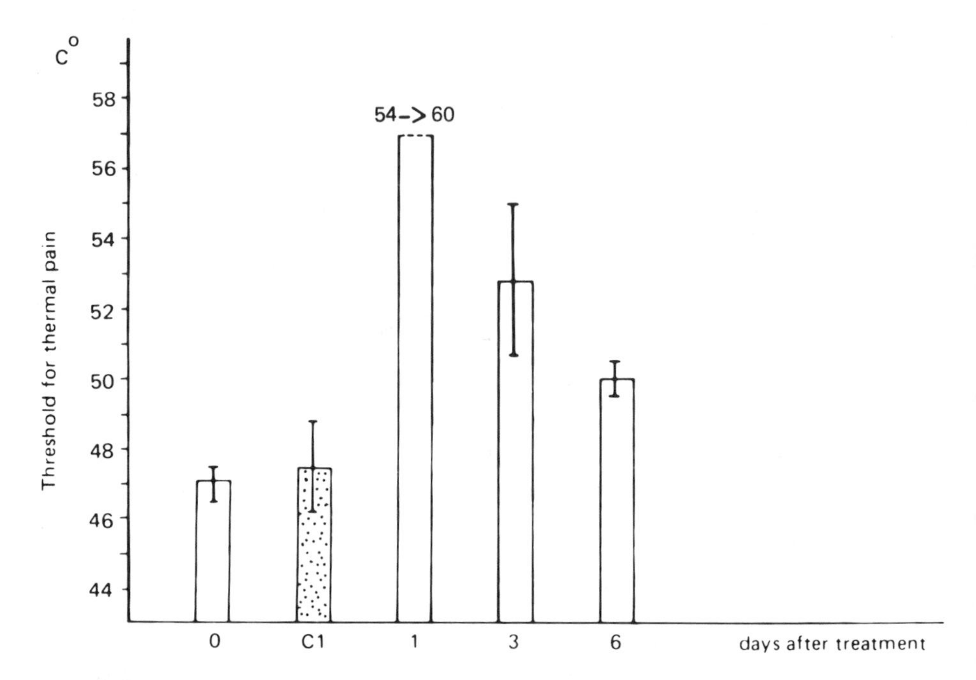


**FIGURE 3** Summary of thermal pain threshold after capsaicin desensitization where ($\bar{x} \pm$ S.E.M. n = 5). Effect of painting the skin of the volar forearm five times with an alcoholic solution of capsaicin (10 mg/ml) on thermal pain is assessed before (0) and on different days after the treatment. $C_1$: effects obtained on the other, solvent-treated side. (Unpublished details from Szolcsányi et al., 1985.)

anonociception was tested by using a probe with four polished needles (tip area: 0.03 $mm^2$) situated in a square area (3 × 3 mm). Slowly increasing pressure (20-60 mN sec) was applied to the probe alternately on the two volar areas until the subject reported pain sensation.

## III. SINGLE UNIT RECORDINGS FROM CUTANEOUS NERVES

### A. Adult Animals

On the basis of multifiber recordings from the saphenous nerve of rats and cats, it was suggested more than 10 years ago that capsaicin in the skin selectively excites polymodal nociceptors and warm receptors (Szolcsànyi, 1977). Table 2 summarizes the results of subsequent single-unit studies. The following conclusions can be deduced from these data.

**TABLE 2** Stimulatory and Desensitizing Effects of Capsaicin on Single Cutaneous Sensory Units

| Type of unit | (1) Rabbit ear i.art | | (2) Cat leg blister base | | (3) Rat paw topical | | (4) Rat paw i.art | (5) Rat paw syst. | (6) Rat paw perineural | (7) Rat paw i.dermal | (8) Human skin topical |
|---|---|---|---|---|---|---|---|---|---|---|---|
| | Stim. | Des. | Stim. | Des. | Stim. | Des | Stim. | Des. | Des. | Stim. | Stim. |
| C-MH | ⊕ | ⊕ | ⊕ | ? | ⊕ | ⊕ | ⊕ | ⊕ | ⊕ | ⊕ | ⊕ |
| HTM | ∅ | | ∅ | | | | ∅ | ∅ | ∅ | | |
| LTM | ∅ | | ∅ | | ∅ | | ∅ | ∅ | ∅ | | |
| Cold | ∅ | ∅ | | | ? | ∅ | ∅ | ∅ | ∅ | | |
| Warm | /+/ | /+/ | /+/ | /+/ | ? | | | | | | |
| "Freezing" | | | | | ∅ | | | | | | |
| Mcold | | | | | | | | | | ? | |
| A$\delta$D-hair | ∅ | ∅ | | | ∅ | | | | | | |
| MH | | | | ∅ | | | ⊕ | /+/ | | | |
| HTM | ∅ | ∅ | | | ∅ | | ∅ | | | ∅ | |
| A$\beta$G-hair | ∅ | | | | ∅ | | | | | | |
| Field | ∅ | | | | | | | | | | |
| SA I | | | ∅ | | ∅ | | ∅ | | | | |
| SA II | | | | | ∅ | | ∅ | | | | |
| "RA" | | | ∅ | | | | | | | | |

MH, polymodal nociceptor; HTM, high-threshold mechanoreceptor; LTM, low-threshold mechanoreceptor; Mcold, mechanocold receptor; SA, slowly adapting mechanoreceptor; RA, rapidly adapting mechanoreceptor; Stim., excitation; Des., desensitization; Syst., systemic.

*References*: 1, Szolcsànyi, 1980, 1983, 1987a; 2, Foster, Ramage, 1981; 3, Kenins, 1982; 4, Szolcsànyi, 1985, Szolcsànyi et al., 1987; 5, Lynn et al., 1984, Szolcsànyi, 1985, Szolcsànyi et al., 1987; 6, Petsche et al., 1983, Handwerker et al., 1984, Lynn et al., 1984; 7, Martin et al., 1987; 8, Konietzny and Hensel, 1983.

1. Capsaicin has a highly selective excitatory and blocking action among the physiologically classified types of cutaneous units. The polymodal nociceptors are affected by the agent irrespective of whether their fibers have conduction velocities in the C- or A-delta range.

2. C-fiber polymodal nociceptors have been described as unusually sensitive to irritant chemicals applied to unbroken skin (Perl, 1984). This group of sensory end organs was preferentially excited also when capsaicin reached the receptors from deeper tissues (intraarterial injection). Consequently, under these conditions a selective excitation cannot be due to a presumably more superficial position of the receptors within the skin.

3. Selective action has been obtained not only in the rat but in two other animal species. In the case of the rat and rabbit where quantitative data are available, the species differences with respect to threshold doses for nocifensor reaction and excitability of polymodal nociceptors by capsaicin parallel one another.

4. When capsaicin diffuses into the tissues from the bloodstream, the primary site of its blocking action is on the receptors and not on the conducting axon. In the rabbit most of the polymodal nociceptors desensitized by intraarterial injections still responded either to chemical, mechanical, or noxious heat stimuli when one or two types of these stimuli became ineffective.

5. The selective site of action was verified over a 100-fold dose range in studies where intraarterial injections were made. In the rat two out of eight SA mechanoreceptors were activated after a latency of several seconds by extremely high doses of capsaicin. In these cases the response was probably due to vasodilation and plasma extravasation induced by the agent.

Recordings from only two warm receptors have been reported; these receptors seem to be relatively rare in the skin. Nevertheless their participation in the thermoregulatory and sensory effects of capsaicin is strongly supported by indirect evidence (Szolcsànyi, 1982, 1983a).

According to a recent hypothesis, novel classes of cutaneous C nociceptors exist in the monkey which are different from C-polymodal nociceptors with respect to their insensitivity to mechanical stimuli. These units responded readily to intracutaneous injection of capsaicin (La Motte et al., 1987). In a series of experiments on the rabbit ear (Szolcsànyi, 1987), all electrically excitable single C-fiber units were tested by intraarterial injection. Excluding the cold receptors, only two mechanically insensitive unclassified C fibers were found. Their responsiveness to intraarterial capsaicin did not differ (number of spikes, time course) from that of the polymodal nociceptors. It is tempting to assume that these mechanically insensitive units are situated deeper in the cutaneous layers and are more easily accessible to chemical agents through diffusion when irritants are injected directly into the skin, but not when a more even distribution is achieved through the microcirculation.

6. Perineural or systemic capsaicin pretreatments induce selective long-term functional and neurotoxic impairments of polymodal nociceptors. After perineural application of capsaicin (33 mM) to the sciatic nerve, the increase in noxious heat threshold (3.5 °C) of the polymodal nociceptors (Lynn et al., 1984) was similar to that obtained in behavioral studies (Table 1). Furthermore, 4 weeks after pretreatment with a total systemic dose of 400 mg/kg, rats showed no increase in heat threshold when measured either behaviorally or via single-unit recordings. The fact that these two types of treatments resulted in a similar loss of polymodal nociceptive units (32-40%) underlies the importance of functional impairment of the desensitized primary afferent neurons in the antinociceptive effect of capsaicin.

**B. Neonatal Treatment**

There were high expectations for the use of this treatment in single-unit studies, since the overwhelming majority of cutaneous C afferents are polymodal nociceptors and neonatal pretreatment of rats has been claimed to induce a selective degeneration of chemosensitive primary sensory neurons (Jancsò, 1977, 1987). The drastic dose-dependent reduction (over 50%) of C- and A-delta afferents (Nagy et al., 1983), however, resulted in an indiscriminate loss of functionally identified sensory C fibers from the skin (Handwerker et al., 1984; Lynn et al., 1984; Welk et al., 1984). Therefore, for evaluation of the functional consequences of neonatal pretreatment in the rat, the following considerations based on the above findings should be kept in mind.

1. In the neonatal rat capsaicin apparently has no selective action on polymodal nociceptors or on chemosensitive primary afferent neurons.

2. Polymodal nociceptive neurons are destroyed by neonatal treatment in similar proportion to other C afferents. In the adult rat the surviving units are quite normal in their properties. Figure 4 shows that three polymodal nociceptive units (one in the middle and two in the lower records) from the saphenous nerve of the rat pretreated at the neonatal age with a dose of 50 mg/kg responded similarly to intraarterial capsaicin as the polymodal nociceptor of an adult untreated rat (upper record). The effects to repeated doses were reproducible. Thus, the capsaicin sensitivity of the units is retained.

3. The spectrum of damaged primary afferent neurons depends on the dose (Nagy et al., 1983). Selective loss of C afferents has been observed after 20-30 mg/kg, but the "optimal dose" of 50 mg/kg (Jancsò et al., 1987) already induced degeneration of A-delta fibers. After 85 mg/kg not only the B type of small dark cells but a portion of A-type sensory neurons were destroyed, although 10-15% of the afferent C fibers were still present (Lawson

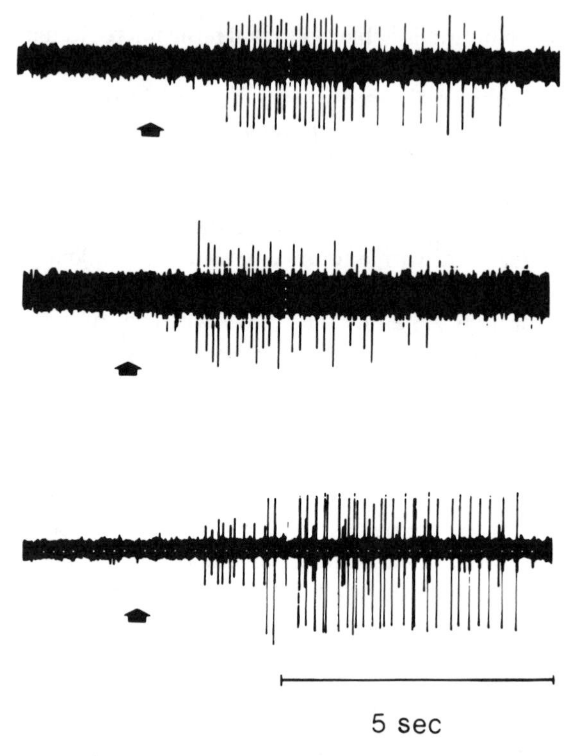


**FIGURE 4** Single unit recordings from C-polymodal nociceptive units (two on the lower tracing) of the rat's saphenus nerve. Capsaicin was injected (0.5 $\mu$g) close arterially into the femoral artery. Upper recording is from a control rat, lower recordings are from two rats which were pretreated in the neonatal period with 50 mg/kg capsaicin. (Unpublished results from Szolcsànyi et al., 1987.)

and Harper, 1984). This dose-dependent shift in selectivity makes it difficult to relate observed biological effects to any well-defined group of primary afferent neurons.

On the other hand, the drug still exhibits an important aspect of selectivity in immature newborn rats. A variety of evidence suggests that efferent neurons and their processes are not eliminated or altered after this treatment (Cervero and McRitchie, 1982). Consequently, it has been used successfully to determine the sensory nature of fibers in which various neuropeptides are localized.

4. We introduced the term "capsaicin-sensitive" sensory neurons or fibers to designate afferent neurons or fibers which are selectively activated

and impaired in function by capsaicin (Szolcsànyi and Barthò, 1978; Szolcsànyi, 1982, 1984b). This label has been accepted and is used frequently even to describe the role of sensory neurons destroyed by capsaicin in newborn animals. After the neonatal treatment, however, the experiments are made in the adult when, say, C mechanoreceptors and some other enteroreceptors damaged by capsaicin in the immature state are no longer sensitive to the drug. Thus, although this terminology turned out to be very useful, it could also be a source of confusion. Therefore, it is suggested that two different acronyms be used: CSA for *c*apsaicin-*s*ensitive afferents in the *a*dult animal and CSB for *c*apsaicin-*s*ensitive afferents in the new*b*orn animal. Furthermore, pretreated adult rats can be labeled as CAP-A rats, and rats treated with capsaicin in the first days of their life as CAP-B rats. The latter abbreviations can also be used to indicate the differences between treatments.

5. An advantage or disadvantage of this method, depending on the scope of investigations, resides in the fact that after the loss of a considerable portion of predominantly slowly conducting afferentation, profound reorganization occurs in the somatosensory afferent pathway within the central nervous system (Réthelyi et al., 1986; Saporta, 1986). Similar secondary changes in the peripheral tissues deprived from the sensory efferent influence of CSA fibers from birth should also be considered.

It is important to note that lower resting values of blood pressure and various changes in cardiorespiratory reflexes have been reported in CAP-B rats. Similar alterations were not observed in CAP-A rats (up to a dose of 500 mg/kg) or CAP-A guinea pigs, with the only exceptions being that in WKY and spontaneously hypertensive rats slightly lower blood pressures were noted (for reference, see Maggi and Meli, 1988).

## IV. EFFECTS ON SENSORY NEURONS OF THE TRIGEMINAL, NODOSAL, DORSAL ROOT (DRG) GANGLIA

### A. Electrophysiological Recordings and Ion Flux Measurements In Vitro

Microelectrode and voltage clamp studies performed in vitro provided further evidence that a distinct subpopulation of sensory neurons has the unique characteristic of being sensitive to the excitatory and long-lasting blocking effects of capsaicin.

Depolarization with conductance increase in C-type neurons of dorsal root and nodosal ganglia (Heyman and Rang, 1985; Bevan et al., 1987; Marsh et al., 1987) and in a subpopulation of DRG neurons in culture (Baccaglini and Hogan, 1983) was observed in response to nanomolar concentrations of capsaicin. A-type cells, or an unclassified capsaicin-resistant subpopulation in tissue culture, sympathetic neurons of the rat, and DRG neurons from the chicken were not depolarized by high concentrations of the agent (10 μM). Depolarization of C-type cells was attributed to an increase in calcium, sodium, and potassium conductances, while the chloride conductance remained unchanged (Wood et al., 1988; Winter, 1987). Mitochondria rather than the endoplasmatic reticulum are the probable destination of the accumulated calcium because ruthenium red inhibits calcium uptake to the DRG cells (Wood et al., 1988) and it also decreases the effectiveness of capsaicin at the peripheral processes (Maggi et al., 1988). These specific actions on cation conductances were not inhibited by tetrodotoxin or calcium channel blockers. Schwann cells, sympathetic neurons, and a number of neuronal cell lines failed to respond to capsaicin by calcium uptake.

Voltage clamp studies on nodosal ganglion cells are in agreement with these findings. Application of capsaicin (0.3 μM) induced an inward current, often followed by an outward current (Marsh et al., 1987). It is interesting to note, however, that no inward current to capsaicin was observed when the intracellular perfusion technique was used on isolated DRG cells of rats, guinea pigs, or chicken (Petersen et al., 1987; Taylor et al., 1984; Szolcsànyi et al., 1984). On DRG cells of all three species the most pronounced effect was a reversible reduction of outward (potassium) current when capsaicin was applied in high concentration (30 μM) either to the outer or inner surface of the cell membrane. This was characteristic also for those DRG neurons which were "chemosensitive" in the sense that they responded with inward current to ATP or protons (Krishtal et al., 1983). Figure 5 shows examples of a reduction of outward currents in response to capsaicin. It can be seen that the threshold concentration for this effect is above 3.3 μM. Thus some membrane effects induced by high (1-200 μM) in vitro concentrations of capsaicin, although they are selective in some respects (Erdélyi and Such, 1985), seem not to be related to the irritant and desensitizing effect induced by capsaicin in vivo in mammals (Szolcsànyi et al., 1984; Bevan et al., 1987). Nevertheless, these results provide further evidence that the neurotoxic effects of capsaicin in high concentrations cannot be attributed simply to damage of the lipoprotein cell membrane of sensory neurons. It seems that well-preserved intracellular structures, e.g., mitochondria (which are certainly altered during intracellular perfusion), are necessary for reproducing the in vivo cellular responses to capsaicin under in vitro conditions.

### B. Ultrastructural Results

#### 1. Adult animals

Systemic treatment of rats results in severe ultrastructural changes in about 50% of the small, dark, B-type neurons in the nodosal, trigeminal, and dorsal root ganglia. (This group comprises the sensory neurons with unmyelinated

DRG - neurone

PH and ATP sensitive

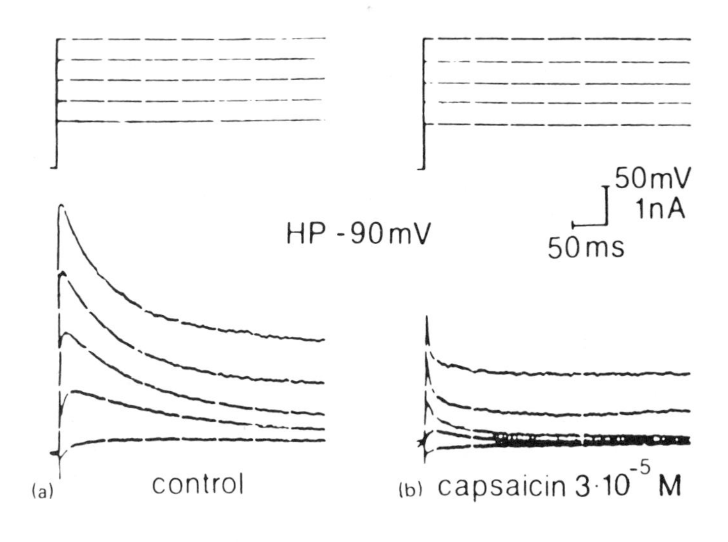


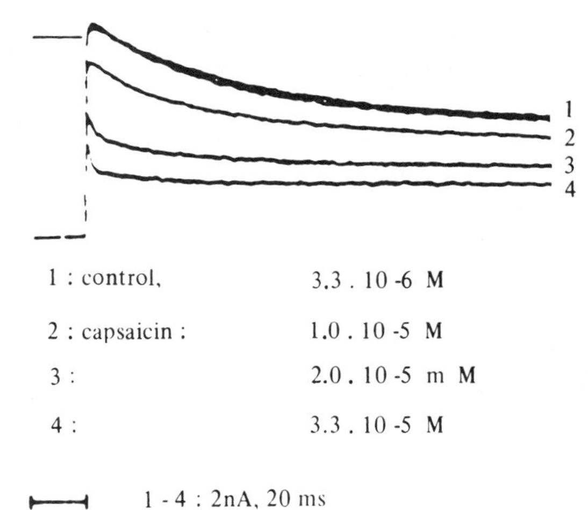


Plus + 90 m V

HP – 100 m V

(c)

**FIGURE 5** Effect of capsaicin on outward currents of two dorsal root ganglion cells. Voltage clamp recordings, intracellular perfusion technique. (Unpublished details from Taylor et al., 1984; Szolcsànyi et al., 1984.)

C fibers denoted as C-type cells by electrophysiological characterization.) A cells remain unaffected.

One of the most striking features of this neurotoxic response is that the mitochondrial damage (swelling, disorganization of christae, dense amorphous material in the matrix) and an irregularly outlined nucleus are both still present 42-60 days after the treatment (Joò et al., 1969; Szolcsànyi et al., 1975, Chiba et al., 1986). These changes, as well as dilation of endoplasmatic reticulum and Golgi apparatus, are due to a direct action of capsaicin on a major subpopulation of B-type neurons, since it can be reproduced by exposing the nodosal ganglia in vitro to 10 μM capsaicin for 30 min (Marsh et al., 1987).

Besides the above neurotoxic features, increased nuclear and cytoplasmic basophilia and vacuolization of the perikaryon were also observed in vivo within the first 6 hr after injection of capsaicin. Using these signs as an indication of cell death, the authors claim that after an exceptionally high first dose of 100 mg/kg, 17% of the total neural population underwent degeneration (Jancsò et al., 1985, 1987). The affected neurons again belonged to the B-type DRG cells. Previous studies revealed that the majority of these cells seemed to survive with damaged mitochondria for months. Furthermore, the analgesic effect of capsaicin is more related to the shift in threshold of the polymodal nociceptors than to the percent loss of fibers (see Table 1 and discussion of single-unit data). Therefore it is an oversimplification to relate all long-term effects of capsaicin to cell death, particularly when it is not verified by morphological evidence of phagocytic activity in the DRG.

Convincing evidence suggests, however, that after systemic capsaicin treatment peripheral and central terminals and unmyelinated axons in rats and guinea pigs do degenerate (Palermo et al., 1981; Hoyes and Barber, 1981; Papka et al., 1984; Chung et al., 1985; Jancsò et al.,1 985, 1987). In addition, in vitro exposure of specimens of rat trachea to capsaicin (660 μM for 5 min) elicited degeneration of intraepithelial axons (Hoyes et al., 1981). No degeneration, only ultrastructural changes lasting for 24 hr, occurred after instillation of a 1% capsaicin solution into the eyes of rats (Szolcsànyi et al., 1975).

The profound, long-lasting ultrastructural changes which are also recognizable under the light microscope provide a unique means of relating the function to the neuropeptide content of sensory neurons (Chiba et al., 1986). Furthermore, since the selective calcium uptake can be visualized by cobalt staining (Winter, 1987; Wood et al., 1988), this approach might even be utilized to develop a selective staining for capsaicin-sensitive nerve fibers in different tissues. The nonlethal neurotoxic effect of capsaicin on sensory neurons and their processes is associated with a pronounced, nonselective increase in cation permeability (Wood et al., 1988). Therefore it seems likely that staining the neurons and terminals of capsaicin-treated animals with silver ions (Eager, 1970) according to the Fink-Heimer technique is not necessarily a sign of Wallerian degeneration.

**2. Neonatal animals**

Systemic treatment of the newborn rat, mouse, and dog results in a dose-dependent reduction in the number of sensory neurons and fibers in the adult animal. Smaller doses (15-30 mg/kg in the rat) seem to affect only B-type neurons with unmyelinated fibers. After higher doses A-type cells and A-delta and A-beta fibers also degenerate. This dose-dependent change in selectivity (Nagy et al., 1983; Lawson, 1987) is in striking contrast to the adult treatment, where a slight fall in A-delta fiber population observed after a dose of 100 mg/kg could not be increased further by increasing the capsaicin dose (Jancsò et al., 1987). A detailed description and interpretation of these data—notably with a summary of quantitative changes reported by different authors (Lawson, 1987)—can be found in the following recent papers (Lawson and Harper, 1984; Jancsò et al., 1985, 1987; Lawson, 1987; Hiura and Sakamoto, 1987).

## V. SENSORY-EFFERENT FUNCTION OF CAPSAICIN-SENSITIVE SENSORY NEURONS

Capsaicin applied to the exteroceptive mucosal areas or injected under the skin not only elicits afferent impulses from the capsaicin-sensitive receptors, but also induces an efferent response of these nerve terminals, which manifests itself in neurogenic inflammation and vasodilation (Jancsò et al., 1967, 1968; Jancsò-Gàbor and Szolcsànyi, 1972; Szolcsànyi, 1984a,b, 1988). The phenomenon is identical to the classical antidromic vasodilation response described by Stricker, Bayliss, Langley, and Lewis (for reference, see Bayliss, 1923; Jancsò et al., 1967, 1968; Holzer, 1988). However, the theory that some receptors have an efferent function was put forward for the first time and supported by experimental data 20 years ago (Jancsò et al., 1968). Previously the exclusive explanation for this reaction had been the axon reflex. As Bayliss stated, "I pointed out that one possibility might be that the dorsal root fibres divide near their peripheral terminations, one branch supplying the sensory end-organ in skin, muscle, etc., while the other ends as an efferent, inhibitory end-organ on the muscular coat of the arterioles . . . which is similar in some ways to that involved in Langley's 'axon-reflex' . . ." Further, Bayliss concluded after describing Ninian Bruce's "evidence in its favour": "Vasodilatation is brought about by impulses from skin receptors, these impulses passing back from the nerve fibres by branches distributed to the inhibitory terminations in the arterioles. It seems desirable, nevertheless, that the experiments should be repeated" (Bayliss, 1923).

Ninian Bruce's experiments concerning the abolition of neurogenic inflammation by topical application of local anesthetics was invalidated by us under the direction of the late Nicholas Jancsò. During the last two decades further evidence has been obtained which enabled me to formulate the concept that polymodal nociceptors and other capsaicin-sensitive mucosal and other enteroceptors have the common characteristic of serving a dual sensory-efferent function (cf. Szolcsànyi, 1982, 1984a,b, 1988; Maggi and Meli, 1988; Holzer, 1988).

During the last 10 years an impressive body of evidence has accumulated throughout the world that the principle mediators of the capsaicin-sensitive efferent responses (inflammation, vasodilation, bronchoconstriction, increase in mucociliary movement in respiratory airways, improvement of peristalsis in gastrointestinal tract, constriction of iris, urinary bladder etc.) are the tachykinins (substance P, neurokinin A) and calcitonin gene-related peptide (for reference, see Lundblad, 1984; Maggi and Meli, 1988; Holzer, 1988). It should be kept in mind, however, that even the CSA population of sensory neurons is comparible in number of neurons to the whole sympathetic adrenergic system. Furthermore, it seems that the neuropeptide repertoire of a functionally homogeneous subpopulation of sensory neurons is variable. Capsaicin still will be a useful tool to clarify the function of neuropeptides, but it is important to realize that a considerable proportion of CSB sensory neurons are in fact not capsaicin-sensitive in adult animals. It should be mentioned that neonatal pretreatment depletes, for example, the vasoactive intestinal polypeptide and bombesin from the sensory neurons, although neither is depleted or released in adult animals (for reference, see Maggi and Meli, 1988; Holzer, 1988).

Another point which deserves comment is related to the effectiveness of this efferent response. Antidromic stimulation of the dorsal roots in the rat and transcutaneous electrical stimulation in the human skin revealed that already one or two impulses are sufficient to evoke vasodilation (Magerl et al., 1987; Szolcsànyi, 1988; Szolcsànyi et al., 1988). Furthermore, the extremely low frequency (0.025-0.2 Hz) of dorsal stimulation that is optimum for the vasodilatory response suggests that the efferent response takes place at low intensities of stimulation which are insufficient to evoke nociception or pain sensation (Szolcsànyi et al., 1988). The mechanical threshold of polymodal nociceptors is moderately high, e.g., it can be activated by pressing a toothbrush to the skin or by knocking on the door, practices which, although they do not evoke pain, still elicit discharges of these capsaicin-sensitive receptors. It is conceivable, therefore, that low-level, nonpainful activity in these receptors might participate in physiological regulation of cutaneous microcirculation. Stronger stimuli (chemical, mechanical, or thermal) elicit pain and neurogenic inflammation. These two steps, which both have bidirectional sensory-efferent aspects, form the basis of the theory we have called "multiple functions of capsaicin-sensitive receptors" (Szolcsànyi, 1988; Szolcsànyi et al., 1988). Preliminary experiments already support the notion that cutaneous blood flux is diminished in skin areas deprived of the majority of capsaicin-sensitive innervation (Sann et al., 1988).

## VI. CONCLUSIONS

In order to facilitate the appropriate use of capsaicin in further research, the present state of knowledge about the site and mode of action at the single-

**TABLE 3** Response Stages of the Capsaicin-Sensitive Primary Afferents (CSA) to the Irritant

*Stage 1. Excitation*

1. Depolarization of the peripheral, central terminals or the cell body by increasing the cation permeability through a tetrodotoxin-insensitive process.
2. Local depolarization initiates regenerative spike potentials at the receptors and maybe at other parts of the neuron.
3. Calcium-dependent release of substance P, other tachykinins, and calcitonin gene-related peptide.
4. Time course: seconds; recovery: seconds or minutes (topical or in vitro application). Effect of systemic application lasts for hours.

*Stage 2. Sensory neuron blocking effect*

1. Desensitization of the sensory receptors to one or more types of stimuli.
2. No measurable depletion of substance P in the peripheral terminals.
3. Ultrastructural changes may occur, but there is no degeneration of the fibers.
4. Start: immediately after stage 1; recovery: hours or days.

*Stage 3. Long-term selective neurotoxic impairment*

1. Sensory receptors are unresponsive to stimuli.
2. Inhibition of axonal transport and its antagonization by nerve growth factor.
3. Depletion of the neuropeptide content from the whole neuron.
4. The cell body is preserved with swollen mitochondria, but their axonal processes might degenerate.
5. Start: within minutes; duration: several weeks and some alterations are irreversible.

*Stage 4. Irreversible cell destructin*

1. Karyolysis within 20 min in neonatal rats.
2. Dose-dependent indiscriminate loss of all types of C afferents and, to a lesser extent, myelinated fibers and even A-type cells of the neonatal rat.
3. Receptors of CSA neurons with unchanged sensitivity to stimuli including capsaicin are still present.

receptor and neural level was critically surveyed. On the basis of these results a distinction between the terms for capsaicin-sensitive neurons in the adult (CSA) and neonate (CSB) has been underlined. The importance of functional and ultrastructural changes in the long-term antinociceptive effect of capsaicin has been reemphasized. Results obtained during the last 3 years and discussed here in detail support the usefulness of differentiating four stages in the action of capsaicin on CSA neurons, which depend on concentration, dose, and time course. These stages have been reported elsewhere (Szolcsànyi, 1985) and a simplified version without the list of references is shown in Table 3.

It was already pointed out (Szolcsànyi, 1982, 1984b) that these stages resemble the effects of guanethidine on adrenergic neurons. In this sense stage 1 would correspond to the tyramine-like effect; stage 2 to the bretylium-like effect; stage 3 to adrenergic neuron blockade, when depletion of noradrenaline is present; and stage 4 to the incomplete sympathectomy which can be produced by guanethidine in the rat (particularly in the neonate). This analogy might help in understanding the output of the cellular response under different experimental conditions; but it should be kept in mind that the intracellular mechanism of these two drugs seems to be entirely different.

From a therapeutic point of view, the stage 1 response to capsaicin has been used for centuries to achieve a counterirritant effect in low back pain and arthritic diseases. Stage 2 seems to be a target of drug development for having capsaicin congeners that have a negligible or no stage 1 and stage 3 effects. Stage 3 could also be utilized for therapeutic purposes by topical application (e.g., in vasomotor rhinitis—Maggi and Saria preliminary observations—or around the nerves or dorsal roots in intractable pain syndromes). The first steps have already been taken in this line. The field for further research seems not only fascinating but also promising in the fight against pain and suffering, and in the attempt to find remedies for various inflammatory, asthmatic, gastrointestinal, and skin diseases.

## ACKNOWLEDGMENT

This work was supported by research grants OTKA-84 and TKT 287. The author is indebted to his coworkers listed as coauthors in the references.

## REFERENCES

Baccaglini, P. J., and Hogan, P. G. (1983). Some rat sensory neurons in culture express characteristic of differentiated pain sensory cells. *Proc. Natl. Acad. Sci. USA* 80: 594-598.

Baranowski, R., Lynn, B., and Pini, A. (1986). The effect of locally applied capsaicin on conduction in cutaneous nerves in four mammalian species. *Br. J. Pharmacol.* 89: 267-277.

Barnes, P. J., Chung, K. F., Lammers, J.-W. J., McCusker M., and Minette, P. (1988). Non-adrenergic bronchodilator mechanism in man. *J. Physiol.* 396: 179.

Bayliss, W. M. (1923). *The Vaso-motor System*, Longmans Green, London.

Bevan, S. J., James, I. F., Rang, H. P., Winter, J., and Wood, J. N. (1987). The mechanism of action of capsaicin: A sensory neurotoxin. In: *Neurotoxins and Their Pharmacological Implications.* P. Jenner (Ed.). Raven Press, New York, pp. 261-277.

Buck, S. H., and Burks, T. F. (1986). The neuropharmacology of capsaicin: Review of some recent observations. *Pharmacol. Rev.* 38: 179-226.

Buck, S. H., Walsh, J. H., Davis, T. P., Brown, M. R., Yamamura, H. I., and Burks, T. F. (1983). Characterization of the peptide and sensory neurotoxic effects of capsaicin in the guinea pig. *J. Neurosci.* 3: 2064-2074.

Carpenter, S. E., and Lynn, B. (1981). Vascular and sensory responses of human skin to mild injury after topical treatment with capsaicin. *Br. J. Pharmacol.* 73: 755-758.

Cervero, F., and McRitchie, A. (1981). Neonatal capsaicin and thermal nociception: A paradox. *Brain Res.* 215: 414-418.

Cervero, F., and McRitchie, H. A. (1982). Neonatal capsaicin does not affect unmyelinated efferent fibers of the autonomic nervous system: Functional evidence. *Brain Res.* 239: 283-288. *Neurosci. Lett.* 53: 221-226.

Chiba, T., Masuko, S., and Kawano, H. (1986). Correlation of mitochondrial swelling after capsaicin treatment and substance P and someatostatin immunoreactivity in small neurons of dorsal root ganglion in the rat. *Neurosci. Lett.* 64: 311-316.

Chung, K., Schwen, R. J., and Coggeshall, R. E. (1985). Ureteral axon damage following subcutaneous administration of capsaicin in adult rats. *Neurosci. Lett.* 53: 221-226.

Coderre, T. J., Abbott, F. V., and Melzack, R. (1984). Behavioral evidence in rats for a peptidergic-noradrenergic interaction in cutaneous sensory and vascular function. *Neurosci. Lett.* 47: 113-118.

Donnerer, J., and Lembeck, F. (1982). Analysis of the effect of intravenously injected capsaicin in the rat. *Naunyn-Schmiedeberg's Arch. Pharmacol.* 320: 54-57.

Eager, R. P. (1970). Selective staining of degenerating axons in the central nervous system by a simplified silver method: Spinal cord projections to external cuneate and inferior olivary nuclei in the cat. *Brain Res.* 22: 137-141.

Eimerl, D., and Papir-Kricheli, D. (1987). Epidural capsaicin produces prolonged segmental analgesia in the rat. *Exp. Neurol.* 97: 169-178.

Erdélyi, L., and Such, Gy. (1985). The effects of capsaicin on action potential and outward potassium currents in a bursting neuron of the snail, *Helix pomatia* L. *Neurosci. Lett.* 55: 71-76.

Fitzgerald, M. (1983). Capsaicin and sensory neurones—a review. *Pain* 15: 109-130.

Foster, R., and Ramage, A. G. (1981). The action of some chemical irritants on somatosensory receptors of the cat. *Neuropharmacology* 20: 191-198.

Fuller, R. W., Dixon, C. M. S., and Barnes, P. J. (1985). Bronchoconstrictor response to inhaled capsaicin in humans. *J. Appl. Physiol.* 58: 1080-1084.

Gamse, R., Jancsò, G., and Kiràly, E. (1984). Intracisternal capsaicin: A novel approach for studying nociceptive sensory neurons. In: *Antidromic Vasodilatation and Neurogenic Inflammation.* L. A. Chahl, J. Szolcsànyi, and F. Lembeck (Eds.). Akadémiai Kiadò, Budapest. pp. 93-110.

Gamse, R., Wax, A., Zigmond, R. E., and Leeman, S. E. (1981). Immunoreactive substance P in sympathetic ganglia: Distribution and sensitivity towards capsaicin. *Neuroscience* 6: 437-441.

Geppetti, P., Fusco, B. M., Marabini, S., Maggi, C. A., Fanciullacci, M., and Sicuteri, F. (1988). Secretion, pain and sneezing induced by the application of capsaicin to the nasal mucosa in man. *Br. J. Pharmacol.* 93: 509-514.

Green, B. G. (1986). Sensory interactions between capsaicin and temperature in the oral cavity. *Chem. Senses* 11: 371-382.

Handwerker, H. O., Holzer-Petsche, U., Heym, Ch., and Welk, E. (1984). C-fibre functions after topical application of capsaicin to a peripheral nerve and after neonatal capsaicin treatment. In: *Antidromic Vasodilatation and Neurogenic Inflammation.* L. A. Chahl, J. Szolcsànyi, and F. Lembeck (Eds.). Akadémiai Kiadò, Budapest. pp. 57-82.

Heyman, I., and Rang, H. P. (1985). Depolarizing responses to capsaicin in a subpopulation of rat dorsal root ganglion cells. *Neurosci. Lett.* 56: 69-75.

Hiura, A., and Sakamoto, Y. (1987). Quantitative estimation of the effects of capsaicin on the mouse primary sensory neurons. *Neurosci. Lett.* 76: 101-106.

Holzer, P. (1988). Local effector functions of capsaicin-sensitive sensory nerve endings: Involvement of tachykinins, and other neuropeptides. *Neuroscience* 24: 739-768.

Hori, T. (1984). Capsaicin and central control of thermoregulation. *Pharm. Therap.* 26: 389-416.

Hoyes, A. D., and Barber, P. (1981). Degeneration of axons in the ureteric and duodenal nerve plexuses of the adult rat following in vivo treatment with capsaicin. *Neurosci. Lett.* 25: 19-24.

Hoyes, A. D., Barber, P., Jagessar, H. (1981). Effect of capsaicin on intrepithelial axons of the trachea. *Neurosci. Lett.* 26: 329-334.

Jancsò, G., Kiràly, E., and Jancsò-Gàbor, A. (1977). Pharmacologically induced selective degeneration of chemosensitive primary sensory neurone. *Nature* 270: 741-743.

Jancsò, G., Kiràly, E., Joò, I., Such, G., and Nagy, A. (1985). Selective degeneration by capsaicin of a subpopulation of primary sensory neurons in the adult rat. *Neurosci. Lett.* 59: 200-214.

Jancsò, G., Kiràly, E., Such, G., Joò, F., and Nagy, A. (1987). Neurotoxic effect of capsaicin in mammals. *Acta Physiol. Hung.* 69: 295-313.

Jancsò-Gàbor, A., Szolcsànyi, J., and Jancsò, N. (1970). Irreversible impairment of thermoregulation induced by capsaicin and similar pungent substances in rats and guinea-pigs. *J. Physiol.* (Lond) 206: 495-507.

Jancsò-Gàbor, A., Szolcsànyi, J. (1972). Neurogenic inflammatory responses. *J. Dental Res.* 41: 264-269.

Jancsò, N. (1955). *Speicherung. Stoffanreicherung in Reticuloendothel und in der Niere.* Akadémiai Kiadò, Budapest.

Jancsò, N. (1968). Desensitization with capsaicin as a tool for studying the function of pain receptors. In: *Proceedings of the 3rd Int. Pharmacol. Meeting, 1966, Pharmacology of Pain,* Vol. 9. Pergamon Press, Oxford, pp. 33-55.

Jancsò, N., Jancsò-Gàbor, A., and Szolcsànyi, J. (1967). Direct evidence for neurogenic inflammation and its prevention by denervation and by pretreatment with capsaicin. *Br. J. Pharmacol.* 31: 138-151.

Jancsò, N., Jancsò-Gàbor, A., and Szolcsànyi, J. (1968). The role of sensory nerve endings in neurogenic inflammation induced in human skin and in the eye and paw of the rat. *Br. J. Pharmacol.* 33: 32-41.

Jhamandas, K., Yaksh, T. L., Harty, G., Szolcsànyi, J., and Go, V. L. W. (1984). Action of intrathecal capsaicin and its structural analogues on the content and release of spinal substance P: Selective action and relationship to analgesia. *Brain Res.* 306: 215-225.

Joò, F., Szolcsànyi, J., and Jancsò-Gàbor, A. (1969). Mitochondrial alterations in the spinal ganglion cells of the rat accompanying the long-lasting sensory disturbance induced by capsaicin. *Life Sci.* 8: 621-626.

Kenins, P. (1982). Responses of single nerve fibres to capsaicin applied to the skin. *Neurosci. Lett.* 29: 83-88.

Konietzny, F., and Hensel, H. (1983). The effect of capsaicin on the response characteristic of human C-polymodal nociceptors. *J. Therm. Biol.* 8: 213-215.

Krishtal, O. H., Marchenko, S. M., Pidoplichko, V. J. (1983). Receptor ATP in membrane of mammalian sensory neurones. *Neurosci. Lett.* 35: 41-46.

La Motte, R. H., Simone, D. A., Baumann, T. K., Shain, C. N., and Alreja, M. (1987). Hypothesis for novel classes of chemoreceptors mediating chemogenic pain and itch. *Pain. Suppl.* 4: 515.

Lawson, S. N. (1987). The morphological consequences of neonatal treatment with capsaicin on primary afferent neurones in adult rats. *Acta Physiol. Hung.* 69: 315-321.

Lawson, S. N., and Harper, A. A. (1984). Neonatal capsaicin is not a specific neurotoxin for sensory C-fibres or small dark cells of rat dorsal root ganglia. In: *Antidromic Vasodilatation and Neurogenic Inflammation.* L. A. Chahl, J. Szolcsànyi, and F. Lembeck (Eds.). Akadémiai Kiadò, Budapest, pp. 111-118.

Lundberg, J. M., Brodin, E., Hua, X., and Saria, A. (1984). Vascular permeability changes and smooth muscle contraction in relation to capsaicin-sensitive substance P afferents in the guinea-pig. *Acta Physiol. Scand.* 120: 217-227.

Lundblad, L. (1984). Protective reflexes and vascular effects in the nasal mucosa elicited by activation of capsaicin-sensitive substance P-immunoreactive trigeminal neurons. *Acta Physiol. Scand.*, Suppl. 529: 1-42.

Lynn, B., Carpenter, S. E., and Pini, A. (1984). Capsaicin and cutaneous afferents. In: *Antidromic Vasodilatation and Neurogenic Inflammation.* L. A. Chahl, J. Szolcsànyi, and F. Lembeck (Eds.). Akadémiai Kiadò, Budapest, pp. 83-92.

Magerl, V., Szolcsànyi, J., Westerman, R. A., Handwerker, H. O. (1987). Laser Doppler measurements of skin vasodilation elicited by percutaneous electrical stimulation of nociceptors in humans. *Neurosci. Lett.* 82: 349-354.

Maggi, C. A., and Meli, A. (1988). The sensory-efferent function of capsaicin-sensitive sensory neurons. *Gen. Pharmacol.* 19: 1-43.

Maggi, C. A., Patacchini, R., Santicioli, P., Giuliani, S., Geppetti, P., and Meli, A. (1988). Protective action of Ruthenium red toward capsaicin desensitization of sensory fibers. *Neurosci. Lett.* 88: 201-205.

Marsh, S. J., Stansfeld, C. E., Brown, D. A., Davey, R., and Mc Carthy, D. (1987). The mechanism of action of capsaicin on sensory C-type neurons and their axons in vitro. *Neuroscience* 23: 275-289.

Martin, H. A., Basbaum, A. J., Kwiat, G. C., Goetzl, E. J., and Levine, J. D. (1987). Leukotriene and prostaglandin sensitization of cutaneous high-threshold C and A-delta mechanonociceptors in the hairy skin of rat hindlimbs. *Neuroscience* 22: 651-659. 399-406.

Mason, R. J., and Maruniak, J. A. (1983). Behavioral and physiological effects of capsaicin in red-winged blackbirds. *Pharmacol. Biochem. Behav.* 19: 857-862.

Monsereenusorn, Y., Kongsamut, S., and Pezalla, P. D. (1982). Capsaicin—A literature survey. *CRC Critical Rev. Toxicol.* 10: 321-339.

Nagy, J. I. (1982). Capsaicin: A chemical probe for sensory neuron mechanisms. In: *Handbook of Psychopharmacology*, Vol. 15. L. L. Iversen, S. D. Iversen, and S. H. Snyder (Eds.). Plenum Press, New York, pp. 185-235.

Nagy, J. I., Iversen, L. L., Goedert, M., Chapman, D., and Hunt, S. P. (1983). Dose-dependent effect of capsaicin on primary sensory neurons in the neonatal rat. *J. Neurosci.* 3: 399-406.

Paintal, A. S. (1973). Vagal sensory receptors and their reflex effects. *Physiol. Rev.* 53: 159-227.

Palermo, N. N., Brown, H. K., and Schmith, D. L. (1981). Selective neurotoxic action of capsaicin on glomerular C-type terminals in rat substantia gelatinosa. *Brain Res.* 208: 506-510.

Papka, R. E., Furness, J. B., Della, N. G., Murphy, R., and Costa, M. (1984). Time course of effect of capsaicin on ultrastructure and histochemistry of substance P-immunoreactive nerves associated with the cardiovascular system of the guinea-pig. *Neuroscience* 12: 1277-1292.

Perl, E. R. (1984). Pain and nociception. In: *Handbook of Physiology, Nervous System III.* J. M. Brookhart and V. B. Mountcastle (Eds.). Bethesda, pp. 915-975.

Petersen, M., Pierau, Fr-K., and Weyrich, M. (1987). The influence of capsaicin on membrane currents in dorsal root ganglion neurones of guinea-pig and chicken. *Pflügers Arch.* 409: 403-410.

Petsche, U., Fleischer, E., Lembeck, F., and Handwerker, H. O. (1983). The effect of capsaicin application to a peripheral nerve on impulse conduction in functionally identified afferent nerve fibres. *Brain Res.* 265: 233-240.

Pierau Fr.-K., Sann, H., Harti, G., Szolcsànyi, J. (1988). A possible mechanism of the absence of neurogenic inflammation in birds: Inappropriate release of peptides from small afferent nerves. *Agents Actions* 23: 12-13.

Réthelyi, M., Salim, M. Z., and Jancsò, G. (1986). Altered distribution of dorsal root fibers in the rat following neonatal capsaicin treatment. *Neuroscience* 18: 749-762.

Rogers, L. R. (1984). Reactions of free-ranging black bears to capsaicin spray repellent. *Wildl. Soc. Bull.* 12: 59-61.
Rozin, P., Mark, M., and Schiller, D. (1981). The role of desensitization to capsaicin in chili pepper ingestion and preference. *Chem. Senses* 6: 23-31.
Russell, L. C., and Burchiel, K. J. (1984). Neurophysiological effects of capsaicin. *Brain Res. Rev.* 8: 165-176.
Sann, H., Pintér, E., Szolcsànyi, J., Pierau, Fr.-K. (1988). Peptidergic afferents might contribute to the regulation of the skin blood flow. *Agents Actions* 23: 14-15.
Saporta, S. (1986). Loss of spinothalamic tract neurons following neonatal treatment of rats with the neurotoxin capsaicin. *Somatos. Res.* 4: 153-173.
Sizer, F., and Harris, N. (1985). The influence of common food additives and temperature on threshold perception of capsaicin. *Chem. Senses* 10: 279-286.
Szolcsànyi, J. (1977). A pharmacological approach to elucidation of the role of different nerve fibers and receptor endings in mediation of pain. *J. Physiol.* (Paris) 73: 251-259.
Szolcsànyi, J. (1980). Effect of pain-producing chemical agents on the activity of slowly conducting afferent fibres. *Acta Physiol. Acad. Sci. Hung.* 56: 86.
Szolcsànyi, J. (1982). Capsaicin type pungent agents producing pyrexia. In: *Handbook of Experimental Pharmacology*, Vol. 60. *Pyretics and Antipyretics.* A. S. Milton (Ed.). Springer-Verlag, Berlin, pp. 437-478.
Szolcsànyi, J. (1983a). Disturbances of thermoregulation induced by capsaicin. *J. Therm. Biol.* 8: 207-212.
Szolcsànyi, J. (1983b). Tetrodotoxin-resistant non-cholinergic neurogenic contraction evoked by capsaicinoids and piperine on the guinea-pig trachea. *Neurosci. Lett.* 42: 83-88.
Szolcsànyi, J. (1984a). Capsaicin and neurogenic inflammation: History and early findings. In: *Antidromic Vasodilatation and Neurogenic Inflammation.* L. A. Chahl, J. Szolcsànyi, and F. Lembeck (Eds.). Akadémiai Kiadò, Budapest, pp. 7-26.
Szolcsànyi, J. (1984b). Capsaicin-sensitive chemoceptive neural system with dual sensory-efferent function. In: *Antidromic Vasodilatation and Neurogenic Inflammation.* L. A. Chahl, J. Szolcsànyi, and F. Lembeck (Eds.). Akadémiai Kiadò, Budapest, pp. 27-56.
Szolcsànyi, J. (1985). Sensory receptors and the antinociceptive effects of capsaicin. In: *Tachykinin Antagonists.* R. Hakanson and F. Sundler (Eds.). Elsevier, Amsterdam, pp. 45-54.
Szolcsànyi, J. (1987a). Selective responsiveness of polymodal nociceptors of the rabbit ear to capsaicin, bradykinin and ultra-violet irradiation. *J. Physiol.* 388: 9-23.
Szolcsànyi, J. (1987b). Capsaicin and nociception. *Acta Physiol. Hung.* 69: 323-332.
Szolcsànyi, J. (1988). Antidromic vasodilatation and neurogenic inflammation. *Agents Actions* 23: 4-11.
Szolcsànyi, J., Anton, F., Reeh, P. W., and Handwerker, H. O. (1988). Selective excitation by capsaicin of mechano-heat sensitive nociceptors in rat skin. *Brain Res.* 446: 262-268.

Szolcsànyi, J., and Barthò, L. (1978). New type of nerve-mediated cholinergic contractions of the guinea-pig small intestine and its selective blockade by capsaicin. *Neunyn-Schmiedeberg's Arch. Pharmacol.* 305: 83-90.

Szolcsànyi, J., and Jancsò-Gàbor, A. (1976). Sensory effect of capsaicin congeners II. Importance of chemical structure and pungency in desensitizing activity of capsaicin-type compounds. *Arzneim.-Forsch.* (*Drug Res.*) 26: 33-37.

Szolcsànyi, J., Jancsò-Gàbor, A., and Joò, F. (1975). Functional and fine structural characteristics of the sensory neuron blocking effect of capsaicin. *Naunyn-Schmiedeberg's Arch. Pharmacol.* 287: 157-169.

Szolcsànyi, J., Pierau, F., and Krishtal, O. H. (1984). Effect of capsaicin on the ionic currents of isolated primary sensory neurones. *Acta Physiol. Hung.* 63: 284.

Szolcsànyi, J., Sann, H., and Pierau, Fr.-K. (1986). Nociception in pigeons is not impaired by capsaicin. *Pain* 27: 247-260.

Szolcsànyi, J., Sebök, B., and Barthò, L. (1985). Capsaicin, sensation and flare reaction: The concept of bidirectional axon reflex. In: *Substance P Metabolism and Biological Actions.* C. C. Jordan and P. Oehme (Eds.). Taylor and Francis, London, p. 234.

Szolcsànyi, J., Westerman, R. A., Magerl, W., Pintér, E. (1988). Capsaicin-sensitive cutaneous sense organs: Nerve terminals with multiple functions. *Regul. Peptides* 22: 80.

Taylor, D. C. M., Pierau, Fr.-K., Szolcsànyi, J. (1984). Effect of capsaicin on rat sensory neurons. In: *Thermal Physiology.* J. R. S. Hales (Ed.). Raven Press, New York, pp. 23-27.

Tòth-Kàsa, I., Jancsò, G., Bognàr, A., Husz, S., and Obàl, F.Jr. (1986). Capsaicin prevents histamine-induced itching. *Int. J. Clin. Pharm. Res.* 6: 163-169.

Welk, E., Fleischer, E., Petsche, U., and Handwerker, H. O. (1984). Afferent C-fibers in rats after neonatal capsaicin treatment. *Pflügers Arch.* 400: 66-71.

Welk, E., Petsche, U., Fleischer, E., and Handwerker, H. O. (1983). Altered excitability of afferent C-fibres of the rat distal to a nerve site exposed to capsaicin. *Neurosci. Lett.* 38: 245-250.

Winning, A. J., Hamilton, R. D., Shea, S. A., and Guz, A. (1986). Respiratory and cardiovascular effects of central and peripheral intravenous injections of capsaicin in man: Evidence for pulmonary chemosensitivity. *Clin. Sci.* 71: 519-526.

Winter, J. (1987). Characterization of capsaicin-sensitive neurones in adult rat dorsal root ganglion cultures. *Neurosci. Lett.* 80: 134-140.

Wood, J. N., Winter, J., James, I. F., Rang, M. P., Yeats, J., and Bevan, S. (1988). Capsaicin-induced ion fluxes in dorsal root ganglion cells in culture. *J. Neurosci.* 8: 3208-3220.

Yaksh, T. L., Farb, D., Leeman, S., and Jessell, T. (1979). Intrathecal capsaicin depletes substance P in the rat spinal cord and produces prolonged thermal analgesia. *Science* 206: 481-483.

# Chapter 8 Discussion

**Dr. Kobal:** How many of the A-delta fibers are sensitive to capsaicin?

**Dr. Szolcsànyi:** We picked up 5 units and 4 of them were activated by capsaicin and 1 was not. They are relatively rare, which is why other people have not discovered this. It may be more common in the human skin and in the monkey; there are convincing data for monkeys, and Campbell and Meyer have found such mechano-heat sensitive units.

**Dr. Silver:** Do you know if the effect that you saw of heat or temperature on the response to capsaicin plays a role in the mouth?

**Dr. Szolcsànyi:** Yes. Although we have not studied this directly in the mouth, we painted the skin with capsaicin as I told you and we saw the noxious heat threshold drop to lower temperatures. It was around 30 °C or 28 °C when a burning sensation could be felt. It is also very interesting that when capsaicin acts and there is a burning sensation in the skin, if you rub the skin you will notice tactile hyperalgesia. If you cool the hand to 25 °C, the hyperalgesia disappears, along with the burning sensation. You might think that this is only perceptual, but this can be confirmed by recording neural potentials. It could be due to sensitization of polymodal nociceptors to mechanical stimuli.

# 9
# Effects of Thermal, Mechanical, and Chemical Stimulation on the Perception of Oral Irritation

**Barry G. Green**
Monell Chemical Senses Center
Philadelphia, Pennsylvania

## I. INTRODUCTION

When it was first described as a sensory system, the common chemical sense (CCS) was considered to be a unidimensional modality whose function was to signal the presence and amount of noxious chemicals (Parker, 1912). It has become increasingly clear, however, that what is referred to as the CCS is a complex afferent system that is not easily accommodated within the classical definitions of a specific sensory system. In contrast to the other chemical sensitivities of taste and smell, the sensitivity to chemical irritants is not restricted to a single, specialized region of skin; *all* skin is to some extent sensitive to irritants (Keele, 1962; Scheuplein, 1976). Furthermore, there is as yet no clear evidence that a specific chemoreceptor (or family of chemoreceptors) exists that mediates the sensitivity to chemical irritants.

There is instead a close anatomical and functional relationship between the classical skin senses and common chemical sensitivity. Consistent with this fact and initially inspired by the inescapable relationship between intense irritation and pain, the CCS has long been considered by some as part of the sense of pain (Jones, 1954). Recent advances in understanding the properties of cutaneous nociceptors have made it clear that this view is at least partly correct: both c- and A-delta nociceptors, which are also sensitive to extremes of temperature and pressure, respond to irritating chemicals

(Torebjörk, 1974; Adriaensen et al., 1980, 1983; Foster and Ramage, 1981; Kenins, 1982; Konietzny and Hensel, 1983; Szolcsànyi, this volume). Whether or not other types of afferents exist which are *specifically* sensitive to chemical irritants remains to be seen (Simone et al., 1987; Lamotte et al., 1988), as does the extent to which thermal afferents respond to irritants (e.g., Szolcsànyi, 1977; Green and Flammer, 1988).

The obvious yet poorly understood relationship between common chemical sensitivity and the cutaneous senses has guided much of the research on oral chemical irritation in our laboratory. The experiments presented here represent some initial efforts to describe psychophysically the relationship between chemical irritation in the oral cavity and the two most common forms of cutaneous stimulation: temperature and mechanical force. In addition, a study is described which explored the interaction over time of two different chemical irritants. The latter study investigated one of the temporal characteristics of some forms of chemical irritation (i.e., "sensitization") while also investigating the degree of interaction between two different irritants. Together the experiments demonstrate that the sensitivity to chemical irritants in the mouth derives from complex and highly interactive sensory processes. The results also affirm the relationship between chemical irritation and the nociceptive sense, and demonstrate that the sensitivity to irritants often depends on both the nature and intensity of stimulation in the other cutaneous modalities.

## II. THERMAL EFFECTS

Temperature is a variable of interest with respect to chemical irritation for several reasons. First, the sensations produced by some of the most common irritants—e.g., capsaicin, piperine, and ethanol—are often described as warm, burning, or hot (e.g., Rozin et al., 1982; Stevens and Lawless, 1988). Based solely on this qualitative information, it is logical to assume that physical warming or cooling would interact with the sensations generated by the irritants. Second, it was reported over a decade ago that the burning sensation produced by capsaicin could be completely turned off by cooling the skin to 28 °C (Szolcsànyi, 1977). This result implied that the neurochemical effect of capsaicin was strongly temperature-dependent, although it did not provide information about the psychophysical properties of this temperature dependence over the normal range of skin temperatures. Third, as was pointed out above, it is well established that capsaicin and other irritants stimulate c-polymodal nociceptors, which are temperature-sensitive afferents (Bessou and Perl, 1969; Croze et al., 1976; Handwerker and Neher, 1976). The same population of nociceptors have also been shown to increase their rate of firing to endogenous chemical "irritants" when their temperatures are raised (Kumazawa and Mizumura, 1983; Kumazawa et al., 1987).

### A. Capsaicin and Temperature

Although the above evidence leaves little doubt that temperature is important for the perception of chemical irritation, it does not provide much information about the basic characteristics of the temperature effects. For example, it was not known over what range temperature would have an effect on capsaicin's burning sensation or to what degree capsaicin might in turn alter the perception of temperature. Both of these questions were of interest from the standpoint of sensory coding. It was assumed that because capsaicin had been shown to lower the threshold to heat pain (Szolcsànyi, 1977), the intensity of burning sensations produced by the compound might increase monotonically from baseline skin temperature up to the threshold for heat pain. This would imply that capsaicin "sensitized" nociceptors (and perhaps warm fibers) to temperature change over the range of innocuous temperatures. It was also of interest to learn whether capsaicin actually affected perceptions of temperature or merely produced burning sensations that could be differentiated from sensations of warmth and heat. This question was of particular interest over the range of temperatures between neutrality and approximately 40 °C, where warm receptors, not nociceptors, are normally active. An enhancement of perceived warmth throughout this range would constitute indirect evidence that the sensitivity of warm fibers is affected by capsaicin and that they therefore contribute to the sensation of irritation that capsaicin produces.

To investigate these questions a simple experiment was devised in which subjects rated either the burning sensation produced by capsaicin or the thermal sensations produced during exposure to solutions of different temperatures (Green, 1986). The experiment had two conditions: subjects received either deionized water to which capsaicin had been added (producing a 2 ppm solution) or deionized water alone. Both conditions, which were blocked, were run in every testing session and the order of presentation was counterbalanced across subjects. At the beginning of the capsaicin condition subjects sipped and expectorated two samples of the capsaicin solution (warmed to 36 °C) to establish a burning sensation in the mouth. In the test proper the solutions were either warm (34-45 °C) or cold (32-13 °C) and were sipped at the rate of one every 30 sec. Exposure to each solution lasted 6 sec during which the subjects rated (using magnitude estimation) the perceived thermal intensity in both conditions and the perceived intensity of the burning sensation in the capsaicin condition. Warm and cold temperatures were run on separate days.

Figure 1 illustrates that the temperature of the solution had a monotonic effect on the perception of the burning sensation over a wide range of temperatures. At warm temperatures (top of Fig. 1) the burning sensation increased by a factor of about 6:1, and at cool temperatures (bottom of Fig. 1)

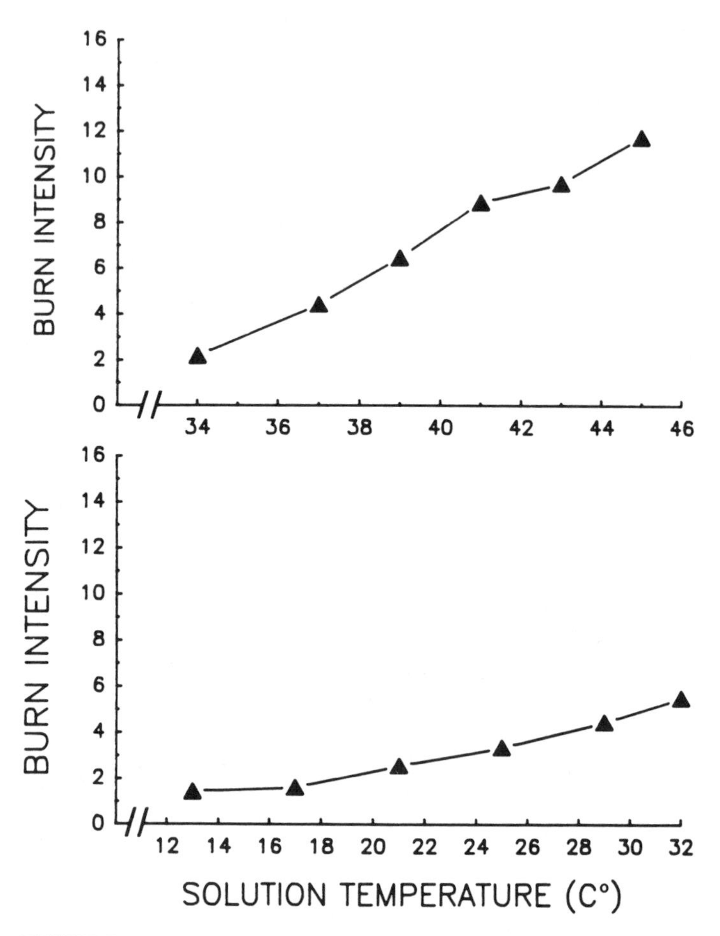


**FIGURE 1** Shown is the effect of temperature on the perceived intensity of the burning sensation produced by capsaicin in the oral cavity. *Top*: solution temperatures at and above normal oral temperature; *bottom*: solution temperatures below normal oral temperature. (Adapted from Green, 1986.)

the burning sensation declined to a low, seemingly asymptotic level. (Note that because the warm and cold temperatures were tested on different days, the numbers reported for the two conditions cannot be directly compared. It is likely that context effects tended to inflate judgments of burn in the presence of cooling relative to judgments of burn in the presence of warming.) The failure of the burning sensation to completely disappear at the

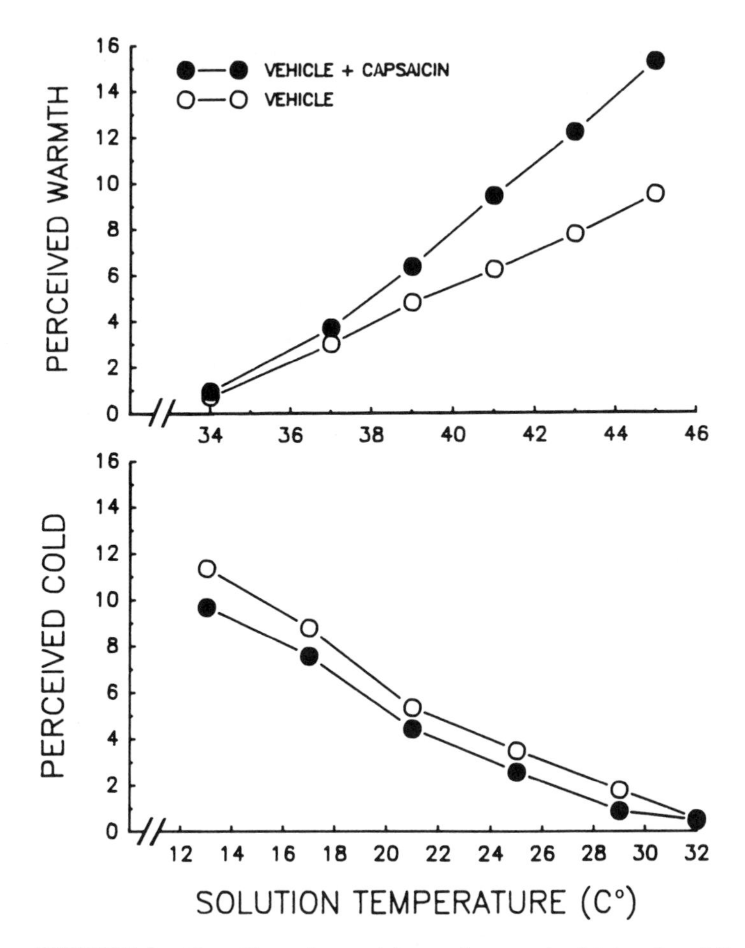


**FIGURE 2** The effect of capsaicin on the perceived intensity of thermal sensations in the oral cavity. *Top*: perceived warmth; *bottom*: perceived cold. (Adapted from Green, 1986.)

colder temperatures was established in an ancillary experiment to be due to insufficient cooling of all of the mucosal areas that were stimulated by capsaicin. Thus, when capsaicin was applied to a confined area of the lip and the skin of the lip was cooled by a contact thermode, the median perceived intensity of the burning sensation fell to zero at a temperature of 24 °C.

Figure 2 shows how capsaicin affected the perception of temperature. Looking first at the data for warmth (top of Fig. 2), it is apparent that the effect of capsaicin is greater at the higher temperatures. The virtual absence of an effect of capsaicin on the perceived intensity of warmth produced by 34 and 37 °C water suggests that capsaicin does not strongly influence the activity of warmth sensors that function at innocuous temperatures. On the other hand, the increase in perceived warmth over the control condition that occurs at the higher temperatures is in agreement with the physiological evidence that capsaicin stimulates or sensitizes polymodal nociceptors. Surprisingly, capsaicin also produced a slight but statistically significant reduction in the perception of cold (bottom of Fig. 2). The source of this effect is not apparent; capsaicin may have a heretofore unobserved inhibitory effect on cold fibers, or some sort of perceptual context effect might have come into play. In any case, we can conclude from these measurements that capsaicin seems to suppress the perception of coolness and enhance the perception of heat while leaving the perception of innocuous warmth unchanged (but see Section III.A).

## B. Temperature and Other Irritants

Because the effect of temperature on the burning sensation produced by capsaicin was so strong, an obvious question that arose was whether thermal lability was a general characteristic of oral chemical irritation. If the effect of temperature were specific to capsaicin it would imply that *at least* two types of afferents (one temperature-sensitive and the other not) mediate oral irritation; if temperature affected the perception of each irritant tested it would imply that chemical irritation is a temperature-dependent phenomenon mediated by one or more types of temperature-sensitive afferents.

An experiment was recently completed which indicated that capsaicin is not the only irritant that exhibits a temperature dependence. Four common irritants were tested in the experiment: ethanol, NaCl, piperine, and capsaicin. Capsaicin was included for purposes of comparison with the other irritants. Piperine was chosen because, on the one hand, its sensory qualities in the oral cavity are generally similar to capsaicin's, but on the other hand differences have been found in such characteristics as the areas of the mouth most affected by the two chemicals and the frequency with which certain adjectives are used to describe the sensations they produce (Stevens and Lawless, 1988). The latter data raise the possibility that the two peppers do not stimulate exactly the same nerve fibers. Ethanol and salt were included because they represent two very different types of oral irritants, one (ethanol) being a potent nasal as well as oral irritant, the other (NaCl) being a potent taste stimulus.

The experiment was run as follows: The irritants were presented to the tip of the tongue on disks of filter paper 0.38 $cm^2$ in area. The nominal concentrations used were 25% (ethanol), 3.0 M (NaCl), 2 ppm (capsaicin), and 100 ppm (piperine), which seemed in informal tests among laboratory personnel to produce roughly equal sensations of irritation. All of the irritants were applied to the filter paper by dipping the paper into solutions of the appropriate concentrations. For NaCl the solvent was deionized water; for capsaicin and piperine it was ethanol. The stimulus disks containing these substances were prepared before each session and allowed to dry. The ethanol stimuli were made up individually at each trial and applied directly to the tongue. So that all stimuli were moist when applied, a drop of deionized water was pipetted onto the filter paper disks that contained the other chemicals immediately before each was placed on the tongue.

Prior to a trial the subject rinsed his or her mouth with 36 °C deionized water. A trial began when the experimenter placed a disk of filter paper on the dorsal aspect of the tip of the subject's tongue. The subject then sat with the mouth closed for 30 sec, after which the experimenter removed the filter paper from the tongue. Fifteen seconds later the subject sipped a 20 ml aliquot of either 24 or 46 °C deionized water. This "temperature rinse" was held without swishing in the anterior of the mouth for 10 sec. After the subject expectorated the rinse he or she responded with a magnitude estimate of the intensity of the irritation perceived on the tip of the tongue at the moment before expectoration. Following the magnitude estimate (and 15 sec after expectorating the first rinse) the subject sipped another sample of water, this time of the opposite temperature as the first sip. The same timing was observed and another magnitude estimate was collected. Thus on each trial the subject was exposed to an irritant, given warm and cold water (in a sequence that was counterbalanced over trials), and asked to rate the intensity of irritation perceived after the tongue had been bathed in the water for 10 sec. A 1-min pause was inserted between trials during which the subject rinsed with 37 °C water to help clear the mouth of the previous stimulus and return the mucosa to its normal temperature.

Figure 3 contains a summary of the data from the experiment. The solid bars show the percentage decline in perceived irritation that resulted from exposing the tongue to 24 °C compared to 46 °C water on trials in which those temperatures were presented first. For example, perceived irritation was 53.9% weaker when the subjects sipped cool water rather than warm water right after exposure to piperine. The open bars show the percentage difference in perceived irritation experienced at the same two temperatures during the second period of exposure (i.e., after the tongue had been exposed to a rinse at the *other* temperature). The generally lower perceived irritation reported for the second rinse apparently reflects both the removal of stimulus from the tongue during the first rinse and any adaptation that might have occurred. Note that the rinsing effect was greatest for ethanol and least for NaCl. The result for ethanol is consistent with the relatively rapid decay in irritation that occurs for that irritant compared to the pepper compounds, and the

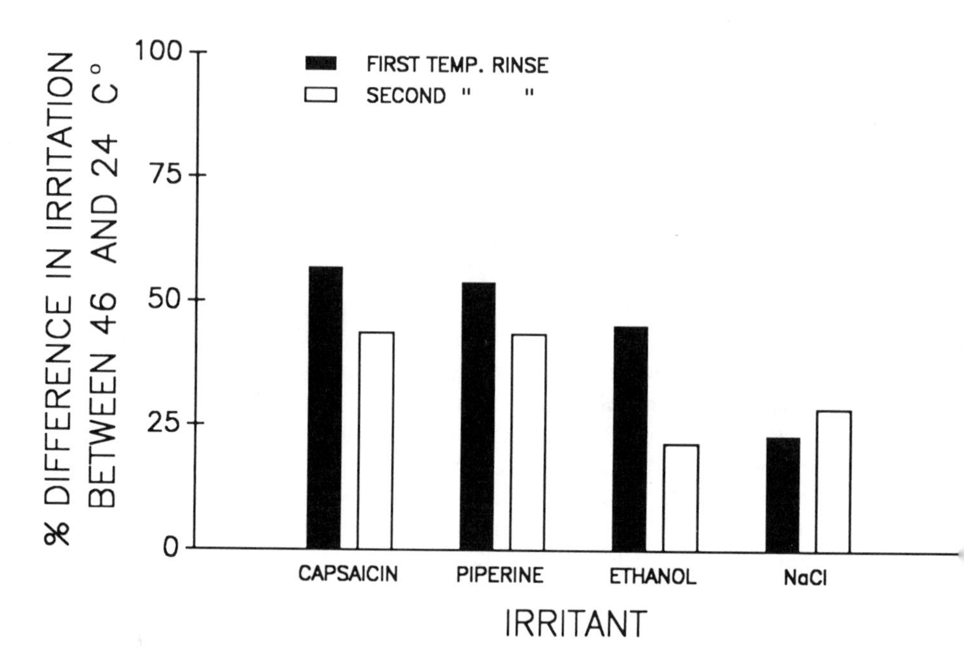


**FIGURE 3** Shown is the difference in perceived irritation reported following rinses of 46 and 24 °C deionized water for four oral irritants (the higher the percentage, the greater was the effect of temperature). Solid bars represent the mean differences obtained after the first of a pair of temperature rinses that were given on each trial; the open bars represent the differences obtained following the second rinse.

result for salt is consistent with the tendency for salt irritation to grow over time (see below).

The general conclusion to be drawn from the data is that for all four irritants, cooling significantly reduced perceived irritation (confirmed by a repeated measures ANOVA and subsequent simple effects analyses). However, the data also imply that the thermal effect is not equal in magnitude for all irritants. In particular, when the results for the first rinse are compared (solid bars), the effect of temperature was only half (or less) as large for salt as it was for the other three irritants when cooling and warming occurred immediately after exposure to the irritant. The smaller effect of temperature on salt irritation could mean that NaCl produces irritation by stimulating at least some afferents that are different from (and less temperature-sensitive than) the afferents that mediate the sensations produced by the other three irritants. This conclusion has intuitive appeal in that the quality of the irritation produced by salt (as well as its temporal properties) seems to differ from the irritation produced by the peppers and by ethanol. However, differences in the perceived intensity of the irritations produced by salt and the other irritants complicated the interpretation of the data. Despite our efforts to equate the level of irritation across compounds, salt produced an irritation that was stronger (about 30%) than the sensations produced by capsaicin and piperine, and slightly stronger than the sensation produced by ethanol. Consequently, the possibility existed that the lesser thermal lability seen for salt owed to the more intense irritation it produced.

We therefore performed another experiment in which two different concentrations of salt were used. In addition to the 3.0 M solution, a 1.5 M solution was tested under the same conditions as the preceding experiment. The 1.5 M solution produced magnitude estimates that were on average 41% lower than the estimates given to the 3.0 M solution, which meant the weaker salt solution was perceived as somewhat less intense than the capsaicin and piperine solutions of the main experiment. The results showed that the difference in perceived irritation between the 24 and 46 °C conditions was 26% for the 3.0 M solution (which was identical to the value found in the main experiment) and 36% for the 1.5 M solution. Because the size of the temperature effect for the two kinds of pepper was near 50%, it appears that salt irritation (as well as ethanol irritation) is indeed relatively resistant to cooling. However, the 10% greater effect that was observed for the 1.5 M solution compared to the 3.0 M solution also supports the notion that the thermal effect depends to some extent on the intensity of irritation. It may well be that as the intensity of stimulation increases, more extreme cooling is required to produce the same percent reduction in irritation. Consistent with this interpretation is the fact that the *absolute* reduction in perceived irritation was nearly identical for the 1.5 and 3.0 M conditions.

A final point of interest regarding the effect of temperature on salt irritation is that the occurrence of the cooling and heating effect contrasts with the absence of an effect of temperature on suprathreshold sensations of salt taste (Green and Frankmann, 1987). Temperature sensitivity is apparently one stimulus dimension on which the afferent systems that mediate salt taste differ from the afferent systems that mediate salt irritation.

## III. MECHANICAL EFFECTS

During the course of our studies we have investigated two possible interactions between mechanical stimulation and chemical irritation in the oral region: the effect of vibration on the sensations produced by capsaicin, and the effect of mechanical pressure on the perceived intensity of irritation produced by ethanol.

### A. Effect of Vibration on Irritation

This experiment was motivated by two observations. The first was an anecdotal one: namely, when questioned informally individuals who ate spicy foods often said that the "heat" imparted by the foods seemed to increase when they stopped eating. One possible explanation of this seeming paradox is that the mechanical stimulation which accompanies chewing acts to mask or otherwise inhibit the irritation produced by the spices. The second observation had come from the study of the effect of temperature on the irritation produced by capsaicin that was discussed above. In addition to the experiment already described, we had run another experiment that enabled us to learn something about the time course of the thermal effect on irritation. In that experiment subjects again received warm and cold capsaicin solutions, but instead of rating the burning sensation only once per trial, they rated it three times: during the sip and 5 and 30 sec after expectorating the solution. The results of this task are shown in Fig. 4.

The data supported the results of the initial experiment but also revealed a puzzling phenomenon. Although we expected to see a slow increase in the

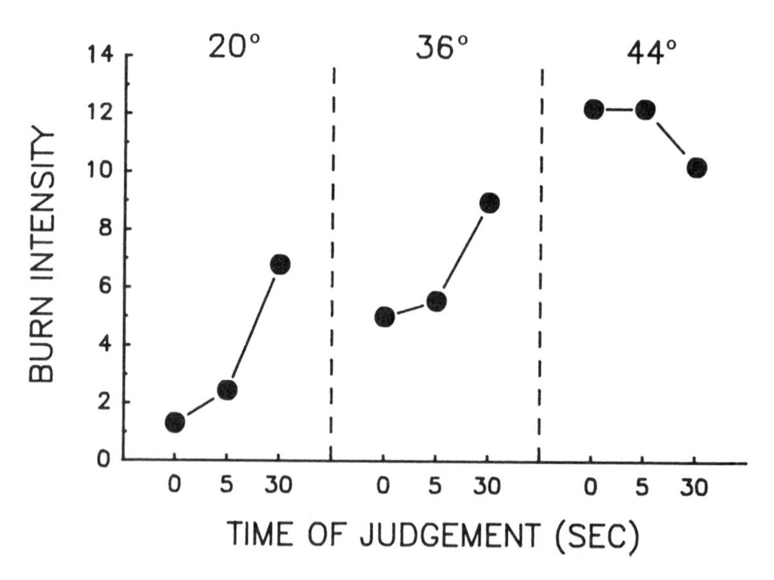


**FIGURE 4** The intensity of sensations of burn is shown as a function of when the judgments were made relative to exposure to water of different temperatures (20, 36, and 44 °C). Judgments at time "zero" were made during the sip, and 5 and 30 sec represent the amount of time postexpectoration.

burning sensation following expectoration of a cold (20 °C) solution as the mucosa returned to its normal temperature (on the left side of the figure), we were surprised to see a similar increase after expectoration of the 36 °C solution (in the middle of the figure). Because 36 °C is so close to normal oral temperature, there should have been little or no "recovery" of the burn after expectoration. One of the few hypotheses we were able to generate to explain this recovery was that the mechanical stimulation associated with sipping and expectorating the solution transiently suppressed the burning sensation. The plausibility of this hypothesis rested on reports that mechanical stimulation in the form of vibration could reduce the perceived intensity of pain (e.g., Melzack et al., 1963; Sullivan, 1968; Ottoson et al., 1981; Bini et al., 1984).

Our initial attempt to test the hypothesis involved having subjects sip a solution containing capsaicin after which they chewed a flavorless gum base. Ratings were to be obtained of the intensity of the burn both during and between bouts of chewing. Pilot work indicated that this approach was compromised by the fact that subjects knew how intense the irritation was when they were not chewing, which made it difficult to give unbiased judgments of the irritation they felt when they were chewing (and vice versa). It concerned us that the tendency for subjects to try to be consistent in their judgments might lead to an erroneous negative finding.

A more rigorous experiment was subsequently designed in which subjects were unaware of the intensity of the irritation produced by capsaicin prior to each trial. The design took advantage of the temperature dependence of capsaicin irritation. Instead of testing in the oral cavity proper, the vermilion border of the lower lip was chosen so that the entire area of skin treated with capsaicin could be cooled, heated, and mechanically vibrated.

The specifics of the design were the following: Either capsaicin (40 ppm) dissolved in light mineral oil or light mineral oil alone was applied to the midline region of the lower lip. (The higher concentration of capsaicin was necessitated by the lower absorption rate of the keratinized skin of the lip and the use of oil rather than ethanol and water as the vehicle. Oil was used because its greater thickness helped ensure that the stimulus would not spread to skin beyond the lip.) The lip was immediately covered with a rectangular-shaped peltier thermoelectric module whose temperature was set at 25 °C, which was sufficiently cold to prevent a burning sensation from appearing (see above). The peltier module was mounted on the shaft of a minishaker vibrator that was clamped in turn to an adjustable stand. Skin temperature was held at 25 °C for 2 min to allow time for the capsaicin to be absorbed into the tissue. The temperature of the thermode was then increased at one of three rates: 0.2, 0.4, or 0.8 °C/sec. The subjects' task was to press a button (a) when a sensation of warmth first appeared, (b) when a burning sensation appeared, and (c) when the sensation became painful. The use of three rates of temperature change prevented subjects from anticipating the various thresholds based on temporal cues and further assured that they would not know whether capsaicin or the vehicle had been applied to the lip (i.e., a strong sensation early in a trial could mean either that capsaicin had been applied or that the rate of temperature change was fast). On half of the trials the vibrator was energized immediately prior to initiating the temperature change, producing a 60 Hz, 1030 $\mu$m (pk-pk) vibration that remained on until the subject pressed the button to indicate that the threshold for pain had been reached. Two minutes was allotted between trials to return the thermode to 25 °C and establish a stable baseline. Either the capsaicin or the vehicle was reapplied after three trials to ensure the chemical stimulus would remain effective throughout the six trials that were run in each session. A total of six subjects served in six sessions each.

The results for the slowest rate of temperature change (0.2/sec) are displayed in Fig. 5 (the results were identical for the other two rates). Graphed

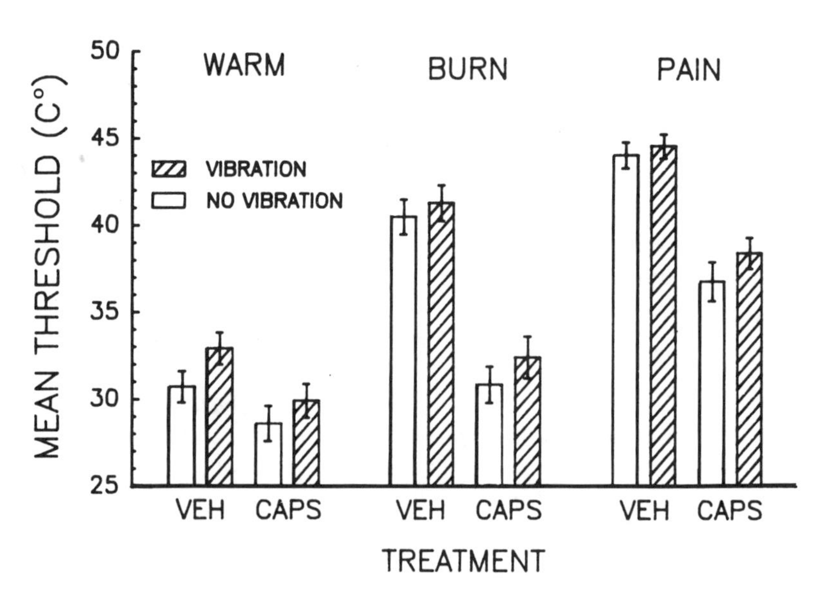


**FIGURE 5** Thresholds for the appearance of sensations of warmth, burn, and pain are shown for two different treatments (vehicle alone or vehicle + capsaicin) when vibration was either present (hatched bars) or absent (empty bars). These data, while representative of the results for all three rates of temperature increase, are for an increase of 0.2 °C/sec.

are the mean thresholds for the sensations of warmth, burn, and pain in the four conditions of the experiment. It is apparent from the figure that vibration reduced sensitivity to all three categories of sensation, whether or not capsaicin was applied to the lip (i.e., all thresholds were higher in the presence of vibration). Although uniformly small (thresholds rose only 1-2 °C), the attenuating effect was statistically significant for each type of threshold (confirmed by a repeated measures ANOVA performed on the pooled data for all three rates of temperature change).

The appearance of an attenuation across all conditions argues against a specific cross-modal inhibitory effect on the nociceptive system. It is more likely that the shifts in threshold were the result of a nonspecific masking or distraction effect. The vibrotactile sensation generated by the 1030 μm stimulus was intense (>40 dB above threshold) and may simply have made it difficult to attend to the other sensations. The magnitude of the effect is in any case too small to explain the apparent suppression of the capsaicin burn observed in Fig. 4.

The absence of a strong or specific suppression of pain sensation is not necessarily inconsistent with previous reports of the reduction of pain with vibration. It has been shown in a number of studies that the analgesic properties of mechanical stimulation may take several minutes to develop fully (Pertovaara, 1979; Lundeberg et al., 1983; Lundeberg, 1984; Hansson and Ekblom, 1986). Presentation of the vibration coincident with the chemical irritation and for less than 2 min each time may not have been optimal for producing a suppressive effect.

Another point of interest in Fig. 5 is the uniform reduction in thresholds in the capsaicin condition. Whereas reductions in the burn and pain thresholds were expected, the lower threshold for warmth was unexpected given the apparent failure of capsaicin to enhance perceived warmth at temperatures near thermal neutrality (see Fig. 2). The present data reopen the question of whether capsaicin stimulates or sensitizes warm fibers in the oral cavity. However, it is possible that in the present experiment subjects identified the "warm" threshold with sensations associated with the onset of activity in nociceptors rather than warm fibers. The occurrence of burning sensations at temperatures in the vicinity of 30 °C indicate that capsaicin can excite nociceptors in that temperature range, and it can only be guessed whether subjects decided that the warmth threshold was the first thermal sensation of any kind they detected (e.g., burning) or maintained a stable criterion for the sensation of warmth.

### B. Mechanical Pressure and Ethanol Irritation

Vibration is, of course, only one form of mechanical stimulation that might interact with chemical irritation. During a study of the sensitivity of the tongue to ethanol we discovered that the simplest form of tactile stimulation—static pressure—had the capacity to greatly *enhance* sensations of irritation produced by ethanol. During pilot work for the experiment one of us happened to press the tip of the tongue, which had been treated with a high concentration of ethanol, against the anterior hard palate. The result was a sharp sensation of pain. Intrigued by this phenomenon, we incorporated a condition into the experiment that enabled us to quantify the relationship between ethanol concentration and the mechanical enhancement that seemed to occur.

The original purpose of the experiment was to assess the growth and decay of ethanol irritation on the tip of the tongue. Ethanol ranging in concentration from 25 to 65% (/vol. in deionized $H_2O$) was applied to the tongue on disks of filter paper (0.38 $cm^2$) and left there for 25 sec. The subjects' task was to rate the intensity of irritation perceived on the tongue tip at six different times: 5, 10, 15, 20, 30, and 40 sec after application of the stimulus. Ratings were made by pressing on a force transducer (using the index finger of the preferred hand) with a force that was proportional to the intensity of the oral irritation. The applied forces were later transformed into values on a psychological scale of intensity by having subjects press on the transducer in response to numbers given to them by the experimenter (i.e., the method of magnitude production).

On half of the trials measurements of the effect of static pressure on the intensity of irritation were made by telling subjects to press the tip of the tongue against the anterior palate at the 35-sec mark of the trial. The instructions were to apply "moderate force" for 5 sec, the duration of which was signaled by a tone. At the end of that time the subjects pressed on the transducer to register their ratings of perceived intensity. In addition, between trials subjects were asked whether the sensation was at any time painfully intense. On the trials in which subjects did not press against the palate they were instructed to rest the tongue tip lightly against the palate to ensure that differences in perceived intensity that might be observed in the "press" condition were not due to mere contact or to a slight warming of the tongue.

A total of 15 subjects served in the experiment, contributing three replicates each for a total of 45 observations per data point. Figure 6 contains the transformed intensity ratings of irritation as a function of the time after application of the stimulus. The parameters are the concentration of ethanol and whether or not the subject pressed the tongue against the palate. A clear concentration effect can be seen and, more important to the present discussion, the ratings of irritation made during the "pressure" trials are all higher than the ratings made when no pressure was applied. Whereas in the no-pressure condition ratings of perceived intensity continued to decline between 30 and 40 sec, in the pressure condition ratings at 40 sec (connected by dashed

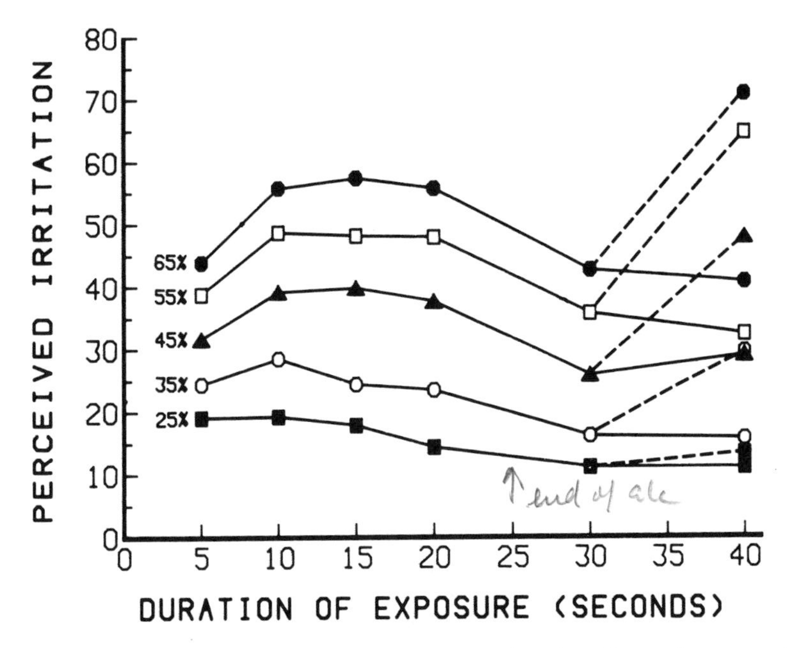


**FIGURE 6** Perceived irritation produced by ethanol applied to the tongue tip as a function of duration of exposure. The parameter is the concentration of ethanol. The data points at 40 sec connected by dashed lines represent the responses given when the tongue tip was pressed against the hard palate.

lines) were always greater than ratings at 30 sec. In addition, the size of the increase in perceived irritation between 30 and 40 sec was directly related to concentration.

The frequencies of pain judgments are shown for the press and no-press conditions in Fig. 7. Pressing the tongue against the palate resulted in a dramatic rise in the number of trials on which pain was encountered at the 40 sec mark. Without pressure, 65% ethanol resulted in pain only 9.0% of the time; with pressure, the sensation was judged painful 48.5% of the time.

The most straightforward interpretation of these data is that the application of pressure heightens the ability of ethanol to stimulate nociceptors. Alternatively, ethanol may be a sensitizing agent which increases the sensitivity of some types of nociceptors to static pressure, i.e., a hyperalgesia to pressure. Whichever view is more correct, the phenomenon can nevertheless be considered a true sensory interaction inasmuch as the responsiveness to both kinds of stimulation is altered in the presence of stimulation in the other modality.

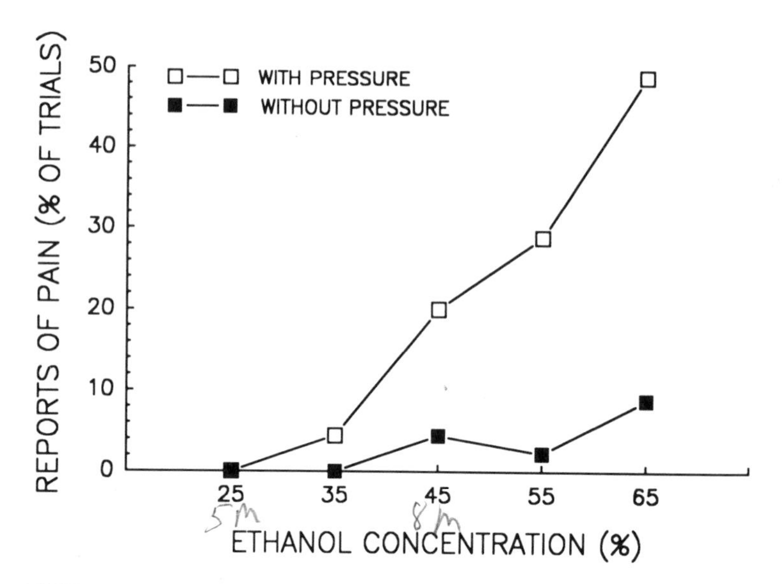


**FIGURE 7** Shown as a function of the concentration of ethanol is the percentage of trials on which subjects reported sensations of pain 40 sec after application of the compound to the tongue. Empty symbols denote trials on which subjects pressed the tongue against the hard palate; solid symbols denote trials on which the tongue touched the palate but was not pressed against it.

While this experiment demonstrated the existence of a mechano-chemical interaction, it provided little information about its basic nature. Additional studies will be required to determine the relationship between the amount of force exerted and the enhancement observed, the temporal characteristics of the enhancement, and the extent to which it occurs with other irritants. It may also be important to study the qualitative features of the sensation (e.g., its quality seemed to some observers to be more like the pain produced by pinching than the pain produced by burning) because of the indirect information those features may give us about sensory coding mechanisms in nociception.

The demonstration of a facilitatory interaction between a chemical irritant and a mechanical stimulus hints at the complexity of common chemical sensitivity. Thus far we have shown that chemical irritation can be either weakly or strongly affected by both temperature and mechanical force. The final experiment demonstrates that similar changes in sensitivity can be caused by preexposure of the skin to either the same or a different chemical irritant.

## IV. CHEMICAL EFFECTS

An obvious characteristic of the sensations produced by many oral irritants is their long decay time. From a practical standpoint the lengthy sensations they often produce make it difficult to study irritants in typical psychophysical paradigms. It was this difficulty that recently led us to investigate the extent to which irritation may grow over repeated exposures.

An initial attempt to evaluate the irritating properties of NaCl by a typical sip-and-spit procedure led us to abandon that approach and to study instead the cumulative effect of repeated exposure to the salt. We quickly realized that the degree of irritation perceived on a given exposure depended on whether a sensation of irritation still lingered from the previous exposure. An experiment was subsequently designed that enabled us to demonstrate the phenomenon of salt sensitization (Green and Gelhard, 1989). In brief, we found that the irritation produced by NaCl grew linearly over a 15 min period when subjects rinsed at the rate of once per minute with a 0.4 or 0.8 M solution (no water rinses were permitted between exposures to salt).

That result led to two immediate questions: first, was this a general characteristic of all oral irritants, and second, would sensitization to one irritant lead to sensitization to another?

### A. Sensitization and Cross-Sensitization of Irritation

NaCl and ethanol were the chemicals chosen for study; NaCl because we had recently quantified its ability to produce sensitization, and ethanol because we had already studied several characteristics of the irritation it produces and had in addition noticed that repeated exposure to ethanol seemed *not* to produce sensitization. The experiment therefore had the potential for giving at least a partial answer to both of the questions we had posed.

The experimental design was relatively complicated because of the problems inherent in studying the time course of irritation. A session began with the subject rinsing with 36 °C deionized water to cleanse the palate and maintain tongue temperature near normal. They then received a stimulus of either 25% ethanol or 3.0 M NaCl (which produced approximately equal irritations for a single presentation) delivered to the tongue tip on a disk of filter paper 0.38 $cm^2$ in area. After 10 sec the subject rated perceived irritation by writing a magnitude estimate on a piece of paper which the experimenter collected. After 30 sec the disk was removed from the tongue and another estimate of irritation was given. Ten seconds later this was repeated. Together these two presentations constituted the preexposure trials. The subject rinsed five times with 36 °C water to wash away as much of the remaining irritant as possible and then waited 60 sec before receiving the first of the "conditioning" stimuli. The conditioning stimuli were either 25% ethanol or 1.5 M

NaCl, again delivered on filter paper but this time on larger disks (1.2 $cm^2$). (A lower concentration of NaCl was used in anticipation of the sensitization that would occur; a 3.0 M stimulus would have eventually produced a painful irritation.) The conditioning stimulus was always different from the preexposure stimulus, so that if a subject received NaCl in the first two trials he or she received ethanol on the conditioning trials. A series of fifteen 30 sec conditioning stimuli were presented with subjects rating irritation just as they had on the preexposure trials. No water rinses were allowed between stimulations, which were separated in time by only 10 sec. The total duration of the conditioning period was therefore approximately 10 min. After the last conditioning stimulus the subject received three postexposure stimuli identical in quality, concentration, and size to the preexposure stimuli. The smaller disks were used to ensure that the postexposure stimuli would fall within the area of the tongue affected by the conditioning stimuli. The timing of the stimuli and the ratings of irritation remained the same.

Only one condition could be tested in a session. Ten subjects participated in a total of six sessions, three with NaCl as the conditioning irritant and three with ethanol as the conditioning irritant. A magnitude-matching paradigm with vibration as the matching modality was used to standardize magnitude estimates across sessions within subjects.

The results of the experiment are shown in Fig. 8. The first point to be taken from the graphs is that the irritation produced by NaCl grew throughout the conditioning period (top of Fig. 8), whereas the irritation produced by ethanol did not (bottom of Fig. 8). Sensitization is apparently not a property of all oral irritants. The other main point evident in Fig. 8 is that NaCl had a cross-sensitizing effect on ethanol sensitivity; the irritation produced by ethanol was significantly stronger after the tongue had been exposed to NaCl than it was before exposure to NaCl. The level of ethanol irritation was approximately four times stronger after the conditioning trials than before, which is virtually identical to the ratio of intensities of NaCl irritations at the beginning vs. the end of the conditioning period. Whether this is mere coincidence or indicates that cross-sensitization is proportional to self-sensitization must be determined in future experiments in which more than one concentration of more than one conditioning stimulus is tested.

It is also notable that the first ratings of ethanol irritation in the postexposure period were higher than the final ratings of salt irritation that preceded them. It is as though the baseline for irritation was raised by sensitization and the additional irritation produced by ethanol was added to that baseline. That the irritation at the end of the conditioning period was greater with the 25% ethanol stimulus than with the 1.5 M salt stimulus is consistent with the fact that ethanol at 25% produces a sensation of about equal intensity to that produced by a 3.0 M salt stimulus (compare the preexposure data

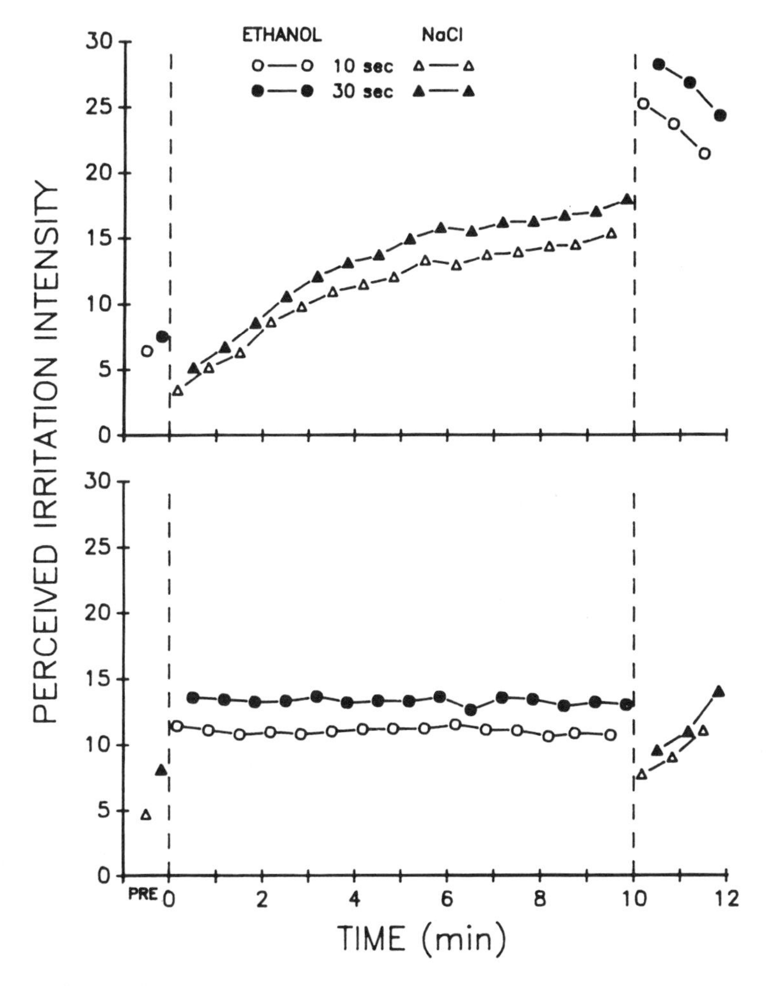


**FIGURE 8** Perceived irritation produced by ethanol and NaCl is shown as a function of time. The parameter is the time after each presentation of the stimulus that the rating of irritation was made. Top half of figure: irritation produced by a single exposure to ethanol (points to the left of zero), to repeated exposures to NaCl (middle), and to subsequent exposures to ethanol (extreme right). Bottom half of figure: irritation produced by a single exposure to NaCl (points to the left of zero), to repeated exposures of ethanol (middle), and to subsequent exposures to NaCl (extreme right).

on the left side of the figure). Thus we cannot conclude from these data that a kind of synergism occurred between ethanol and NaCl; the results seem more in accord with simple additivity. To determine with certainty which phenomenon occurs will require another experiment in which a concentration of ethanol that produces a sensation equal to 1.5 M NaCl is applied after conditioning.

## V. CONCLUSIONS

The data presented here illustrate the complexity of the CCS and its intimate functional relationship with the classical skin senses. In the oral cavity the sensitivity to chemical irritants is clearly vulnerable to stimulation in the other cutaneous modalities. This vulnerability presumably exists because many if not all of the receptive elements of the CCS are either nociceptors or thermoreceptors. By their nature polymodal nociceptors are sensitive to temperature and mechanical stimulation, which makes it inevitable that the sensitivity of such units to chemicals will sometimes be modified by heat and pressure. Even more obvious is the effect temperature should have on the response of thermoreceptors to chemical irritants. The CCS may therefore qualify as the least specific of all of the senses. Indeed, it is arguable as to whether the sensitivity to irritants should be characterized as a separate sense at all. If chemical irritation is actually mediated exclusively by elements of the classical senses of pain and temperature, it may be better to avoid reference to the common chemical sense and refer instead to common chemical sensitivity or the sensitivity to chemical irritants (see Preface to this volume).

It remains possible, however, that a set of cutaneous afferents exist which are sensitive only to chemicals (Simone et al., 1987; Meyer and Campbell, 1988). This hypothesis is currently being explored in electrophysiological studies, and a new technique has been found that no longer relies on traditional methods of locating nociceptors by stimulating with intense mechanical or thermal stimuli (Meyer and Campbell, 1988).

Regardless of what the physiological studies may eventually reveal, it will be important to continue with psychophysical studies if we are to sort out the contributions to sensation that are made by each of the underlying neural elements. This chapter has only hinted at the variety of sensory interactions that are likely to occur when chemical irritants stimulate the cutaneous sensory systems.

## ACKNOWLEDGMENTS

These studies were supported in part by grants NS20577 and NS20616 from the National Institutes of Health. The author thanks Barbara Gelhard and

Mary Ellen Bailey for carrying out the experiments and assisting with the analysis of the data.

## REFERENCES

Adriaensen, H., Gybels, J., Handwerker, H. O., and Van Hees, J. (1980). Latencies of chemically evoked discharges in human cutaneous nociceptors and of the concurrent subjective sensations. *Neurosci. Lett.* 20: 55-59.

Adriaensen, H., Gybels, J., Handwerker, H. O., and Van Hees, J. (1983). Response properties of thin myelinated (A-8) fibers in human skin nerves. *J. Neurophys.* 49: 111-122.

Bessou, P., and Perl, E. R. (1969). Response of cutaneous sensory units with unmyelinated fibers to noxious stimuli. *J. Neurophys.* 32: 1025-1043.

Bini, G., Cruccu, G., Hegbarth, K. E., Schady, W., and Törebjork, E. (1984). Analgesic effects of vibration and cooling on pain induced by intraneural electrical stimulation. *Pain* 18: 239-248.

Croze, S., Duclaux, R., and Kenshalo, D. R. (1976). The thermal sensitivity of the polymodal nociceptors in the monkey. *J. Physiol.* 263: 539-562.

Foster, R. W., and Ramage, A. G. (1981). The action of some chemical irritants on somatosensory receptors of the cat. *Neuropharmacology* 20: 191-198.

Green, B. G. (1986). Sensory interactions between capsaicin and temperature in the oral cavity. *Chem. Senses* 11: 371-382.

Green, B. G., and Gelhard, B. (1989). Salt as an oral irritant. *Chem. Senses* 14: 259-271.

Green, B. G., and Flammer, L. J. (1988). Capsaicin as a cutaneous stimulus: Sensitivity and sensory quality on hairy skin. *Chem. Senses* 13: 367-384.

Green, B. G., and Frankmann, S. P. (1987). The effect of cooling the tongue on the perceived intensity of taste. *Chem. Senses* 12: 609-619.

Handwerker, H. O., and Neher, K. D. (1976). Characteristics of c-fiber receptors in the cat's foot responding to stepwise increases of skin temperature to noxious levels. *Pflugers Arch.* 365: 221-229.

Hanson, P., and Ekblom, A. (1986). Influence of stimulus frequency and probe size on vibration-induced alleviation of acute orofacial pain. *Appl. Neurophysiol.* 49: 155-165.

Jones, M. H. (1954). A study of the common chemical sense. *Am. J. Psychol.* 67: 696-699.

Keele, C. A. (1962). The common chemical sense and its receptors. *Arch. Int. Pharmacodyn. Ther.* 139: 547-557.

Kenins, P. (1982). Responses of single nerve fibers to capsaicin applied to the skin. *Neurosci. Lett.* 29: 83-88.

Konietzny, F., and Hensel, H. (1983). The effect of capsaicin on the response characteristic of human c-polymodal nociceptors. *J. Therm. Biol.* 8: 213-215.

Kumazawa, T., and Mizumura, K. (1983). Temperature dependency of the chemical responses of the polymodal receptor units in vitro. *Brain Res.* 278: 305-307.

Kumazawa, T., Mizumura, K., and Sato, J. (1987). Response properties of polymodal receptors studied using in vitro testis superior spermatic nerve preparations of dogs. *J. Neurophysiol.* 57: 702-711.

Lamotte, R. H., Simone, D. A., Baumann, T. K., Shain, C. N., and Alreja, M. (1988). Hypothesis for novel classes of chemoreceptors mediating chemogenic pain and itch. In: *Proceedings of the Vth World Congress on Pain*. R. Dubner, G. F. Gelhart, and M. R. Bonds (Eds.). Elsevier, New York, pp. 529-535.

Lundeberg, T. A. (1984). Comparative study of the pain alleviating effect of vibratory stimulation, transcutaneous electrical nerve stimulation, electro-acupuncture and placebo. *Am. J. Clin. Med.* 12: 72-79.

Lundeberg, T., Ottoson, D., Hansson, S., and Meyerson, B. A. (1983). Vibratory stimulation for the control of intractable chronic orofacial pain. In: *Advances in Pain Research and Therapy*, Vol. 5. Raven Press, New York, pp. 555-561.

Melzack, R., Wall, P. D., and Weisz, A. S. (1963). Masking and metacontrast phenomena in the skin sensory system. *Exp. Neurol.* 8: 35-46.

Meyer, R. A., and Campbell, J. N. (1988). A novel electrophysiological technique for locating cutaneous nociceptive and chemospecific receptors. *Brain Res.* 441: 81-86.

Ottoson, D., Ekblom, A., and Hansson, P. (1981). Vibratory stimulation for relief of pain of dental origin. *Pain* 10: 37-45.

Parker, G. H. (1912). The relation of smell, taste and the common chemical sense in vertebrates. *J. Acad. Natl. Sci.*, Ser. 2, 15: 221-234.

Pertovaara, A. (1979). Modification of human pain threshold by specific tactile receptors. *Acta Physiol. Scand.* 107: 339-341.

Rozin, P., Ebert, L., and Schull, J. (1982). Some like it hot: A temporal analysis of hedonic responses to chili pepper. *Appetite* 3: 13-22.

Scheuplein, R. J. (1976). Permeability of the skin: A review of major concepts and some new developments. *J. Invest. Dermatol.* 67: 672-676.

Simone, D. A., Ngeow, J. Y. F., Putterman, G. J., and Lamotte, R. H. (1987). Hyperalgesia to heat after intradermal injection of capsaicin. *Brain Res.* 418: 201-203.

Stevens, D. A., and Lawless, H. T. (1988). Responses by humans to oral chemical irritants as a function of locus of stimulation. *Percept. Psychophys.* 43: 72-78.

Sullivan, R. (1968). Effect of different frequencies of vibration on pain threshold detection. *Exp. Neurol.* 20: 1356-142.

Szolcsànyi, J. (1977). A pharmacological approach to elucidation of the role of different nerve fibers and receptor endings in mediation of pain. *J. Physiol.* 73: 251-259.

Torebjörk, H. E. (1974). C units responding to mechanical, thermal and chemical stimuli in human non-glabrous skin. *Acta Physiol. Scand.* 92: 374-390.

# Chapter 9 Discussion

**Dr. Lawless:** I noticed in your experiment on capsaicin and cooling that you plotted solution temperature. Is it possible that the decrease in perceived cooling with capsaicin is an effect of vasodilation in the oral cavity and a warming due to an enhanced peripheral blood flow?

**Dr. Green:** We actually looked at that as best we could. We put thermocouples in the mouth after exposure to capsaicin and saw no rise in temperature. Then we thought about it and decided there really shouldn't be a temperature change because the mouth is at core temperature already. So all you are doing is engorging an already engorged tongue with more 37° blood, so I don't think that would account for it.

**Dr. Cain:** With respect to the last figure you showed, it seemed to me that ethanol shows more effective adaptation than salt. Once and a while I drink alcohol, and last night when I had my martini, I noticed that even Beef Eater's gin, good gin, burns like hell at first—unless you cool it, by the way—but then after you sip two, three, and four times, the burn goes away. If you are getting a lot of adaptation to ethanol, and you have essentially a cross-adaptation, and for sodium chloride you are showing integration over time (this is integration we see all the time that nobody seems to understand or know anything about), it doesn't surprise me that you see a balancing-off of integration with adaptation. That's why you are getting this flat function for ethanol. In the case of sodium chloride, the adaptation has perhaps been started, we don't know the magnitude of it, but it's not as far along. So we still see some temporal integration and it looks to me that if you think about those two processes working one against the other, that you have a simple explanation for it.

**Dr. Green:** I think that if you put it in those terms, I agree you may say it's a simple process. However, another way of looking at this is that a sensitization process is occurring along with adaptation. I use the term sensitization, by the way, somewhat incautiously because that term is really used by physiologists to talk about sensitization of a single receptor or nerve fiber. But it is descriptive of what is happening; you grow more and more sensitive to sodium chloride. With respect to the adaptation to ethanol that you spoke of, when I began this experiment I assumed that the functions for salt and ethanol would go in opposite directions and I'd have a nice comparison. Then, if you present sodium chloride after you've adapted to ethanol, are you also less sensitive to sodium chloride? Basically, I think that's what tends to happen. But subjects had difficulty, I think, with this task. Some showed a little increase over time and some showed adaptation. And what happens in addition to the change in sensitivity is that the quality of the ethanol sensation changes. It goes from the burning and sharpness that you spoke of to something that has a quality of numbness or dullness. But it is still there, and when you have subjects judge irritation, you can't tell them to ignore the numbness or whatever you may call it. I think that some subjects did ignore it and gave lower estimates and other subjects said, boy, there is still a lot going on on my tongue, and gave it higher estimates. But I think the main point is that exposure to a given compound had the same effect on another compound, whether it was adaptation or sensitization.

**Dr. Szolcsànyi:** You have demonstrated a lot of interesting interactions, and I'm wondering if you can find some volunteers who would like to undergo the procedure of desensitization of the tongue. I did it, actually, and it was not too pleasant; it caused sneezing and discharging from the nose when we put a 1% capsaicin solution on the tongue. It was unpleasant for about a half hour, but tolerable. After an hour or so, there was no paresthesia or numbness; it was as if it was not treated at all with capsaicin. What I wanted to mention to you was that when we followed this with mustard oil, which is an irritant, there was no irritation or pain. When we measured the threshold, all we tasted was sweetness. So this experiment came to mind because you are using stimulation to steady these interactions, but you could also see what happens after you have desensitized the tongue.

**Dr. Green:** Yes, I think that's a very good point. I've thought about doing that. I've read your experiments and have great respect for you having done them. It is difficult to imagine going through that initial pain on an area like the tongue. About half of my work now is devoted to looking at some of these same issues on hairy skin, and I've thought of doing similar experiments on the arm. Perhaps there we could try to get desensitization while we cool the skin, so the subject won't experience as much pain and discomfort. I don't think you can control temperature well enough in the oral cavity, and I think it might be difficult to recruit subjects who will agree to let us desensitize the tongue.

**Dr. Kobal:** You could take a local anesthetic in the initial phase, and then when the action of the local anesthetic is gone you have the effect of desensitization.

**Dr. Green:** That's a good suggestion also.

**Dr. Alarie:** I was interested in your comment about there being so many irritants out there that act in different ways and cause different perceptual experiences. About 20 years ago, I accidently exposed myself to an irritant on the skin, and the effect went away and I went home that night and watered the lawn. As I watered the lawn, I got water on my hand. I felt intense pain all night; it just wouldn't go away. I repeated the experiment a few weeks later, by taking that chemical and painting a very small area on my skin and then dipping it in water 2-3 hours later. The intense pain came again. So I'm not surprised at the way you are thinking in terms of the variety of possible interactions.

**Dr. Green:** What chemical was it?

**Dr. Alarie:** I can send you some and you can do the experiment on yourself. It was one of the chemicals we were screening for the army, for tear gas; a very potent irritant.

**Dr. Green:** You've reminded me of a problem in our experiments with capsaicin. Even when we apply it at only moderate levels on external skin, and the subjects don't complain of a pain or strong burn during the 20 min after exposure that we observe them, we get what we call "shower shock" or "the morning-after effect." When they shower the next morning and hot water hits the arm, the stimulated area burns and brightens right up again. There is a long-lasting hyperalgesia to heat. So it is really cumbersome to deal with these compounds; we've had a couple of subjects refuse to continue because they say they can't enjoy their showers anymore.

# 10
# Differences Between and Interactions of Oral Irritants
## Neurophysiological and Perceptual Implications

**Harry T. Lawless**
Cornell University
Ithaca, New York

**David A. Stevens**
Clark University
Worcester, Massachusetts

## I. INTRODUCTION

Several sources of evidence point to the fact that capsaicin and piperine, the irritative components in red and black pepper, respectively, are neither perceptually identical nor the same in their modes of physiological stimulation. The compounds differ in their effective concentration ranges, in their abilities to stimulate different areas of the oral cavity, and possibly in their time intensity relations and interactions with classical taste qualities. Experiments with sequential and simultaneous stimulation with the two compounds provide data consistent with partially independent receptor mechanisms. Although direct tests of perceived quality differences are difficult to perform, there is a growing body of experimental results suggesting differences in the sensation quality elicited by the two compounds as well as different peripheral mechanisms of stimulation that may give rise to such sensation differences.

## II. QUALITATIVE DIFFERENCES AMONG IRRITANTS?

As consumers of spicy food, we have been impressed by the apparently wide array of irritative sensations one experiences in the mouth from different spicy condiments, pepper sauces, and ethnic cuisines. In other words, there appears to be wide qualitative variation in the types of mouth burn one can get from spices in real foods. Is this qualitative variation due to the other sensory characteristics of the food, such as gustatory or olfactory sensation? Or is this variation in perceptual qualities true for single irritant chemicals as well? The question concerns whether all mouth-burning chemicals produce the same kind of heat and pain sensations, or whether there are perceivable differences. Such differences would have implications for the possibility of multiple neurophysiological mechanisms operating at the periphery.

Over the past 5 years, we have documented psychophysical characteristics of irritant chemicals when applied to the mouth under a variety of conditions. The stimuli we have chosen are the chemical irritants commonly occurring in spices. These irritants are shown in Fig. 1. Capsaicin, the active ingredient in red pepper (from plants of the genus *Capsicum*), is actually one member of a chemical family whose members differ in potency and whose structures differ as to chain length and the saturation of the double bond. Vanillyl nonamide, the straight-chain saturated version of capsaicin, is an easily synthesized molecule, is commonly used in research, and has a similar potency to that of capsaicin. Piperine is the active ingredient in black pepper, derived from the seeds of *Piper nigrum*. Capsaicin is about 100 times as potent as piperine (Govindarajan, 1979; Lawless, 1984). Both have limited sensory

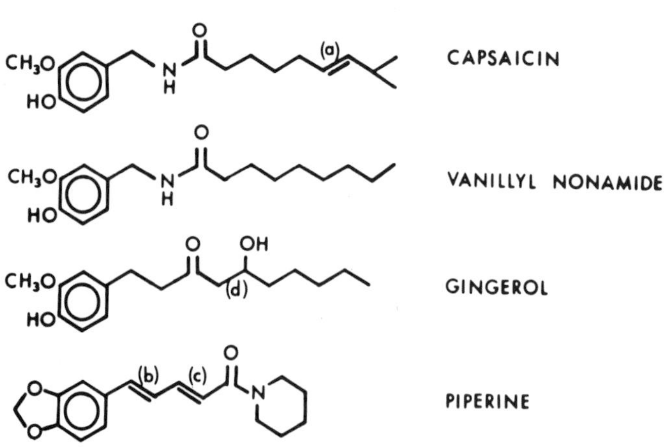

**FIGURE 1** Irritant compounds.

characteristics other than their tactile, thermal, and painful attributes, i.e., in pure form they are relatively free of tastes and odors.

How might irritant chemicals differ in terms of their sensory properties? As mentioned above, they differ in their potency, i.e., the chemicals differ in the concentration which will produce a certain level of sensory intensity. These stimuli also produce long-lasting irritative sensations. After a single application the mouth burn may last 10 min or more. Irritants may differ in their time course regarding the growth and recession of burning sensations. Other possible sensory variations pertain to the areas of the mouth in which the burn is sensed. The spatial pattern of irritation produced by irritant chemicals may not be the same. A fourth possibility is that even though irritant chemicals such as capsaicin and piperine have relatively few sensory qualities other than chemical heat, they may show minor differences in side tastes or odors. Finally, they may differ in irritative quality. By a qualitative difference we mean to imply a discrimination based on quality alone rather than intensity or some other attribute of the stimuli such as perceived area, onset time, decay characteristics, etc. The analogous situation in color vision would be a discrimination based on hue (chroma), all other things being equal.

When first presented with the question of perceptual differences between these two chemicals, one might propose a direct attack on the issue. Quite simply, can people distinguish between irritant chemicals like capsaicin and piperine? In applied sensory evaluation, this would call for a discrimination test, such as the triangle or duo-trio procedure. However, discrimination among these chemicals could be achieved by a number of means other than through a true qualitative distinction as we define it in the restricted sense. As noted above, there are many additional characteristics that could facilitate a sensory discrimination by subjects in addition to the type of irritation perceived. The difficulty arises in equating all of these other characteristics (intensity, temporal pattern, spatial pattern, side tastes) in order to eliminate them as logical possibilities on which a discrimination might have been based.

The logistics of such a discrimination test present additional difficulties. Within a session, subjects become partially desensitized to the stimuli such that later-presented stimuli in a sequence are perceived as less intense than stimuli presented earlier (Lawless and Gillette, 1985). As noted by O'Mahony and Odbert (1985), such sequential intensity effects can have profound influences on the results of discrimination tests. For example, when two strong stimuli and one weak stimulus form the three trials in a triangle procedure, the second presentation of a stronger stimulus may be perceived as weaker, thus leading to the incorrect judgment of the first presentation of that stronger stimulus as the odd sample.

An additional complication arises from individual differences. There are wide variations in the responsiveness of individuals to these irritants due to

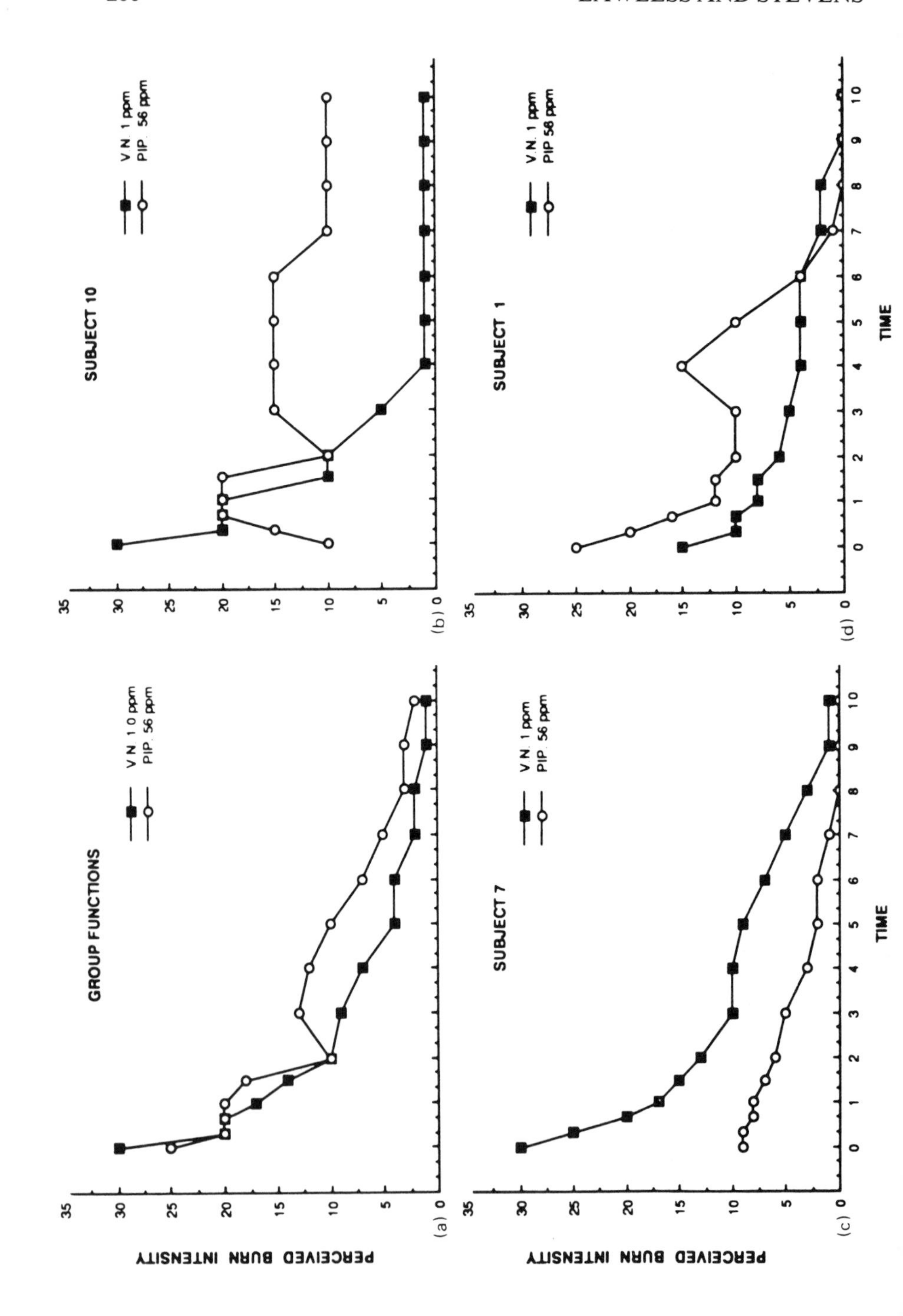
GROUP FUNCTIONS
V N 1 0 ppm
PIP. 56 ppm
SUBJECT 10
V N 1 ppm
PIP. 56 ppm
SUBJECT 7
V.N. 1 ppm
PIP. 56 ppm
SUBJECT 1
V.N. 1 ppm
PIP 56 ppm
PERCEIVED BURN INTENSITY
TIME
(a)
(b)
(c)
(d)

consumption habits and other unknown factors (Lawless et al., 1985). Furthermore, a given individual may be more responsive to capsaicin than piperine, or vice versa. Once concentrations have been set to produce equal perceived intensities (for a group average) through pilot testing, individual subjects may still judge them to be different. Figure 2 shows the intensities of 1 ppm vanillyl nonamide and 56 ppm piperine which are about equal for this group of subjects (replotted from Lawless, 1984). However, subject 1 judges piperine to be more intense while subject 7 judges vanillyl nonamide to be more intense. Subject 10 shows a pronounced flattening in the time function for piperine, producing a dramatic crossover. Such subject-by-compound-by-time interactions are common in our data. They illustrate the difficulty in choosing concentrations of the two irritants that would produce equal burn intensities.

In summary, a traditional discrimination test for differences in perceived quality is bound to give a misleading result insofar as it is difficult to equate two irritants for all of the other perceptual attributes on which a discrimination might be based. Unfortunately, subjects cannot be told simply to ignore all the other properties besides quality—they will use any information at their disposal to get the "correct" answer in such procedures.

## III. IMPLICATIONS FOR PHYSIOLOGICAL CODING

Logically, if some qualitative perceptual distinctions are possible in a sensory modality, some differences in physiological coding mechanisms must exist to give rise to differences in the peripheral afferent information. It is possible, of course, for the brain to interpret two equivalent incoming signals as different, as happens in perceptual illusions. However, such failures of constancy are almost always due to different *contexts* within which the two stimuli are viewed. To the extent that context can be controlled, the logical implications for different coding mechanisms must hold. Like Fechner, we propose that subjective perceptual events have correlates in physiological function. Furthermore, we feel that a comprehensive understanding of a sensory system, modality, or mechanism necessarily involves the correlation of perceptual data (or behavioral, in the case of animal research) with physiological and anatomical information.

Two potential physiological mechanisms that could give rise to quality differences among irritants include, first, different receptor mechanisms and/or different tuning of receptor cells to various irritants, and second,

**FIGURE 2** Perceived intensity of 1 ppm vanillyl nonamide (open circles) 56 ppm piperine (filled squares) over time (min) for the group data (a) and three individual subjects. (Replotted from Lawless, 1984.)

different routes of access of irritant molecules to receptor structures. The anatomical substrate for differences in peripheral mechanisms is present in the oral epithelium insofar as a variety of morphologically different end-organs of trigeminal origin are visible histologically (Grossman and Hattis, 1967). There is much greater diversity than suggested by the common practice of referring to simply "free nerve endings" as the supposed nociceptors. The oral epithelium also contains a diversity of organized endings, both with and without connective tissue capsules, fibrillar extensions into the epithelium from some organized endings in the papillary lamina propria, and afferents with branched terminations that are different from free nerve endings (Grossman and Hattis, 1967).

A further complication in the anatomical picture of the periphery is the large number of trigeminal afferents in contact with taste buds. In the fungiform papillae, trigeminal terminations form a chalice-like structure around the taste bud in the hamster (Whitehead et al., 1985). In the rat, these fibers have been shown to have substance P immunoreactivity (Nagy et al., 1982). The exact role played by these fibers in irritant sensitivity is unclear, although psychophysical data from humans show a strong correlation of responsiveness to irritants with areas of the oral epithelium which are dense in fungiform papillae (Lawless and Stevens, 1988a; Miller, 1987). Whether all irritant compounds have equal access to this apparently sensitive structure is unknown.

A variety of mechanisms could influence the relative access of irritant molecules to receptors. For example, solubility differences might affect the local equilibrium in solution-receptor boundary layers. It is also unknown whether capsaicin and piperine bind to some membrane-bound protein-like receptor, or whether they interact more directly with the cell membrane bilayer.

## IV. PERCEPTUAL DIFFERENCES BETWEEN CAPSAICIN AND PIPERINE

As mentioned above, capsaicin and piperine differ in their average relative potency as well as in their potency from person to person. Another perceptual difference occurs in the sensitivity of different areas of the mouth to each irritant. There is some overall similarity in the spatial pattern of responsiveness to these two chemicals, e.g., in the high degree of responsiveness of tongue tip and edges, areas rich in fungiform papillae. However, there are also some differences between the two irritants in their spatial patterns.

Lawless (1984) requested that subjects report the areas of the mouth in which they felt the burn after whole-mouth rinses. The lips and gum were more frequently reported after stimulation with vanillyl nonamide than after stimulation with piperine. These results were more suggestive than definitive, since the subjects' accuracy in localizing their sensations of oral irritation was unknown.

To avoid this issue, Lawless and Stevens (1988) reported data from stimulation of specific areas of the oral cavity with filter paper swatches impregnated with capsaicin and piperine. The results of this study are shown in Figs. 3 and 4. The stimuli were equated for intensity on the tip of the tongue during the initial report period (30 sec after application). For capsaicin, the perceived intensity on the lip was second only to the tongue tip in magnitude, while for piperine, the lip was fourth in responsiveness after the tongue tip, side of the tongue, and posterior palate.

These figures illustrate a complexity in the pattern of responsiveness with regard to irritant, intensity, location, and time. The interaction of intensity, location, and time points out the difficulty of performing a direct discriminability assessment of perceptual differences between the irritants, while also

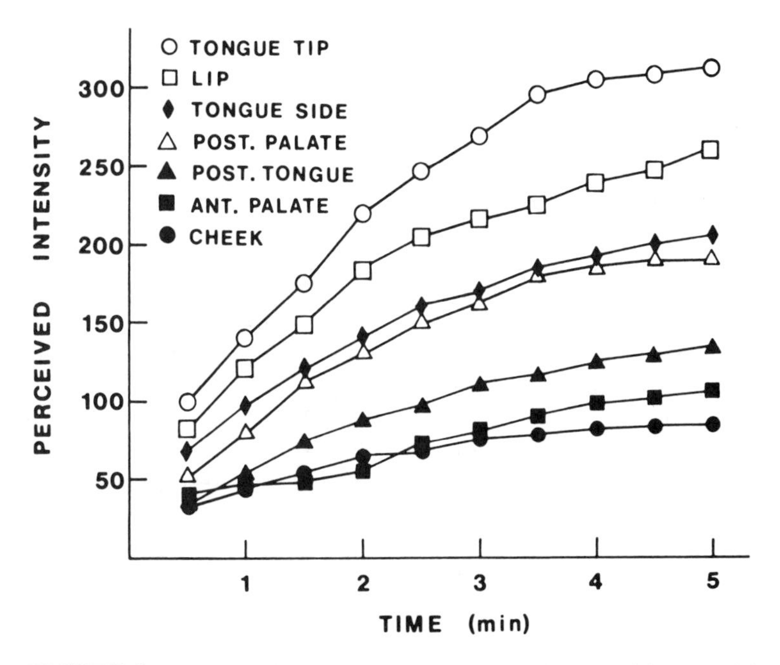


**FIGURE 3** Perceived intensity of filter paper swatches containing capsaicin, plotted over time for seven locations in the mouth. (From Lawless and Stevens, 1988a. Reprinted by permission.)

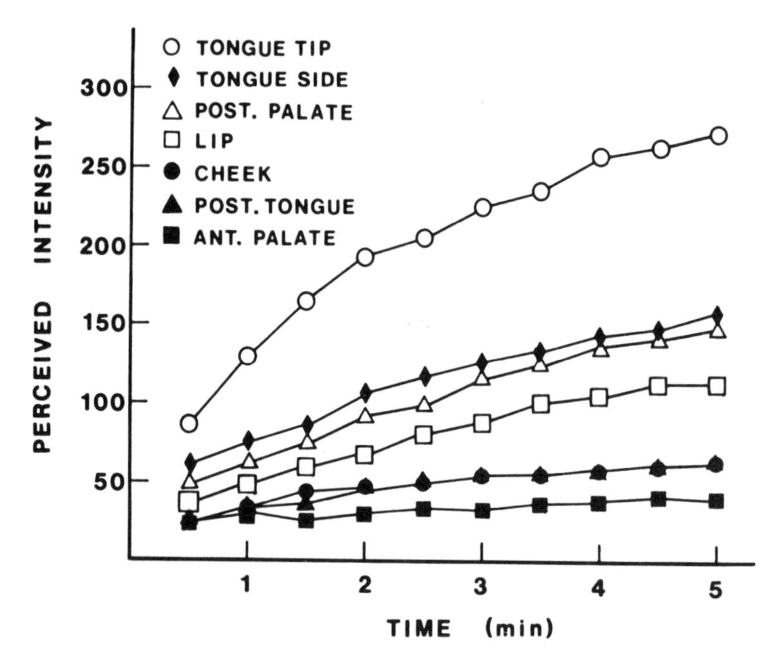


**FIGURE 4** Perceived intensity of filter paper swatches containing piperine, plotted over time for seven locations in the mouth. (From Lawless and Stevens, 1988a. Reprinted by permission.)

ruling out those three variables as the sensation differences on which correct discrimination is based. These results suggest that a demonstration of true quality discrimination can only be done on a particular area of the mouth for which the two irritants have been equated in intensity at the initial report interval. Otherwise, the basis for discrimination is ambiguous.

During the course of these localized stimulation trials, adjectives describing the perceived qualities of the stimuli were probed. Again, some similarities between the two irritants were observed. However, there were also compound-by-descriptor interactions and compound-by-locus-by-descriptor interactions suggesting differences in perceived quality. Capsaicin evoked more responses of *stinging, biting,* and *piercing,* while piperine evoked more choices of the descriptor *itching.* These reports are consistent with the notion of perceptual quality differences among irritants.

Two other observations during the course of our research have suggested differences between capsaicin and piperine. In a study of the interactions of irritation with repeated sequential tastant rinses, the time course of decay of the burn was approximately linear for piperine but exponential for capsaicin (Stevens and Lawless, 1986), an additional perceptual difference in the irritant-time profile. The second interaction concerns the ways in which irritants may inhibit sensations from the classical taste modalities. When intense irritant rinses are interspersed with trials in which tastants are given, capsicum oleoresin was seen to inhibit the sensations fro citric acid and quinine more than sensations from NaCl and sucrose (Lawless and Stevens, 1984; Cowart, 1987). However, the pattern of inhibition from piperine was more uniform across tastants (Lawless and Stevens, 1984).

This difference has implications for mechanisms as well as differences in perceived quality. As stated above, differences in perception logically entail some differences in the way the nervous system and/or peripheral physiology treats stimuli. For example, the differences in local responsiveness could be due to differences in the spatial distributions of receptors types, densities, or depth in the epithelium. Alternatively, the nature of the epithelium itself may affect stimulus access to receptors in some ways that are different for capsaicin and piperine. Conversely, differences in the exponential vs. linear time course of sensation decay could be due to differences in mechanisms of stimulus egress from receptors, or in mechanisms of the metabolism or destruction of molecules inside receptor cells. Whether the mechanism of stimulation involves a membrane-bound protein-like receptor site (Szolcsànyi and Jancsò-Gàbor, 1976), or whether these highly lipophilic molecules interact directly with the membrane phospholipids is also open to question (Nagy, 1982).

## V. IRRITANT INTERACTIONS

Two classical paradigms for the investigation of interactions among stimuli in the chemical senses have concerned stimuli presented simultaneously, i.e., mixture experiments, and stimuli presented sequentially, as in the case of cross-adaptation studies. An interaction of the two irritants and a demonstration of their nonequivalence were reported by Stevens and Lawless (1987). In this experiment, irritants were presented in pairs of successive mouth rinses lasting about a minute each. The intensity of the irritation was increased during the second rinse, and even more so if the second irritant was different than the first (see Fig. 5). This effect is consistent with partially independent populations of receptors for the two irritants, i.e., one parsimonious explanation for such enhancement is the recruitment of an additional population of receptors, receptor cells, and/or fibers with the second (and different) irritant.

A second type of interaction is observed when the irritants are presented simultaneously, i.e., in mixtures (Lawless and Stevens, 1989). The perceived

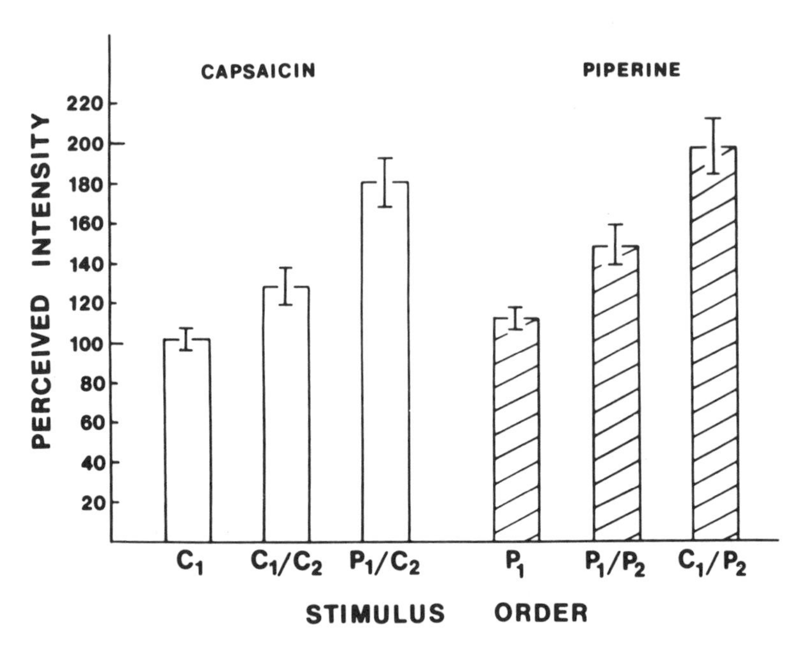


**FIGURE 5** Perceived intensity of oral irritants as a function of stimulus order. Open bars show the intensity of capsaicin and hatched bars the intensity of piperine. $C_1$, intensity of 1 ppm capsaicin after water rinse. $C_1C_2$, intensity of capsaicin after capsaicin. $P_1C_2$, intensity of capsaicin after piperine. $P_1$, intensity of 38 ppm piperine after water rinse. $P_1P_2$, intensity of piperine after piperine. $C_1P_2$, intensity of piperine after capsaicin. (Reprinted from Stevens and Lawless, 1987 by permission.)

intensities of piperine and capsaicin components were approximately matched in intensity at three sensation levels representing weak, moderate, and strong levels of burn. The 50/50 mixtures of these concentrations were compared to the intensities of the components. Figure 6 shows that at the intermediate concentration level, there is a marked synergy between the two irritants, such that the 50/50 mixture produces greater burn intensity than either twice the concentration of the piperine component or twice the concentration of the capsaicin component. This is also consistent with partial specificity of receptors, providing that the degree of specificity is inversely related to stimulus concentration.

This is a reasonable assumption in the chemical senses (e.g., Hanamori et al., 1987) as well as in other sensory modalities. At low levels, relatively specific and nonoverlapping populations of receptors may be stimulated by those irritants to which they are best tuned. As concentrations increase, each

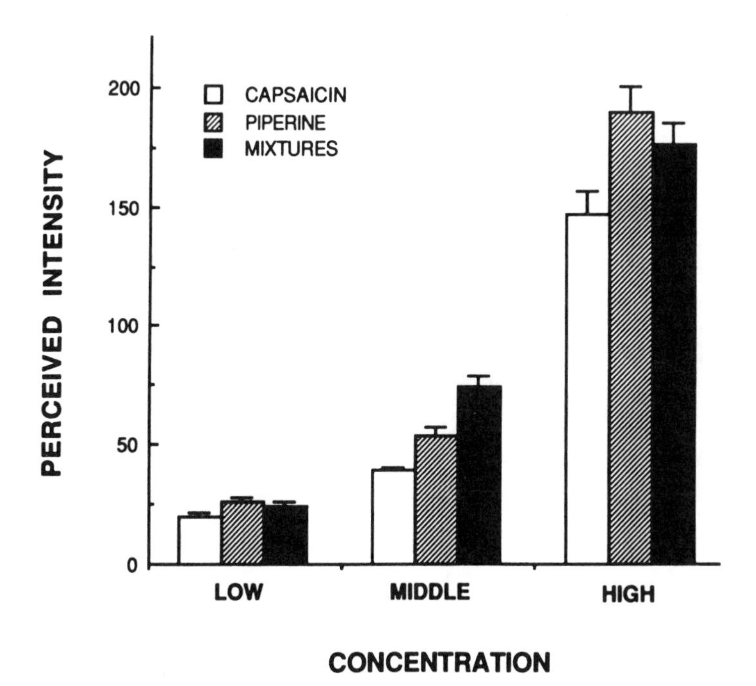


**FIGURE 6** Perceived intensity of capsaicin, piperine, and their 50/50 mixtures at three concentration levels. Capsaicin concentrations were 0.5 (low), 1 (middle), and 2 (high) ppm. Piperine concentrations were 17.5 (low), 35 (middle), and 70 (high) ppm. (Replotted from Lawless and Stevens, 1988b.)

irritant stimulates a wider and wider array of units as the thresholds for more and more units (for that given compound) are surpassed. Concentration-dependent synergy can be explained by assuming a subpopulation of units at or near threshold (Fig. 7). At intermediate levels, there is sufficient overlap in the near-threshold populations such that a redundancy gain occurs (the combined stimulation pushes some of them above threshold), similar to the "facilitation" described by Sherrington in spinal reflexes from compound stimuli.

In contrast to the partial specificity suggested by these experiments, studies of capsaicin desensitization have shown a greater degree of generalization among irritants. When capsaicin is injected into an animal, or applied topically and repeatedly to the skin or cornea, sensitivity to a variety of chemical irritants is greatly diminished (Jancsò, 1960; Szolcsànyi, 1977). This effect is not inconsistent with the cross-enhancement and mixture synergy we observed, providing the two sets of observations have different underlying

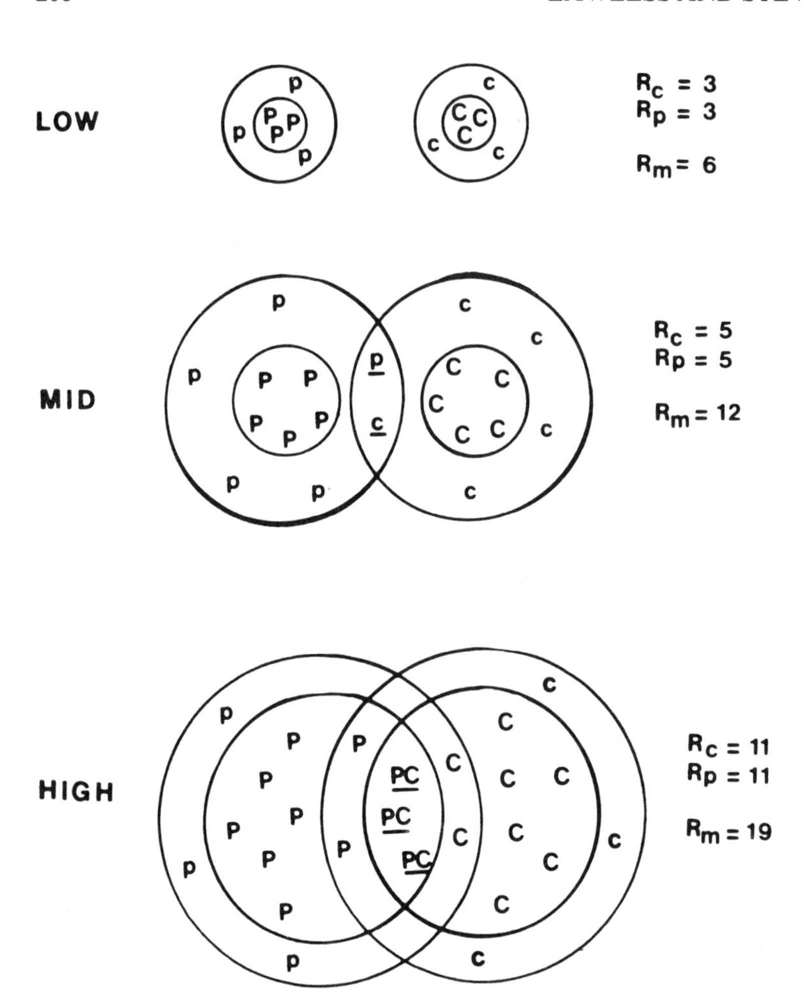


**FIGURE 7** Scheme for explaining concentration-dependent synergy in mixtures. Circles represent populations of neural units. P, units responding to piperine. C, units responding to capsaicin. p, near-threshold units responsive to piperine. c, near-threshold units responsive to capsaicin. Units in overlapping areas are responsive to both chemicals. $R_c$, response to capsaicin. $R_p$, response to piperine. $R_m$, response to mixture. (Reprinted from Lawless and Stevens, 1989 with permission.)

physiological mechanisms. For example, capsaicin desensitization could be the result of neurotransmitter depletion. On the other hand, the effects we observed in rapid sequential stimulation may involve the recruitment of additional receptors before transmitter depletion can occur. As Nagy (1982) pointed out, stimulatory potency and potency as a desensitizing agent are weakly correlated (Szolcsànyi and Jancsò-Gàbor, 1977). Irritants have a variety of physiological effects and a variety of underlying physiological mechanisms.

In summary, while there is no airtight proof that the irritant compounds from red and black pepper have different peripheral modes of stimulation and different perceptual properties, a growing body of consistent information points in that direction. Certainly, there are perceptual similarities among irritant compounds and perhaps in stimulatory mechanisms as well. However, the red and black pepper compounds are not interchangeable, as any chef will tell you.

Three experimental programs present clear opportunities for future research. The first concerns the individual differences in responsiveness noted in Fig. 2. If families of capsaicin-related and piperine-related compounds turn out to be correlated within a family and uncorrelated between a family with regard to individual responsiveness, additional evidence for independent mechanisms would be gained. This is analogous to the low correlation of responsiveness to quinine and phenylthiocarbamide in bitter taste. This approach might involve factor analysis to define functionally intercorrelated groups of stimuli.

A second area of research concerns the extrapolation of our sequences paradigm to longer time intervals and perhaps more intense oral stimulation such that cross-desensitization can be examined. A lack of cross-desensitization would not be expected from the pharmacological literature but if obtained would suggest separate peripheral mechanisms and separate neural pathways for the piperine and capsaicin. A result showing cross-desensitization would be consistent with partially independent peripheral mechanisms feeding into a higher common pathway and would explain some of the perceptual similarity between the two compounds in the face of possible differences in receptors or in receptor access.

Finally, the language of chemical heat, like the language of pain, is impoverished. One convincing demonstration of perceptual differences among irritants would be the training of a descriptive panel to distinguish among the irritants and to apply a specialized vocabulary to the different sensations elicited by the compounds in a reliable fashion. Unfortunately, such descriptive analysis training is both time- and labor-intensive. Furthermore, training a descriptive analysis panel for chemical heat would be doubly difficult due to the limitations imposed by the long time course of sensation for each stimulus and by such carryover effects as desensitization. Perhaps this could be achieved in an industrial setting in which standing panels of volunteers are available for testing and training on a daily basis for periods of years.

## REFERENCES


Cowart, B. J. (1987). Oral chemical irritation: Does it reduce perceived taste intensity? *Chem. Senses* 12: 467-479.

Govindarajan, V. S. (1979). Pungency: The stimuli and their evaluation. In: *Food Taste Chemistry*. J. C. Boudreau (Ed.). American Chemical Society, Washington, D.C., pp. 53-92.

Govindarajan, V. S. (1987). Capsicum production, technology, chemistry and quality. Part I: History, botany, cultivation, and primary processing. *CRC Crit. Rev. Food Sci. Nutr.* 22: 109-175.

Grossman, R. C., and Hattis, B. F. (1967). Oral mucosal sensory innervation and sensory experience: A review. In: *Symposium on Oral Sensation and Perception*. J. F. Bosma (Ed.). Charles C Thomas, Springfield, IL., pp. 5-62.

Hanamori, T., Miller, I. J., and Smith, D. V. (1987). Taste responsiveness of hamster glossopharyngeal nerve fibers. *Ann. N. Y. Acad. Sci.* 510: 338-341.

Jancsò, N. (1960). Role of the nerve terminals in the mechanism of inflammatory reactions. *Bull. Millard Fillmore Hosp. Buffalo* 7: 53-77.

Lawless, H. (1984). Oral chemical irritation: Psychophysical properties. *Chem. Senses* 9: 143-155.

Lawless, H., and Gillette, M. (1985). Sensory responses to oral chemical heat. In: *Characterization and Measurement of Flavor Compounds*. D. D. Bills and C. J. Mussinan (Eds.). American Chemical Society, Washington, D.D., pp. 26-42.

Lawless, H. T., and Stevens, D. A. (1988). Responses by humans to oral chemical irritants as a function of locus of stimulation. *Percep. Psychophys.* 43: 72-78.

Lawless, H. T., and Stevens, D. A. (1989). Mixtures of oral chemical irritants. In: *Perception of Complex Smells and Tastes*. D. G. Laing (Ed.). Academic Press, Sydney, pp. 297-309.

Lawless, H., Rozin, P., and Shenker, J. (1985). Effects of oral capsaicin on gustatory, olfactory and irritant sensations and flavor identification in humans who regularly or rarely consume chili pepper. *Chem. Senses* 10: 579-589.

Miller, I. J. (1987). Human fungiform taste bud density and distribution. *Ann. N. Y. Acad. Sci.* 510: 501-503.

Nagy, J. I. (1982). Capsaicin: A chemical probe for sensory neuron mechanisms. In: *Handbook of Psychopharmacology*, Vol. 15. L. L. Iverson, S. D. Iverson, and S. M. Snyder (Eds.). Plenum Press, New York, pp. 185-235.

Nagy, J. I., Goedert, M., Hunt, S. P., and Bond, A. (1982). The nature of the substance P-containing nerve fibers in taste papillae of the rat tongue. *Neuroscience* 7: 3137-3151.

O'Mahony, M. A., and Odbert, N. (1985). A comparison of sensory difference testing procedures: Sequential sensitivity analysis and aspects of taste adaptation. *J. Food Sci.* 50: 1055-1058.

Stevens, D. A., and Lawless, H. T. (1986). Putting out the fire: Effects of tastants on oral chemical irritation. *Percep. Psychophys.* 39: 346-350.

Stevens, D. A., and Lawless, H. T. (1987). Enhancement of responses to sequential presentation of oral chemical irritants. *Physiol. Behav.* 39: 63-65.

Szolcsànyi, J., and Jancsò-Gàbor, A. (1976). Sensory effects of capsaicin congeners. Part I. *Arzneim.-Forsch.* 25: 1155-1158.

Szolcsànyi, J., and Jancsò-Gàbor, A. (1977). Sensory effects of capsaicin congeners. Part II: Importance of chemical structure and pungency in desensitizing activity of capsaicin-type compounds. *Arzneim.-Forsch.* 26: 33-37.

Szolcsànyi, J. (1977). A pharmacological approach to elucidation of the role of different nerve fibers and receptor endings in mediation of pain. *J. Physiol.* (Paris) 73: 251-259.

Whitehead, M. C., Beeman, C. S., and Kinsella, B. A. (1985). Distribution of taste and general sensory nerve endings in fungiform papillae of the hamster. *Am. J. Anat.* 173: 185-201.

# Chapter 10 Discussion

**Dr. Walker:** Have you considered looking at the question of qualitative differences between capsaicin and piperine in a discrimination paradigm?

**Dr. Lawless:** I'm going to be forced to do that experiment. My hestitation has always been that someone can come back to me and say, well for that subject, the intensities really weren't equated, or they perceived the two stimuli in a different part of the mouth. So my search for a qualitative difference would always be open to some criticism. I think I'll do the experiment anyhow, though, because although it is not a sufficient condition to demonstrate perceptual differences, it is a necessary condition.

**Dr. Silver:** Just a quick comment about the recordings from the lingual nerve. Several of us have tried to record from the lingual nerve while stimulating with capsaicin, including myself and Gary Schwartz here at Monell. We have not been very successful. Harata in Japan has recorded from the lingual nerve and has gotten single-unit responses to capsaicin, but those units, according to him, are very, very rare. For some reason, it's difficult to find responses to capsaicin in the lingual nerve.

**Dr. Rozin:** I think that obviously one needs neurophysiological evidence to converge on an explanation, but I think historically in the senses, neurophysiological evidence has not led directly to an understanding of how to divide the senses. Color vision is probably the best advanced field, and neurophysiology has trailed way behind the psychophysics. In fact, it was the psychophysicists who told the neurophysiologists what to look at, and when they told them the wrong thing the neurophysiologists found the wrong thing. That's because it's hard to do a quick analysis of what's going on in a single unit. I think in taste as well it would be fair to say that the problem of how to divide the tastes into their components or standards has not been totally clarified by studying neuroscience. We do need multiple routes of attack and it seems to me that historically the first approach has been to get some sense of how the human organism divides the stimuli up. That gives you a framework with which to look at the nervous system instead of just going in there and hoping to find something.

**Dr. Cain:** One of the first-order questions I think you should have listed there, and I consider it to be the most important thing to try to understand, is how these compounds distribute themselves in the oral cavity and what they adsorb to. All other things are going to flow from that. I don't know if fluorescence can do it for you, for instance. I don't know what kind of sensitivity you would have, but if the biophysical properties of the mouth vis-à-vis the stimuli are such that you are going to get them distributed in different places, you will see different rules of mixture interaction, different rules of adaptation and cross-adaptation, than if they were all restricted to one area. And I think that if you do the kind of experiment you reported here restricted to one spot on the mucosa, it would help you determine to what degree these are sharing a common mechanism vs. having different mechanisms. I would encourage you to do that. But I think that coming to know the distribution of these things in the mouth is crucial and fluorescence might help you there; or perhaps some kinds of dyes. It might allow you to find out why things take a long time to go off, and in that context I often wonder if some detergents or ethanol or things like that might give you some insight into what's going on.

**Dr. Lawless:** You wouldn't be concerned with nonspecific binding?

**Dr. Cain:** Yes, but you don't even known, for instance, to what degree things that are bound nonspecifically are serving as reservoirs of stimuli for neurons. I don't mean to imply that you will learn everything from this, but if you find that piperine and capsaicin affect very different areas, even if you are looking at nonspecific binding, then it seems to me that you have learned something important. And restricting it to one area might give you more insight with respect to things like mixtures and adaptation.

**Dr. Green:** I'll jump into this messy area of labeling sensations. It seems to me that to call sensations like sting and burn tactile sensations is problematic, because tactile really means touch, and historically that has been related to mechanical stimulation. So I think to throw those all into one basket as you have done is somewhat misleading.

**Dr. Lawless:** Are you concerned about the label on that one slide that said tactile? That was just a convenient way of categorizing things for subjects to relate to. They had a ballot with four rows of adjectives in front of them, some of them thermal, some of them painful, some of them were called tactile, some of them were taste or flavor; these were just general categories. I didn't mean to imply anything mechanistic or really categoric by that label. It was mainly for the understanding of the subject.

**Dr. Green:** But when these things get into the literature people sometimes go on to use the terms incorrectly.

**Dr. Lawless:** Yes, I appreciate your concern about that.

**Dr. Green:** The question I had relates to the enhancement that you obtained with sequential stimulation. You mentioned that you did this before desensitization began to occur. In my experience with capsaicin, at, say, 2 ppm, which are the concentrations you were using, I really don't see any desensitization with repeated exposure in the oral cavity. You see a growth in sensation over the first few exposures, and if you keep hitting the oral cavity with capsaicin, it pretty much levels out at a moderate level rather than adapting. I know you did some work here at Monell on that issue, so I don't think I'm really telling you anything. But my point is, do you think you would have gotten the same enhancement effect if you had produced a steady-state burn with capsaicin and then followed with piperine? Do you think you would then see the enhanced response, or do you think it's specific to the period when sensation is building over the first couple of trials?

**Dr. Lawless:** I think the case where we will see desensitization is when we have an extended mouth rinse, we let the burn build, then let it fall off for 15, 20 min or a half hour, before we apply the next stimulus. When I've analyzed sequences of repeated stimulations within an experimental session, things have been less intense if they followed a strong trial than if they appeared first in that day's session. Regarding timing relative to the steady-state burn, it's difficult to determine where the crossover between enhancement and desensitization would occur. At some point enhancement should be negated by desensitization. One could investigate that in a parametric study.

**Dr. Szolcsànyi:** Just a short remark because it seems to be rather discouraging that in the tongue, people have found very few fibers that respond to capsaicin. This might simply be for technical reasons. For a long time we could not find any "pain receptors" of any type because they are thin and therefore difficult to dissect. So I think there actually are a lot of receptors in the tongue. In the skin we have a rather complete picture of the types of receptors, and the tongue was one of the first areas from which single units were recorded by Zotterman. But I don't know of any more recent, careful, single-unit studies of the type that could hope to detect all of the varieties of receptors that innervate the tongue. I think that this would be a very worthwhile undertaking. A lot of problems raised in the psychophysical studies could be addressed by studies of single units in the tongue.

# 11
# Personality Variables in the Perception of Oral Irritation and Flavor

**David A. Stevens**
Clark University
Worcester, Massachusetts

Individual differences in psychophysical studies, if not accounted for, can produce serious problems. If the sources of such variance are not identified, the variance will typically be included in the error estimate used in statistical analyses. This inflated error term increases the type II error rate; no effects, or weaker effects are more likely to be inferred than if the error rate was smaller.

Further, a failure to identify systematic individual variation results in generalizations being made from the date under the assumption that a simpler model accounts for the data than is required. For example, bi- or multimodal populations might be erroneously treated as unimodal ones. Then measures of central tendency, e.g., means and medians, would be incorrectly used to describe those populations.

Increasing the size of the sample is not a solution if there are systematic individual differences. A larger number of subjects will indeed increase reliability but will not ensure that the inferences made are valid. For example, a mean is as invalid a measure of a large bimodal population as of a smaller bimodal population.

The solution to this problem is the identification of sources of individual variation and their control by an appropriate method.

That there are a number of sources of individual differences is well known. For example, Pangborn (1981) lists subject variables that define genetic and biological differences, intellectual differences, semantic differences, and personality differences as known sources of individual differences. Here I will present evidence to support the argument that individual differences in responses to taste, smell, and flavor stimuli can be identified from a theoretical position of self-perception in personality.

Self-perception theory (Bem, 1967, 1972) asserts that we know our own feelings, moods, abilities, and so forth by the same means by which we know them in others: by inference from observed actions and their context. This thinking follows directly from the theory of emotion published by William James (1884). Since one's perception of one's own behavior is critical to the experience of emotions, feelings, and moods, the theory is called a self-perception theory.

Recent research indicates that individuals differ consistently in the types of information they use in making inferences about themselves. Two types of cues have been identified: (a) those that arise from the individual's actions and personal properties, including visceral responses, expressive behavior, and instrumental actions and the consequences of those actions, and (b) those that come from the environment, including normative expectations about what most people would feel in a particular situation. The former are called self-produced cues, and the latter, situational cues (Laird and Berglas, 1975).

All persons seem to use situational cues, but there are reliable differences among individuals in the extent to which they are affected by manipulations of self-produced cues. For example, when people are induced to perform the muscle movements associated with smiling and frowning, which produce self-produced cues, some subjects report strong emotional fluctuations corresponding to the manipulation of facial expression and some do not.

Those individuals who are affected by experimental manipulation of facial expression are called self-produced cuers. Research has shown that they respond to self-produced cues in other situations as well. They change their impression of themselves as a result of changes in appearance (Kellerman and Laird, 1982), feel more romantic love after gazing into their partner's eyes (Kellerman et al., 1989), and show reverse placebo effects—they feel *more* fear rather than less after taking placebos identified as relaxers, and *less* aroused rather than more after taking placebos identified as arousers (Duncan and Laird, 1980). Presumably, when self-produced cuers experiencing fear are given a placebo identified as a relaxer, the realize that those responses that produce fear have not changed. Accordingly, they conclude that they must be even more fearful than they thought since the pill did not work; a reverse placebo effect is found.

Situational cuers (people who do not respond to facial manipulation) will accept more readily an experimenter's suggestion about what they should feel (Kellerman and Laird, 1982) and show the usual positive placebo effects (Duncan and Laird, 1980). Being less sensitive to those responses that are interpreted as fear, they respond to the situational cue of the placebo.

Taste, smell, and flavor perception, and one's hedonic responses to those sensory experiences, are the result of both self-produced and situational cues. Self-produced cues result in part from mouth and facial movements, salivation, swallowing, inhalation, and other behaviors that accompany tasting and smelling. Situational cues are the normative expectancies produced by the situation, e.g., that roses will have a floral odor and that it is pleasant, or that a maple syrup will taste sweet.

If these two kinds of cues are used in the evaluation of taste, smell, and flavor, and if there are individual differences in the extent to which people use them, it follows that one should be able to find systematic differences in flavor experiences that vary with the types of cues people use.

Since the consumption of pepper produces a variety of responses, including expressive, autonomic, and instrumental ones, and people seem to differ widely in their responses to peppers, I reasoned that capsaicin and piperine would be ideal stimuli with which to test the notion that individual differences in flavor perception would be related to individual differences in self-produced cueing.

Harry Lawless and I have been jointly conducting a series of studies on the sensory effects of capsaicin and piperine, some of which are summarized in Chapter 10. All of our subjects were given private body consciousness (PBC) tests (Miller et al., 1981) so that I could investigate the possibility that sensitivity to bodily changes are a source of individual differences in response to oral irritants. People were classified as having low or high PBC by their answers to questions on how sensitive they are to changes in body temperature, internal tensions, heart rate, dryness of the mouth and throat, and hunger contractions. While this test does not determine the extent to which people respond to self-produced cues per se, it does measure the extent to which they are sensitive to bodily changes such as those produced by caffeine. The PBC test was appended to a food preferences survey.

In one study we were interested in the effect of tastants on oral irritation. All of the subjects rinsed their mouths with a solution of one of the irritants, which produced the burning sensations associated with the eating of peppers. Then they tasted samples of citric acid, sodium chloride, quinine hydrochloride, sucrose, water, or tasted nothing. The subjects reported the intensity of the burn using magnitude estimation.

Figure 1 shows the results when piperine (100 ppm) was the irritant and judgements were made while the tastant was held in the mouth (Stevens and

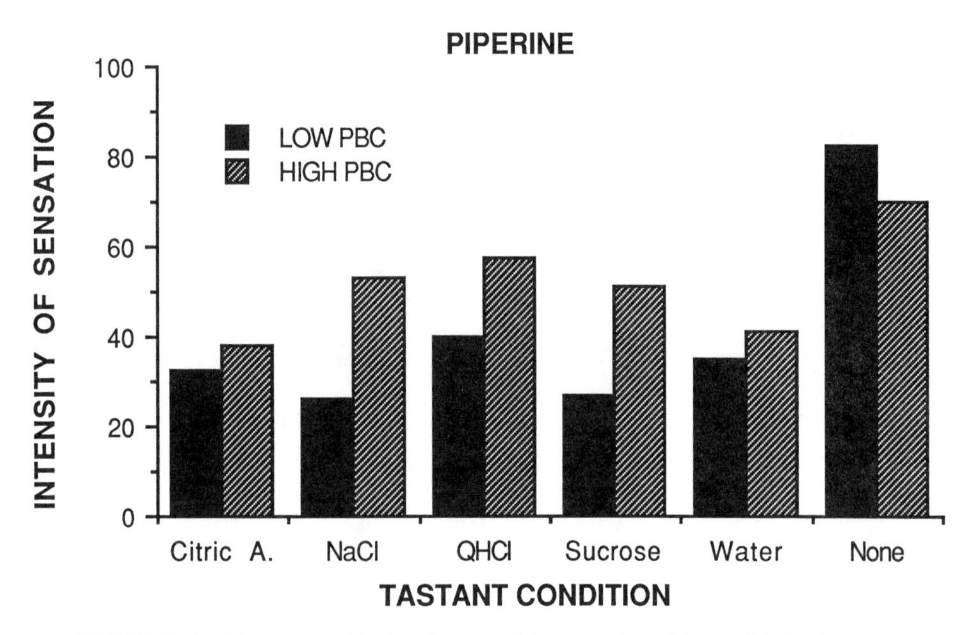


**FIGURE 1** Mean normalized estimates of the intensity of the oral irritation produced by piperine with solutions of various tastants or no tastant in the mouth by subjects classified as having high and low private body consciousness.

Lawless, 1986, exp. 1). There were clear differences between high and low body consciousness people in the effects of the tastants on the burn piperine.* The effects are not simply due to low PBC people giving generally lower estimates than do high PBC people, since these data were normalized prior to analysis using estimates of a standard solution of NaCl obtained before the subjects rinsed their mouths with the irritant. Further, in other studies, the low PBC subjects gave higher estimates than high PBC subjects in some conditions.

In another study (unpublished) we stimulated various parts of the mouth with filter paper treated with capsaicin (3.2 $\mu$g) or piperine (320 $\mu$g) and the subjects (N = 20) reported the intensity of the burning sensations. Figure 2 shows the intensities of irritation over time reported by low and high PBC subjects for capsaicin, averaged over different parts of the mouth. The function is considerably flatter for the low PBC subjects—the result you would expect if they were less sensitive to the self-produced cues from the irritants.

*This and other effects reported here are statistically significant; $ps < 0.05$.

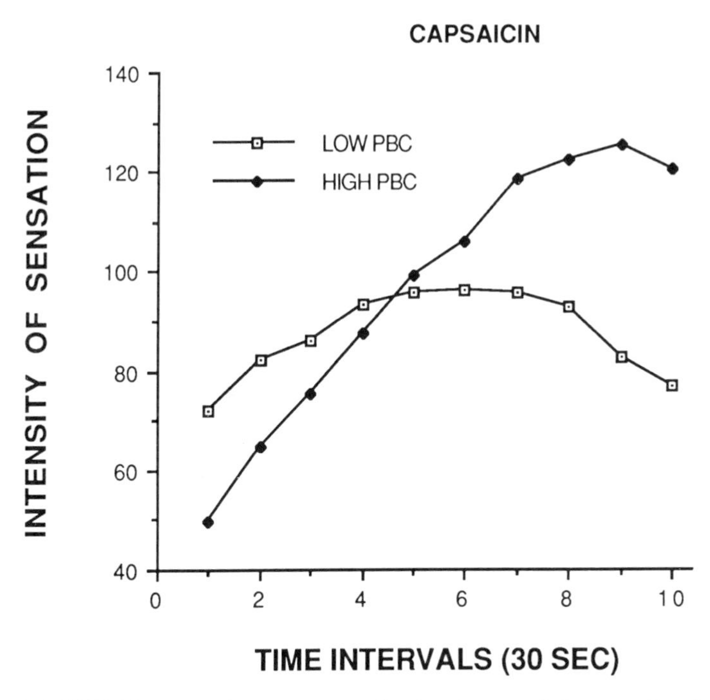


**FIGURE 2** Periodic mean normalized estimates of the intensity of the oral irritation produced by capsaicin over 5 min by subjects classified as having high or low private body consciousness.

Figures 3 and 4 show the responses of those same subjects for the various loci which were stimulated by capsaicin- and piperine-treated paper, respectively. Differences attributable to cue type were found for both irritants. The effect was greatest at the tip of the tongue where the highest intensity of burn was reported; subjects with higher PBC scores reported more intense sensations than did those with lower PBC scores.

In these studies I did not deliberately manipulate situational cues. Accordingly, in the next study both situational cues and stimuli which should affect self-produced cues were varied. This was done in a factorial experiment in which subjects identified as being more or less sensitive to bodily responses evaluated soup samples. These samples differed orthogonally in sodium chloride concentration (a likely stimulus for self-produced cues) and in the verbal information given about the flavor (a situational cue).

The subjects were given a facial manipulation procedure similar to that used by Duncan and Laird (1980) for classifying subjects as self-produced or situational cuers, the test for PBC, and the food preferences survey.

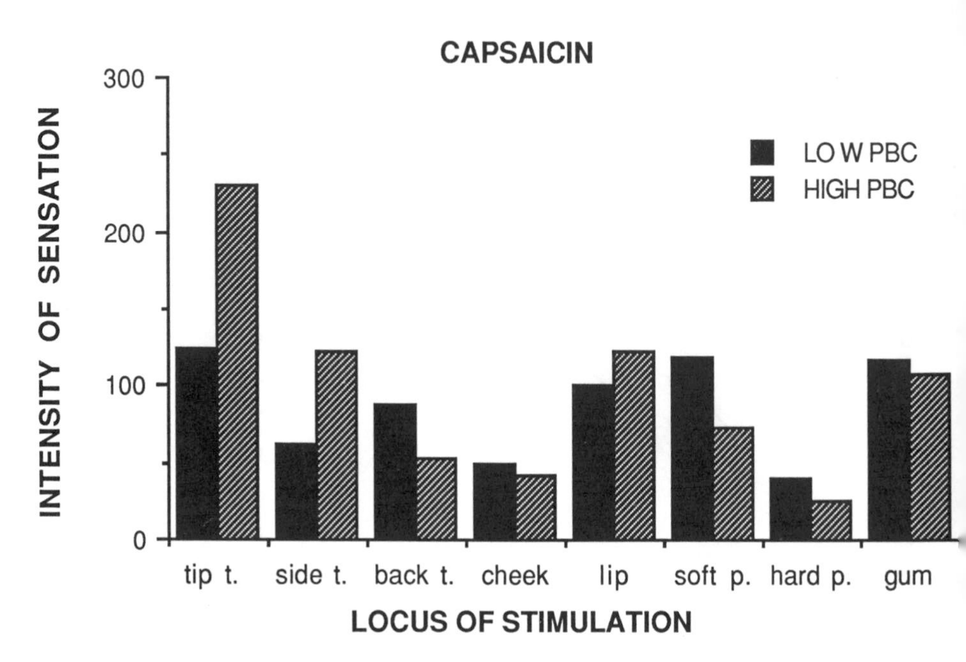


**FIGURE 3** Mean normalized estimates of the intensity of the oral irritation produced by capsaicin applied to tip of tongue, side of tongue, back of tongue, inside of cheek, inside of lower lip, soft palate, hard palate, and lower gum, by subjects classified as having high or low private body consciousness.

For the facial manipulation test, the subjects were given a cover story to account for their being asked to manipulate their facial muscles. Then they were told to contract the muscles near the corners of their mouth and extend the eyebrow muscles slightly, or to contract the muscles located above their eyebrows by drawing them down and together and to contract the muscles at the corners of the jaw by clenching the teeth. The first manipulation made a smile; the latter a frown. While posed in these ways, the subjects viewed patterns of geometric forms identified as abstract art and having visible titles (situational cues) with an emotional tone the opposite of that associated with the facial expression, e.g., one title used for a smile trial was ''rage''. Then they completed a mood adjective check list derived from the Nowlis-Green Adjective Check List (Nowlis, 1965). The subjects were scored on the basis of the extent to which they changed their mood in the direction consistent with the facial manipulation.

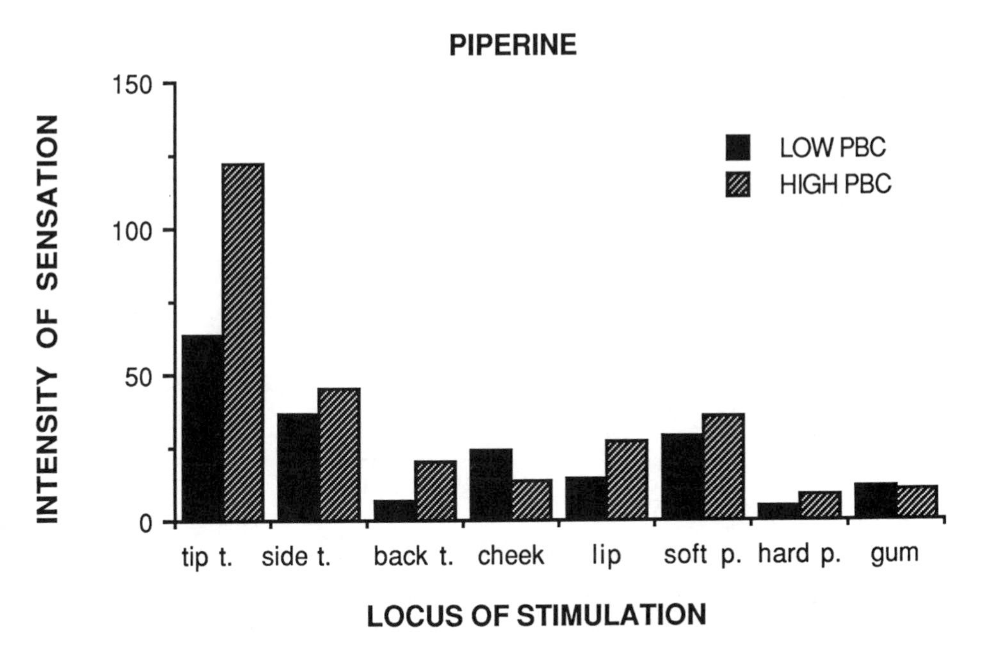


**FIGURE 4** Mean normalized estimates of the intensity of the oral irritation produced by piperine applied to eight oral loci by subjects classified as having high or low private body consciousness.

Subjects falling in the upper and lower thirds of the distributions of facial manipulation results and PBC scores were classified as having high (N = 32) or low (N = 28) private body consciousness and being self-produced (N = 29) or situational (N = 31) cuers, respectively. Unexpectedly, there was no correlation between the two classification systems; how much people think they respond to internal stimuli is not related to how their mood is affected by facial muscle tensions.

All nine combinations of samples of chicken soup having three levels of sodium concentration (normal, 0.276%; high, 0.420%, or very high, 0.564% w/v), three kinds of statements about the soup (told the soup had "less than normal" flavoring, told it had "more than normal" flavoring, or told nothing about flavoring) were presented to the subjects. They tasted the samples and rated them using line scales on several attributes including the intensity of overall flavor, saltiness, chicken flavor, oiliness, and hedonic quality.

With the subjects classified as to the type of cuer, there was an interaction between classification and ratings of overall intensity of flavor for the three sodium concentrations. The self-produced cuers were more sensitive

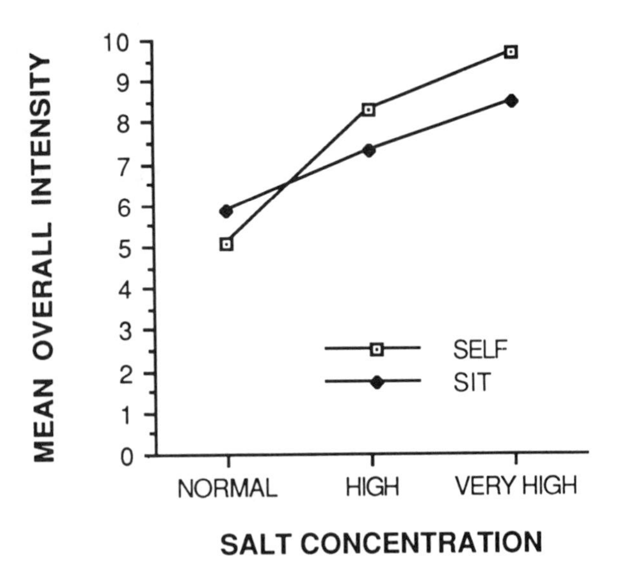


**FIGURE 5** Mean estimates of overall intensity of flavor of soups having three concentrations of salt by subjects classified as being self-produced or situational cuers. Adapted from Stevens, et al. (1988).

to the changes in sodium concentration than the situational cuers. The psychophysical slope for flavor intensity was 0.92 for the former and 0.52 for the latter. The functions are shown in Fig. 5. A reliable difference in functions for estimates of saltiness was not found.

When the subjects were classified by PBC, an interaction was found between classification and ratings of saltiness, but not overall flavor, for the three sodium concentrations shown in Fig. 6. The high PBC group had a steeper psychophysical slope (1.17) than the low PBC group (0.85); people with high PBC were more sensitive to changes in the physical qualities of the soup than those with low PBC.

With regard to hedonic ratings, across all three sodium concentrations the low PBC group consistently liked samples identified as having flavor added more than those identified as having flavor reduced (a positive placebo effect), whereas there was no consistent relationship between verbal information about flavor and hedonic ratings for the high PBC group across sodium concentrations.

The results found with the normal salt concentration samples are shown in Fig. 7. They are similar to those found in studies comparing responses by people of different cue types and PBC scores to placebo treatments (Duncan and Laird, 1980, Brockner and Swap, 1983). In those studies, the situational cuers and low PBC subjects showed positive placebo effects, while the self-produced cuers and high PBC subjects showed reverse placebo effects.

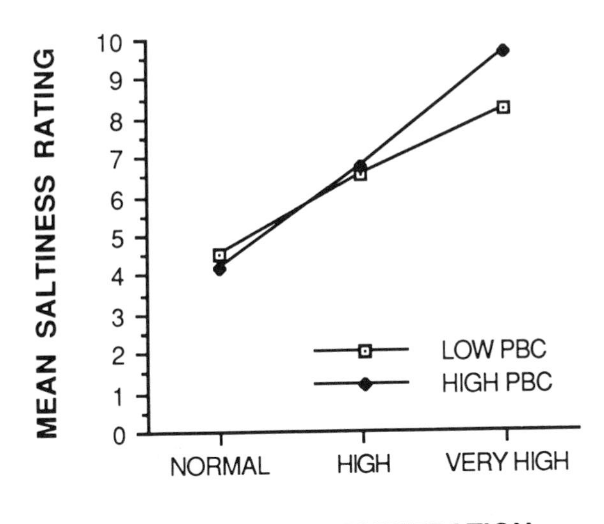


**FIGURE 6** Mean estimates of saltiness of soups having three concentrations of salt by subjects classified as having high or low private body consciousness.

The present high PBC subjects also showed a reverse placebo effect; their relative hedonic judgments were in the opposite direction to that suggested by the verbal information. Presumably the subjects found the soup sample having less than the optimal level of flavor, heard it described as having flavor added, and concluded that a soup tasting like that after having flavor added cannot be a very good soup. They assigned it a low hedonic rating. However, the same soup described as having flavor reduced was given a higher hedonic rating. Presumably, the subjects concluded that a soup having that much flavor left after flavor reduction must have redeeming qualities. This reverse placebo effect was not found at the higher sodium concentrations, and this is consistent with placebo effects in general. As the sensory event being manipulated by a placebo becomes stronger, the placebo becomes less effective.

The no-information condition was included as a control for information, but on reflection it is obvious that it was not an adequate one; no information in a context where information is more often provided is not a neutral event. Thus it is difficult to interpret the effects of that condition.

The food preferences survey to which the PBC test was appended included items in which the subjects indicated how frequently they ate a hot or spicy food, added hot sauce to Mexican food, chose hot and spicy Chinese and Indian food, and added hot and spicy oil to Chinese food. I hypothesized that there would be a correlation between PBC scores and the extent to which

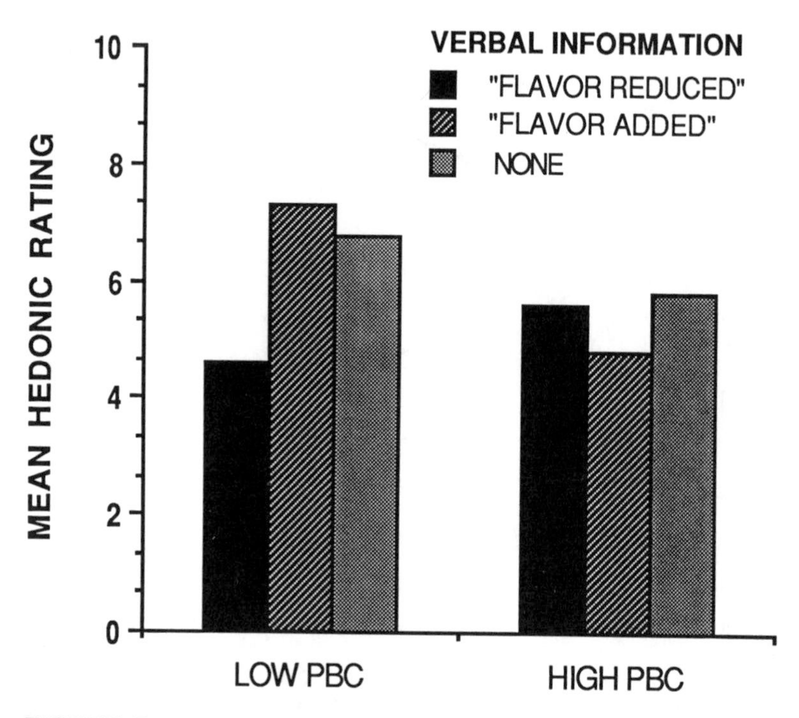


**FIGURE 7** Mean hedonic ratings of soup having normal salt concentration by subjects classified as having high or low private body consciousness when given three kinds of verbal information. Adapted from Stevens, et al. (1988).

people eat spicy food in general and add hot sauces to it when served. This idea was tested by analyzing the responses from 227 people who had volunteered as subjects for research in my laboratory. These were mostly undergraduate students; the remaining were other members of the Clark University community.

The vast majority of these people were infrequent consumers of hot and spicy food, and no general relationships were found between personality types and preferences for that food. However, when I examined only the responses from frequent consumers, those 13 subjects reporting consumption of spicy food more than four times a week, significant inverse relationships were found. The higher the PBC score, the less likely people were to add hot sauce to Mexican food ($r = 0.67$) or to choose spicy Chinese or Indian dishes ($r = 0.65$). Thus, while the personality classifications did not predict the use of hot and spicy foods in general, private body consciousness scores predicted the extent to which frequent users added hot condiments and chose the hotter dishes; the higher the PBC score, i.e., the more people are sensitive to bodily changes, the less likely they are to add hot pepper to food. Rozin and Schiller (1980) noted that the motivation for choosing to eat chili peppers may include personality, social, and associative factors as well as sensory ones. The present finding is consistent with this conclusion. Among those who chose to consume hot and spicy foods frequently, more heat is not necessarily better. Predictable individual differences exist.

The present work provides good evidence that there are interactions between sensory variables and personality variables that determine the kinds of cues people use in evaluating sensory events. Classification of subjects on the basis of how sensitive they are to their own bodily changes and the extent to which their mood is affected by manipulation of facial muscle tension was associated with differences in responses to flavor stimuli. However, as expected from initial studies, a number of questions were raised. For example, why was there no apparent relationship between private body consciousness and cue type? Why did the PBC test identify individual differences in estimates of overall flavor of soup but not saltiness, and the cue-type test identify individual differences in estimates of saltiness but not flavor intensity? The PBC test was validated using sensitivity of responses to caffeine (Miller et al., 1981), the types of responses which self-produced cuers are likely to use.

Self-perception theory asserts that self-produced cuers differ from situational cuers not in the intensity of sensations but in differences in the extent to which cues resulting from their actions are used. Thus stimuli for which differences in self-produced and situational cuers are found in flavor perception should be stimuli that elicit mouth movements, salivation, etc., and those stimuli with which personality differences are not found should be those which do not elicit the cues that self-producers can use. This relation remains to be established, but there is support for it. In the soup study there were no differences attributable to personality in ratings of intensity of chicken flavor or oiliness, qualities which should not have been affected by differences in salt concentration. Salt does not produce responses associated with the flavor of chicken per se nor the mouth feel of oil.

Working from the viewpoint of a self-perception theory of personality I hypothesized that classifying people by their level of private body consciousness and the extent to which they use self-produced cues in making self attributions would identify sources of systematic individual differences in responses to oral chemical stimulation. The results of the several studies reported here suggest that this hypothesis is valid. Certainly the measures used here to classify people are not in themselves error-free, but they do provide consistent evidence for self-perception playing an important role in psychophysical judgments.

## ACKNOWLEDGMENTS

The advice and assistance of James D. Laird, Charles Bresler, Robin Damrad, Dale A. Dooley, and Debra Thomason, and partial support by NIH grant NS-20616 and Thomas J. Lipton, Inc. are greatfully acknowledged.

## REFERENCES

Bem, D. J. (1967). Self-perception: An alternative interpretation of cognitive dissonance phenomena. *Psychol. Rev.* 74: 183-200.

Bem, D. J. (1972). Self-perception theory. In *Advances in Experimental Social Psychology, Vol. 6.* L. Berkowitz (Ed.). Academic Press, New York, pp. 1-62.

Brockner, J., and Swap, W. C. (1983). Resolving the relationships between placebos, misattribution, and insomnia: An individual-differences perspective. *J. Pers. Soc. Psychol.* 45: 32-42.

Duncan, J. W., and Laird, J. D. (1980). Positive and reverse placebo effects as a function of differences in cues used in self-perception. *J. Pers. Soc. Psychol.* 39: 1024-1036.

James, W. (1884). What is an emotion? *Mind* 9: 188-205.

Kellerman, J., and Laird, J. D. (1982). The effect of appearance on self-perception. *J. Pers.* 50: 296-315.

Kellerman, J., Lewis, J., and Laird, J. D. (1989) Looking and loving: The effects of mutual gaze on feelings of romantic love. *J. Res. Pers.* 23:145-161.

Laird, J. D., and Berglas, S. (1975). Individual differences in the effects of engaging in counter-attitudinal behavior. *J. Pers.* 43: 286-304.

Miller, L. C., Murphy, R., and Buss, A. H. (1981). Consciousness of body: Private and public. *J. Pers. Soc. Psychol.* 41: 397-406.

Nowlis, V. (1965). Research with the mood adjective check list. In *Affect, Cognition and Personality.* S. S. Tomkins and C. E. Izard (Eds.). Springer, New York, pp. 352-389.

Pangborn, R. (1981). Individuality in responses to sensory stimuli. In *Criteria of Food Acceptance: How Man Chooses What He Eats.* J. Solms and R. L. Hall (Eds.). Forster Verlag, Zurich, pp. 177-219.

Rozin, P., and Schiller, D. (1980). The nature and acquisition of a preference for chilli pepper by humans. *Motivat. Emotion* 4: 77-100.

Stevens, D. A., and Lawless, H. L. (1986). Putting out the fire: Effects of tastants on oral chemical irritation. *Percep. Psychophys.* 39: 346-350.

Stevens, D. A., Dooley, D. A., and Laird, J. D. (1988). Explaining individual differences in flavor perception and food acceptance. In *Food Acceptability.* D. M. H. Thomson (Ed.), Elsevier Science, London, pp. 173-180.

# Chapter 11 Discussion

**Dr. Walker:** Would it be better if you are trying to be psychophysical to test only self-produced cue, high private body consciousness people? Is that a reasonable conclusion?

**Dr. Stevens:** Yes, but you might only be talking about 11% of the population.

**Dr. Walker:** But you are interested in doing psychophysical experiments that elucidate peripheral mechanisms; wouldn't it be better to use those people who seem to be less situationally manipulated?

**Dr. Stevens:** It depends on the questions you want to answer. It's the same sort of thing, should I use a highly trained panel? If you're using the panel for quality control, then do it; if you're using the panel to generalize to the world, I think it's better to sample from the world.... There are some theories of olfaction based on the use of a highly selected panel, and we come up with odor classification theories, but it's a kind of self-selecting, evolving thing. You change your panel to conform to your theory. You're asking me what to sample, and I'm saying it depends on what kind of generalizations you want to make.

**Dr. Rozin:** I think that this's really interesting. I just want to ask two questions. Taking your introduction into account, I can't resist asking you what percent of the variation in preference for any of these things is accounted for by using this measure? Personality variables usually account for very little of the variance, so I don't expect to hear a big number. Secondly, how do you compare it to sensation seeking, which also has a low correlation with liking for spicy foods?

**Dr. Stevens:** The second one is easy. I haven't done anything on sensation-seeking. I'm aware of the idea and hypothesis. But I haven't collected any data and I haven't looked at any relationships. I don't remember the percents of the variance accounted for; they are typically small. It doesn't bother me for a number of reasons. For reasons more traditional than rational, journal editors and yourselves like p values smaller than 0.05, and sometimes a relatively small amout of change in error variance is the difference between the scientific public hearing what you say and not hearing what you say. So 2 or 3% sometimes can buy you into the archives. As I understand it, the reason we are people and monkeys are monkeys is the result of very small but reliable changes over time. You don't have to have big reliable differences for some factor to have an impact. One or 2% of the variance will get this plant growing and that one dying. I'm not bothered by the fact that right now I'm accounting for relatively small differences and I'm delighted that there seems to be reliability and stability.

**Dr. Walker:** Just to follow up on that, how would you relate, for example, this sort of personality variable as a covariate to what you might get if you used Eysenk's extroversion/introversion, which seems to be fairly closely tied to, for example, brain recordings. Again, trying to get a little more physiological, you can test people without, for example, having them classify themselves in terms of personality? Should that also be included?

**Dr. Stevens:** Yes, but what we are going to end up with is a 400 page battery of things to classify people with. You will hear people say, now what do we do—give everybody facial manipulation, give everybody MMPI, run them through this, run them through that? The answer is yes, but it's not practical. I don't know how to respond. I can't say do my tests but don't do someone else's.

# 12
# Getting to Like the Burn of Chili Pepper
## Biological, Psychological, and Cultural Perspectives

**Paul Rozin**
University of Pennsylvania
Philadelphia, Pennsylvania

### I. INTRODUCTION

We can only presume that there was some astonishment in the mind of Christopher Columbus when he entered into his journal on Jan. 15, 1493, with respect to the ingestion of fiery hot chili peppers by American natives, that "toda la gente no come sin ella, que la halla muy sana" (they eat nothing without it, and deem it very wholesome) (Columbus, 1493/1986, p. 176). DeCuneo (1495; cited in Govindarajan, 1985a), accompanying Columbus says "those Caribes and Indians eat that fruit like we eat apples." Bernardo de las Casas (1909, p. 436) comments, with respect to the recently discovered Mexican indians, that "without chile, they do not think they are eating" (p. 436). Indeed, according to Diaz del Castillo (1963), the Aztecs practiced their cannibalistic practices by eating their victims with chimole sauce: chili, tomato, onion, and salt.

This unlikely enthusiastic and widespread preference was soon exported to the rest of the world. It wasn't just the Mexicans and Caribes who were to be so fond of the fiery stuff. In one of the most remarkable culinary progressions in history, hot chile peppers became a mainstay of cuisine in diverse parts of the world, such that today about one-quarter of the adults in the world eat them on a daily basis. Chile pepper is a basic, recurrent, almost ubiquitous flavoring ingredient in many world cuisines. It is part of what

Elisabeth Rozin (1982, 1983) calls a flavor principle in the cuisines of Korea, India, southwestern China, Indonesia, west and east Africa, Hungary, Spain, and, of course, Mexico. In all of these cuisines, chili pepper is used on a meal-to-meal basis; it is not an occasional condiment but rather a fundamental part of the daily diet. In Mexico, home of the most commonly consumed species of *Capsicum,* chili pepper is eaten at all meals. It is consumed in three forms: as a principal component of a set of sauces (salsas), along with tomatoes and a few minor ingredients, that is widely used on tortillas and other foods; cooked as a component of main course stews; and raw, often in slices, on foods. It is also used as a principal flavoring in snacks (reviewed in Rozin and Schiller, 1980). It tends not to be used in beverages and in sweets, but there are exceptions.

In spite of its New World origins, India and China are currently the biggest producers of chili peppers, and China is the biggest exporter (Govindarajan, 1985b). Korea has perhaps the hottest of all cuisines, with the highest intake of chile pepper (9 g/person/day) (Govindarajan, et al., 1987). Black pepper is the only spice with a bigger trade volume in the world, in terms of dollars, but there is much more chili pepper consumed; it is much less expensive, and a much larger proportion of the amount produced is consumed locally (Govindarajan, 1985b). Chili is a basic constituent in most of the tropical cuisines of the world and many temperate cuisines as well; it is not only the most consumed spice in the world, but, because of its association with low-meat diets and its low cost, the spice of the underprivileged (Rosengarten, 1969).

In a survey of a few hundred tropical and semitropical cultures designed to determine the best constituents for universal emergency rations, Moore (1970) concludes: "Let us imagine that you wished to invite the entire populations of East, South and Southeast Asia, the Middle East, Africa, Latin America and Oceania to dinner. What to serve that will be acceptable to all and reasonably balanced nutritionally? This is the menu: chicken, rice, squash and chili sauce with tea for a beverage and a banana for dessert."

This extraordinarily popular item is the same one whose smoke when burned, according to sixteenth century chroniclers, was used as a toxic gas in wars by South American Indians against the Spanish invaders. It is the same food that these chroniclers claimed the Spanish found too strong, at least for awhile (Heiser, 1969). Indeed, the genus name *Capsicum* and the name of the pungent chemical capsaicin may well derive from the Greek root *kapto* ("I bite") (Heiser, 1969). The handling of this preferred food is described in a popular cookbook featuring recipes with chili pepper as follows: "They require special handling. Their volatile oils may make your skin tingle and your eyes burn. Wear rubber gloves if you can and be careful not to touch your face or eyes while working with the chilies. . . . After handling hot chilies, wash your hands thoroughly with soap and warm water" (Time-Life, 1970).

Strange, indeed, that in the great Columbian exchange (Crosby, 1972) of foods between the New and Old Worlds, in which chile peppers, tomatoes, potatoes, chocolate, squash, peanuts, vanilla, turkey, and other foods made their first contacts with the Old World, chile pepper, surely the least palatable on first tasting, became one of the most widely accepted ingredients.

## II. BACKGROUND

### A. Botany

All current chili peppers trace their origin to the New World, and were unknown in the Old World until the time of Columbus. The genus *Capsicum* contains many species, five of which have been domesticated. One of these, *C. annuum,* originating in Mexico, accounts for the great majority of existing peppers (Andrews, 1984; Heiser, 1985; Govindarajan, 1985a,b). There is much dispute on the number of varieties of domesticated peppers, with estimates varying from 30 to 150 (Heiser, 1985). The wild form of the chili peppers is small and highly pungent; it is through domestication that mild forms, such as the bell pepper, have been created. The pungency is controlled by a single dominant allele (Heiser, 1969). The pungency results from a family of chemicals, called capsaicins, which are localized primarily in the placenta, the inner ribs of the peppers to which the seeds attach (Govindarajan, 1985a,b). Capsaicin levels for most peppers vary between 0.1 and 1%.

There is evidence for five different natural capsaicins, all decylenic acids of vanillylamide (Table 2) (Szolcsanyi and Jancso-Gabor, 1975; Todd, et al., 1977; Govindarajan, 1986). These have somewhat different sensory properties (Todd et al., 1977) (see below) and vary in concentration from variety to variety, and as a function of ripeness. In general, the amount of capsaicins in peppers increases with ripening (Govindarajan, 1985a).

Chili peppers are high in volatiles and have characteristic flavors and aromas which vary with maturity and variety. The appeal of the peppers results from both the volatiles and the capsaicins as well as their striking and saturated colors. The colors are produced to a large extent by carotenoids, which also account for one of the primary nutritional components of the peppers—high levels of vitamin A or its precursors. The other most significant nutritional component is vitamin C (Govindarajan, 1985b).

### B. Cultural History

In Mexico, there is evidence for consumption of chili peppers dating back to about 7000 B.C., and there is evidence for cultivation at some time in the period between 5200 and 3400 B.C. (Heiser, 1969; Pickersgill, 1969). Domestication probably began in about 1500 B.C. Peppers were brought back to Europe by Columbus and subsequent explorers, and subsequently spread to many parts of Europe, Africa, the Middle East, and Asia (Halasz, 1963). Some of the spread in Europe may have been motivated by the attraction of peppers as ornamental plants (Halasz, 1963), and there were also medicinal (e.g., fever prevention, antimalarial) as well as food uses for the plant (Halasz, 1963). Present day Hungary became a center for chili pepper cultivation and was the location where sweet (non-pungent) paprikas were developed in the early twentieth century (Halasz, 1963). By 1650, peppers were cultivated in Europe, Africa, and Asia (Rosengarten, 1969); pepper use was widespread in India by the eighteenth century (Crosby, 1972). The widespread acceptance in the Old World may have been prompted by previous acceptance and the strong desire for black pepper (Andrews, 1984); chile peppers provided a similar "burn" and were much cheaper. The similarity in sensations is indicated by the similar names provided to black and chili pepper in Spanish (pimienta and pimiento, respectively).

### C. Pharmacology of Capsaicin

Capsaicin is a pharmacologically active substance (see Jancsó, 1960; Jancsó-Gábor and Szolcsányi, 1969; Maga, 1975; Rozin, 1978; Buck and Burks, 1986) for reviews. For this reason, it has a history of medicinal use. The Mayans used it to treat cramps, diarrhea, aching gums, and sore muscles (Rosengarten, 1969; Schweid, 1980), and in the Old World it was used as a skin irritant, carminative (inducing gas expulsion), stimulant, and as a treatment for rheumatism, gastritis, and throat irritations (Heiser, 1969; Parry, 1969).

The effects of capsaicin depend very much on the route of delivery: skin contact, oral contact, ingestion, and systematic effects (via injection). The skin effects involve stimulation of circulation at the site of contact, and mobilization of an inflammation response (Jancsó, 1960, 1968; Jancsó-Gábor and Szolcsányi, 1969). So far as we know, the capsaicin acts as a mimic in the sense that it does not directly produce harm; the body responds to it as if it were a harmful agent.

Capsaicin acts as an irritant to the oral and gastrointestinal membranes, activating both the defensive and digestive systems. It produces copious salivation and gastric acid secretion (Solanke, 1973). Oral capsaicin also induces sweating on the face, neck and the front of the chest; this effect seems to be a reflexive response to oral stimulation, since it does not occur if the mouth is anesthetized (Lee, 1954).

So far as I know, very little capsaicin is absorbed. On the basis of its molecular structure, one might expect modest levels of absorption of intact capsaicin. Certainly, at least some capsaicin passes through the gut untouched, accounting for the Hungarian saying, "Paprika burns twice," or the

Indian saying, "You feel chili on the way out." There is a report from an animal study indicating that capsaicin lowers blood pressure (Porszasz, et al., 1955). Experimental work on the effects of capsaicin emanated originally from the Hungarian laboratories of Jancsó (1960, 1968), Jancsó-Gábor, and Szolcsányi (Jancsó-Gábor and Szolcsányi, 1969; Szolcsányi and Jancsó-Gábor, 1973); these laboratories remain centers for research on this subject. Large amounts of capsaicin in the blood stream of animals seem to have two prominent effects: there are massive effects on thermoregulation with damage to the hypothalamic centers for temperature control, and there is permanent, system-wide damage to chemical irritant receptors (Jancsó, 1960; Szolcsányi and Jancsó-Gabor, 1973; Szolcsányi, 1977). The cells (receptors) in the anterior hypothalamus that mediate the response to heat and chemical irritant noniceptors in and on the body are destroyed. These effects have become of central interest in neuropharmacology, both for their own sake and because they provide a powerful tool for the investigation of neurochemical phenomena (see chapter 8 in this volume). A review of the neuropharmacology of capsaicin in 1986 cites 283 references (Buck and Burks, 1986).

The site of action of capsaicin is primarily a subset of afferent endings (principally nociceptive C fibers) that subserve detection of chemical irritation in various parts of the body. The stimulatory effect and the inflammation or desensitization that may follow seem to be mediated by release and subsequent depletion of substance P (Buck and Burks, 1986).

Many of the effects of capsaicin resemble general cholinergic effects, particularly activation of the gut and lowering of blood pressure. There is some evidence from animals and humans that chili ingestion is associated with exacerbation of, and perhaps increased incidence of, ulcers of the stomach or intestine (Bergsma, 1931; Nopanitaya, 1974). Such an effect would be an expected consequence of the great amount of acid secretion that capsaicin produces. However, at low levels the inflammatory response that capsaicin stimulates may have a protective value against erosion by acid (Szolcsanyi and Bartho, 1980).

The data on ulcers aside, it is surprising how little evidence there is that chronic consumption of this active substance has any effect, positive or negative, on health (but see Lee, 1963). There is the potential for massive cross-cultural comparisons, since many cultures consume very high levels of capsaicin, but the appropriate controls for correlated factors are daunting enough to have discouraged systematic study. If capsaicin was even moderately toxic, we would certainly know this by now; and it would be surprising, given that there are many millions of elderly Mexicans, Koreans, Thais, and Asian Indians.

So far as is known, the noncapsaicin constituents of chili pepper are also not toxic; this is strikingly illustrated by a case from Germany (Tokay, 1932). A 20-year-old woman consumed an excessive amount of paprika (at one point, 2 kg over 2 weeks, probably non-pungent). The resulting syndrome,

"Kapsizimus," was characterized by reddish color of the skin, especially hands and feet. There was some loss of appetite but no other untoward symptoms.

### D. Why Is Chili Pepper Irritating? Why Is Anything Irritating?

Capsaicin is a natural substance. Since birds participate in the dissemination of pepper seeds, it should not come as a surprise that they are not deterred by (and presumably do not sense) the presence of capsaicin (see Mason, this volume). We do not know whether the strong aversion that mammals show to capsaicin is a result of evolution of a defensive response by plants that capitalized on a pre-existing irritation-sensing system in mammals or is an accident. Since capsaicin does not appear to be harmful, the strength of negative response to it is particularly noteworthy. In fact, the major sources of irritation in the world seem to be a few harmless plant products (chili pepper, black pepper, ginger, horseradish, and plant sources of menthol) and a variety of manmade aerial pollutants, including industrial byproducts, petroleum products, and tobacco smoke. It is indeed puzzling that this highly elaborated irritation detection system seems to be detecting primarily harmless plant components or chemicals that have entered our environment since the industrial revolution. Where does the irritation sense come from, and what were the selection pressures that promoted its evolution?

### E. Preferences for Chili Pepper in Animals

In subsequent sections, we will address the question of why humans consume and like chili pepper. Since some of the proposed mechanisms are uniquely human, and others have more general applicability it is of special importance to determine whether acquired likes for chili pepper can occur in nonhuman mammals. The evidence on this point is mixed. Birds readily consume hot peppers; indeed, some types of chili peppers are called bird peppers (Heiser, 1985). I know of no reports of consumption, outside of conditions of great privation, by mammals in nature. Two studies directed at fostering a preference for spicy foods in laboratory rats by extended exposure to these foods failed to establish a preference (Hilker et al., 1967; Rozin et al., 1979). In different experiments, Rozin et al. (1979) continued exposure for a year, used gradually increasing levels of pepper, and associated the pepper with recovery from deficiency. None of these manuevers led to a preference. However, a recent study with rats (Dib, 1989) revealed strong preferences after minimal exposure. I cannot explain this disparity in results. In a yet more recent study, Galef (1989) reports substantial and enduring preferences for an otherwise mildly unpalatable pepper diet by rats exposed on a few occasions to a "demonstrator" rat that had just eaten this diet.

Humans are exposed to chili pepper as a flavoring in a set of dishes that are much more varied than the rat chow that has served as the vehicle in the laboratory studies. Perhaps this is a factor in the human preference. In Mexican villages, domesticated animals, including chickens, pigs, and dogs, are steadily exposed to chili since they eat meal leftovers that are almost invariably piquant. A survey of residents of a Mexican village in Oaxaca about the preferences of the local animals, and tests of a number of dogs and pigs, revealed not a single case of a clear preference for food (usually a tortilla seasoned with hot sauce, as opposed to the same food without much seasoning, offered in a simultaneous choice) (Rozin and Kennel, 1983). In contrast, I have now identified five cases of a clear preference for hot foods in mammals. One is a pet dog in the United States who developed a taste for hot foods from table scraps offered by her mistress. In direct preference tests, "Moose" showed a clear preference for seasoned foods (Rozin and Kennel, 1983). Dua-Sharma and Sharma (1980) fed two pet Indian macaque monkeys their daily Indian cuisine over a period of time. The monkeys showed a clear preference for the seasoned food. Finally, Rozin and Kennel (1983) carried out an experiment on two young captive chimpanzees. The chimps were offered pairs of identical crackers, except that the crackers differed in color and the crackers of one color contained capsaicin oleoresin (an oil extract of chili peppers which is the commercial vehicle for capsaicin). Both chimpanzees strongly preferred the nonpiquant chip on initial testing. The chimps were then offered mildly piquant chips (100 Scoville units, or SU) with a distinctive color, by their trainer over a period of weeks. Both chimps reversed preferences, one after 25 samplings of mildly piquant chips and the other after 125 samplings. This effect, with a small amount of additional exposure, extended to crackers of the same color with piquancy strengths of 200 and 400 SU. The effect was still present 2 months later. Furthermore, when a new pair of crackers of very different composition from the original pair was offered in two different colors, one of which was paired with piquancy, the chimps promptly preferred the hot cracker.

With the exception of the recent, anomalous finding of Dib (1989), all cases of animal piquant preferences arose in the context of social interaction, i.e., rats served as social facilitators in Galef's work (1989); in all other cases humans were the social facilitators. These findings support the idea that there is something uniquely human about the acquired liking for chili pepper in the sense that it may involve human mediation and a social/affective relation to humans.

## III. EXPLAINING THE LIKING FOR CHILI PEPPER

What might an explanation for the liking of chili pepper look like? That depends on the type of explanation one is looking for. Confusion about the type of explananation often causes unnecessary debate about alternative explanations that are not comparable. One type of explanation is adaptive.

It accounts for a behavior in terms of its survival value. For the case of chili pepper, an example of such an explanation would be that chili is eaten to supply vitamin A. Associated with adaptive explanations are evolutionary explanations, which give a historical account of the biological or cultural evolution of a practice. A second type of explanation deals with immediate causation, i.e., what factors operating at the current time cause the behavior? For the case of chili pepper, such an explanation might be that the burn produces pleasure or, at a physiological level, that it causes secretion of endorphins in the brain. A third type, developmental explanations, account for a current behavior in terms of the past history of the organism. For chili pepper, such an explanation might be that chili liking results from past association of chili with eating with the family.

These different types of explanation are all consistent with one another; the adaptive value or developmental origins of a behavior do not necessarily account for, or even constrain, the current motivation to perform it. For this reason, we will separately consider adaptive/evolutionary, current motivation, and development explanations for the liking for chili pepper. Most attention will be focused on developmental explanations, since this is where most of the fundamental issues and most of the research resides.

## A. Adaptive/Evolutionary Explanations of Chili Preference

As we have indicated, there is no convincing evidence that the regular consumption of chili pepper causes any negative effects (Nutrition Reviews, 1986). However, especially since the taste is innately negative, there is a challenge to discover the adaptive value and evolutionary history of pepper consumption. The history and motives that might account for the origin of pepper ingestion in the New World are buried in the preliterate past. However, because chili pepper was introduced to the Old World in the sixteenth century, in literate times, there is the unfulfilled possibility that the causes of its successful adoption could be uncovered. The existing preferences for black pepper, which produces a corresponding burning sensation but a very different flavor, was certainly of importance in encouraging acceptance of chili pepper (Andrews, 1984). It was a much cheaper way of getting a for-some-reason desired sensation; indeed, from an Old World perspective, chili pepper was discovered by Columbus in a search for spices, black pepper being foremost among them.

Adaptive/evolutionary explanations for chili use are commonly offered, but uncommonly evidenced. A particularly popular view is that chili helps to preserve food, a valuable characteristic in prerefrigeration, premodern Europe. There is no historical or biological evidence for this claim, so far as I know (see Sass, 1981, for a discussion of motivations for using spices in premodern Europe). Although capsaicin seems like it should be able to kill anything, with respect to microbes as well as the oral mucosa, its bark is much worse than its bite. There is no evidence for an antibacterial effect (Dold and Knapp, 1948; Govindarajan et al., 1987) although there is some evidence for an antioxidant effect (Govindarajan et al., 1987).

A second, related claim is that the pepper helps to cover the sensory evidence of spoilage and was sought for that reason. Again, there is neither historical evidence for this (Sass, 1981), nor is there current evidence that it serves that function. In fact, both the original New World setting for chili and in the areas where it is currently most widely used, the diet is primarily of vegetable matter, and it is employed to a large extent on grain dishes that are rather resistant to spoilage.

A third claim is that the sweating/cooling effect elicited by capsaicin provided relief from the high temperatures of the tropics. There is no evidence that this is a significant function of chili pepper as it is used today. Chili pepper is popular in many moderate climates, including Korea and the Mexican plateau, the very place of its origin.

Although any of these three factors might have a minor role in both the cultural adoption and adaptive value of chili pepper, the answer to the adaptive evolutionary question cannot lie here. Indeed, this argument requires the establishment of two facts: (1) that there is a particular adaptive value (not yet demonstrated for spoilage retardation or masking) and (2) that this value had some role in cultural adoption. Four other hypotheses seem more promising.

Chili peppers are extraordinarily rich sources of both vitamins A and C. These vitamins are often in short supply in the tropical and semitropical grain-dominated cuisines in which chili pepper is most commonly consumed. For example, chilies are estimated to provide 33% of the vitamin A in the rural Mexican diet (Anderson, et al., 1946). Furthermore, rats have been shown to grow better when fed a South Indian rice diet supplemented with chili and tamarind than with the same diet without the supplement (Krishnamurthy et al, 1948). This may well result from vitamin A supplementaion. The peppers therefore fill a nutritional niche. It is hard to imagine how people discovered this nutritional advantage, if they ever did, since both vitamin C and A deficiencies develop and recover slowly and insidiously; indeed, scurvy decimated European sailors on long voyages for centuries before the simple fresh fruit cure was appreciated. In any event, the vitamins provide a clear adaptive value for chili pepper.

Chili pepper activates the gut and produces extensive salivation, presumably facilitating the processing and digestion of food. Digestion of the bland, high complex carbohydrate diets characteristic of the cultures that use chili as a flavor principle may well be improved with the capsaicin stimulant. It is surely true that mastication of these often dry and mealy diets is facilitated by the copious flow of saliva induced by chili pepper. In this sense, the pepper may also improve the flavor of the diet.

Finally, to move to more "psychological" adaptive values, chili peppers add a meat quality, a mouth fullness to the otherwise bland diets with which they are associated (Rozin, 1978). They also add variety to the diet because of the wide variety of colors, aromas, flavors, and burn intensities and patterns that are produced by different combinations of peppers (see next section).

Finally, judging by the history of spice use in Europe (Sass, 1981), the initial adoption of chili pepper was medicinally (e.g., Holasz, 1963) and socially motivated. In medieval Europe, the use of spices, including black pepper, was mainly limited to the upper classes; consumption of spices was a sign of elevated social status, and was no doubt sought by those not privileged as an indicator of social advancement (Sass, 1981). However, the establishment of chili pepper as a common food in Hungary seems to have taken a course opposite to what normally occurs; chili was initially a food of the lower classes, and moved into the upper classes (Halasz, 1963). There is much to be said, in general, for the powerful role of social forces in the establishment of cultural traditions (see, for example, Simoons, 1961, for a detailed consideration of Old World meat taboos in this context).

I have presented a list of possible adaptive values. In at least some cases, such as induction of salivation and vitamin content, there is little question of the existence of an adaptive value. However, the argument that a particular value or set of values actually supported the adoption of peppers needs more than a demonstration of adaptive value (see Rozin, 1982, for a more extended discussion). In this regard, the effects that are most perceptible, such as salivation and flavor enhancement, seem most likely to have been discovered and exploited.

### B. Explaining the Current Liking for Chili Pepper

Chili peppers are not simply vehicles for the conveyance of capsaicin. They have attractive colors, aromas, and flavors, and all of these factors contribute to their appeal. Commercially, in both Mexico and other countries, piquancy is conveyed to foods using either fresh or dried peppers, or capsaicin oleoresin, an oil extract of peppers that contains the capsaicin along with much of the coloring and aroma. Indeed, neither capsaicin (as an additive to produce an isolated burn) nor bell peppers (aroma/flavor/color without capsaicin) are particularly popular in Mexico. The color and aroma/flavor are not only attractive, but they provide richness and variability in the diet. Figure 1 shows one of many dried chili pepper stalls in a market in Oaxaca, Mexico (fresh peppers are sold at different stalls). This particular stall offers many different types of peppers, each with distinctive shapes, colors, aromas, flavors, and burn intensity and pattern. Thus, while chilies add a familiar burn to almost all staple Mexican foods, they also provide a subtle variety, in the mixture of different peppers in different dishes, and the distinctive taste, smell and irritation

**FIGURE 1** A typical stall selling dried chilies in the market in Oaxaco, Mexico. Note the wide variety of dried peppers on sale.

of each pepper type. We (Rozin and Rozin, 1981) have described this persistent use of a basic flavor principle, but varied in many subtle ways, as culinary themes and variations. The complexity of the aroma/flavor part of the stimulus is illustrated in Table 1, showing identified aromatics in three varieties of chili peppers.

Furthermore, it is a mistake to think of capsaicin as a monolithic stimulus. Five natural capsaicins have been isolated, and they have different sensory properties (Szolcsányi and Jancsó-Gábor, 1975, 1976; Todd et al., 1977; Govindarajan, 1986; Govindarajan et al., 1987). According to Todd et al. (1977), capsaicin, dihydrocapsaicin, and nordihydrocapsaicin have a more rapid bite and produce irritation more toward the back of the mouth than do homo- or homodihydrocapsaicin (Table 2). Since these capsaicins are present in different amounts in different peppers, there is considerable variety in the intensity and burn locus in different dishes (Govindarajan, Rajalakshmi, and Chand, 1987).

The burn produced by capsaicin is a major contributor to the liking for chili pepper. One hundred twenty adults (57 from a university community in the United States and 63 villagers in Oaxaco, Mexico) were asked why they

**TABLE 1** Compounds Identified in Different *Capsicum* Volatiles by GC-MS

| Alcohols | In volatiles of | | |
|---|---|---|---|
| Cyclopentanol | O | — | — |
| 2-Methylbutanol | — | E | — |
| 3-Methylbutan-2-ol | — | E | — |
| 1-Pentanol | O | — | — |
| 2,3 Butandiol | — | E | — |
| *trans*-2-Hexen-1-ol | O | — | — |
| *cis*-3-Hexen-1-ol | O | — | F |
| 2-Methylpentan-2-ol | — | E | — |
| 3-Methylpentan-3-ol | — | E | — |
| 4-Methylpentan-1-ol | O | — | F |
| 1-Hexanol | O | E | — |
| 2-Hexanol | O | E | — |
| 3-Hexanol | — | E | — |
| Linalool | O | E | F |
| Terpinen-4-ol | O | E | — |
| *alpha*-Terpineol | — | E | F |
| Carbonyls | | | |
| 2-Butanone | — | E | — |
| 2-Methylbutanol | O | E | — |
| Cyclohexanone | — | E | — |
| 4-Methyl-3-penten-2-one | O | E | — |
| *n*-Hexanol | O | E | — |
| 2-Hexanone | O | E | — |
| Benzaldehyde | O | — | — |
| 2-Acetylfuran-E | O | E | — |
| 5-Methyl-2-furfural | O | E | — |
| 2-Heptanone | O | — | — |
| 2-Octanone | O | — | — |

| Alcohols | In volatiles of | | |
|---|---|---|---|
| Esters (continued) | | | |
| Ethyl-3-methylbutyrate | — | E | — |
| Methylheptanoate | O | — | — |
| Methylphenylacetate | O | — | — |
| Methyloctanoate | O | — | — |
| $\beta$-Phenylethylacetate | O | — | — |
| Methyl-$\beta$-phenylpropionate | O | — | — |
| Ethyloctanoate | O | — | — |
| Methylnonanoate | O | — | — |
| Methyl-8-methyl-6-nonenoate | O | — | — |
| 4-Methyl-1-pentyl-2-methylbutyrate | O | — | F |
| 4-Methylpentyl-2-methylbutyrate | O | — | F |
| Methyl-8-methylnonanoate | O | — | — |
| Methyldecanoate | O | — | — |
| Methyldodecanoate | O | — | — |
| Ethyldodecanoate | O | — | — |
| Methyltetradecanoate | O | E | — |
| Methyltetradecanoate | O | — | — |
| Methylhexadecanoate | O | — | — |
| Methyloctadecanoate | O | — | — |
| Pyrazines | | | |
| 2,3-Dimethylpyrazine | — | E | — |
| 2,3,5-Trimethylpyrazine | — | E | — |
| 2-Methyl-5-ethylpyrazine | — | E | — |
| 2,3-Dimethyl-5-ethylpyrazine | — | E | — |
| Tetramethylpyrazine | — | E | — |
| 2-Methoxy-3-isobutylpyrazine | O | E | — |

| | | | |
|---|---|---|---|
| *para*-Methylacetophe-none | O | — | — |
| Carvone | O | — | — |
| Camphor | O | — | — |
| Thujone | O | — | — |
| *iso*-Thujone | O | — | — |
| 2-Undecanone | O | — | — |
| $\beta$-Ionone | — | — | F |
| Geranylacetone | O | — | — |
| Carboxylic acids | | | |
| Acetic | O | — | — |
| 2-Methylpropionic | — | E | — |
| 2-Methylbutyric | O | E | — |
| 3-Methylbutyric | — | E | — |
| Pantanoic | O | — | — |
| 4-Methylpentanoic | O | E | — |
| Hexanoic | O | E | — |
| Heptanoic | O | — | — |
| 2-Octanoic | O | — | — |
| Octanoic | O | E | — |
| 7-Methyloctanoic | O | — | — |
| Nonanoic | O | — | — |
| 2-Decenoic | O | — | — |
| 8-Methylnonanoic | O | — | — |
| Esters | | | |
| Methylpentanoate | O | — | — |
| Methylhexanoate | O | — | — |
| Terpene Hydrocarbons | | | |
| *para*-Cymene | — | E | — |
| Camphene | — | E | — |
| $\delta$-3-Carene | — | E | — |
| Limonene | O | E | — |
| Myrcene | O | E | — |
| $\alpha$-Pheilandrene | O | E | — |
| $\alpha$-Pinene | — | E | — |
| $\beta$-Pinene | O | E | — |
| Sabinene | — | E | — |
| $\gamma$-Terpinene | — | E | — |
| Terpinolene | — | E | — |
| $\alpha$-Thujene | — | E | — |
| Caryophyllene | O | E | — |
| $\alpha$-Copacne | — | E | — |
| Miscellaneous | | | |
| Toluene | — | E | — |
| *para*-Xylene | O | E | — |
| Octane | — | E | — |
| 2-Pentylfuran | O | E | — |
| 2-Pentylpyridine | — | E | — |
| 1,8-Cineole | O | — | — |
| Eugenol | — | E | — |
| Pentadecane | O | — | — |
| Hexadecane | O | — | — |
| Heptadecane | O | E | — |

*Note:* O — Oleoresin African chilies, *C. frutescens;* E — Miscella, mild chili, *C. annuum;* F — Fresh jalapeno, *C. annuum.*

**TABLE 2** Sensory Properties Of Capsaicins

| Capsaicinoid | Pungency | Time-Course | Location | Structure |
|---|---|---|---|---|
| Capsaicin | High | Rapid | Back | 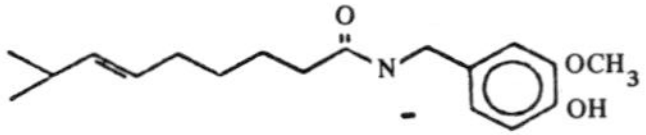 O N $OCH_3$ OH Capsaicin (C) |
| Nordihydrocapsaicin | Mod. | Rapid | Back | 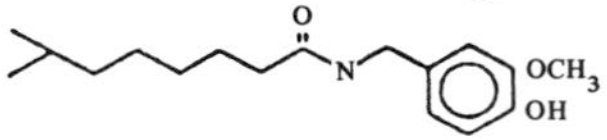 O N $OCH_3$ OH Nordihydrocapsaicin (NDC) |
| Dihydrocapsaicin | High | Rapid | Back | 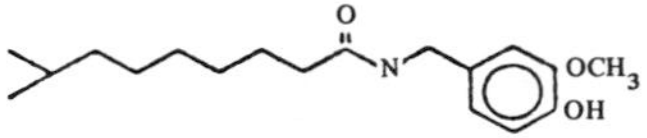 O N $OCH_3$ OH oDihydrocapsaicin (DC) |
| Homocapsaicin | Low | Slow, long | Mid | 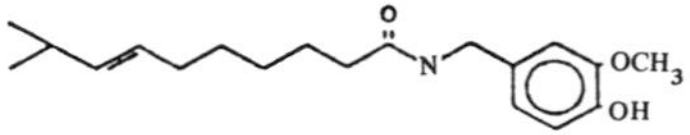 O N $OCH_3$ OH Homocapsaicin (HC) |
| Homodihydrocapsaicin | Low | Slow, long | Mid | 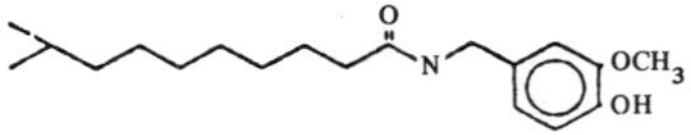 O N $OCH_3$ OH Homodihydrocapsaicin (HDC) |

*Source:* Adapted from Todd, P. H. Jr., Bensinger, M. S., and Biftu, T., *Journal of Food Science*, 42, 660-665.

ate chili pepper (Rozin and Schiller, 1980). These adults generated 125 reasons for ingestion; 106 of these referred explicitly to the flavor, i.e., the enhancement of flavor or piquancy. The term flavor is used to include piquancy: 28 of the Mexicans who responded with the word flavor were asked if that included the burn; 26 answered affirmatively. The modal Mexican response was: "da sabor a la comida" (adds flavor to food). Only 15% of the responses dealt with the consequences of eating chili, such as improving health, giving strength, or cooling the body. Mexicans and Americans were also given a checklist of 19 possible reasons for consuming chili, including all reasons that the investigators could think of. The only two reasons that were subscribed to by at least 50% of both Mexicans and Americans were "It tastes good" and "I like the burning or tingling feeling." It appears that for the great majority of chili eaters, the burn produced by capsaicin is pleasant. That is, a sensation that is initially negative becomes positive. There is a hedonic reversal. The source of pleasure in consuming chili is well captured by the Indian writer, Kamala Markandaya (1954, p. 57), who, in Nectar in a Sieve, a novel about life in rural India, says ". .when the tongue rebels against plain boiled rice, desiring ghee and salt and spices which one cannot afford, the sharp bite of a chillie renders even plain rice palatable."

### C. Liking the Burn: What Is Liked?

If there is a liking for the burn of capsaicin, then those who like chili pepper should also like other foods that produce a burn. Two common examples are ginger, which contains the irritant zinzerone, and black pepper, with the irritant piperine (see Szolcsányi & Jancsó-Gábor, 1975; Stevens & Lawless, 1987, for comparisons of the sensory properties of the different irritant substances). Other common irritant-containing foods or substances are raw onion, tobacco, mustard, horseradish, alcohol, and menthol. Table 3 presents correlations for preferences among these substances, gathered from three studies on students that I have carried out over recent years. As indicated in the table, there are substantial positive correlations between liking for chili and liking for all irritants except menthol; in contrast, the correlation between chili preference and liking for a bitter taste (a totally different innately unpalatable taste) is essentially zero (quinine in Table 3).

### D. Locus Specificity of Liking

Those who like chili pepper seem to like a burning irritation on their oral mucosa. Do they like this burn only on those surfaces that have actually contacted chili pepper? If they had, under experimental conditions, only experienced chili pepper on their left lip, would they only like it on their left lip? Normal mastication guarantees that the pepper will be distributed throughout

**TABLE 3** Relation Between Liking for Chili Pepper and Liking for Other Innately Unpalatable Substances

| Food/substance | Correlation with chili preference |
|---|---|
| Black pepper | 0.44 |
| Horse radish | 0.40 |
| Ginger | 0.24 |
| Raw onion | 0.20 |
| Cigarettes | 0.18 |
| Quinine water | 0.02 |
| Menthol | 0.02 |

*Note:* Data from college students (N = 110-300, depending on item. Correlations are Pearson or Spearman coefficients.) These findings come from three different studies. When more than one study included the same correlation, the values were averaged. Some data appear in Rozin andShenker, 1989.

the oropharynx, so that we cannot address this question. However, we can compare liking for irritation on widely separated surfaces. Generally, chili likers do not seem to relish it in their noses, and they seem to dislike it in their eyes; indeed, exposure of the eyes to the smoke of burning peppers was a punishment practiced by the ancient Mexicans, as illustrated in the accompanying figure from the *Codex Mendoza*, a sixteenth century Aztec manuscript (*Codex Mendoza*, 1978) (Fig. 2).

Chili pepper is an occasional component in skin linaments, such as Icy-Hot (menthol is a more common ingredient). The burning sensations resulting from application of these linaments may become pleasant. Sandy Koufax (1966), the great Los Angeles Dodge pitcher, regularly used capolin hot ointment on his arm. He reports: "I also had discovered, very early in my career, that I enjoy the feeling of warmth on my arm. After all these years, a mixture that might parboil someone else's arm is no more warm to me" (p. 238). These linaments produce a burning sensation which seems related to the oral sensations. Because menthol dominates as the irritant in skin creams and linaments, we can examine the mouth-to-body generalization hypothesis with menthol. There is a weak relation between liking for the burn of menthol in the mouth and on the body (the spearman $\rho$ is 0.24; Rozin and Shenker, 1989). Correlations are higher when menthol burns in closer areas are compared (mouth and face, $\rho = 0.50$). For chili pepper, the only comparisons we can make are between the burn of chili pepper in the mouth and horseradish in the nose ($\rho = 0.33$) or the hot burn sensation on first stepping into a hot bath ($\rho = 0.05$) (Rozin and Shenker, 1989). Again, it seems that there is generalization to neighboring but not to distant surfaces.

**FIGURE 2** Drawing from the Codex Mendoza, a post-Columbian document created by Aztecs for the Spanish, depicting aspects of Aztec life. Shown is child being punished by being held over the smoke of burning chilies. (from Codex Mendoza, 1978, p. 80.)

### E. Do Chili-Experienced and Chili-Naive People Experience the Same Sensations?

Dislikers of chili often complain that the burn of chili pepper makes it difficult to experience other flavors, whereas likers often talk about chili pepper as enhancing flavor. If we accept these reports at face value, the explanation could occur at a sensory or cognitive level. Lawless, et al. (1985) report that the presence of capsaicin reduces magnitude estimations for solutions of basic tastants or odorants, but that this effect appears in roughly equal amounts for chili likers and dislikers. Cowart (1987), using a different procedure (see chapter by Cowart), finds no masking effect in either likers or dislikers. Neither study reports a difference in the masking power of the burn in likers vs. dislikers.

Independent of masking effects, one can ask whether there is a difference in sensitivity to capsaicin in likers and dislikers. Given the phenomenon of desensitization, which can occur topically (Jancsó, 1960, 1968; Jancsó-Gábor and Szolcsányi, 1969; Szolcsányi, 1977), a desensitization might well occur in frequent users. Before exploring this issue, it is important to realize that even if desensitization occurs in users, it could not account for the liking for chili pepper. At most, desensitization by itself would make an aversive sensation less aversive. This point is emphasized by research on chili preference and desensitization in rats (Rozin et al., 1979). Although almost a year on a very piquant diet did not cause a significant reduction in aversion to that diet (in comparison to the same diet without chili pepper), desensitization of rats to capsaicin by systematic injection caused them to be indifferent to the chili-flavored diet. This study indicates both that desensitization could only account for a reduction in aversion and that chronic oral exposure to piquant foods does not seem to have a long-term desensitizing effect. Nonetheless, data from humans would be most relevant and are indeed available. The evidence indicates at most a slight desensitization from chronic use.

1. Mexicans eating chili regularly show a burn detection threshold for chili pepper, measured by exposure to increasing levels in a palatable cracker base, that is only about 1 log(2) unit higher than the threshold for American subjects (Rozin and Schiller, 1980).
2. American chili likers show a 0.6 log(2) unit increase in detection threshold, using the same type of measure as above, in comparison to Americans who are neutral to chili or dislike it (Rozin and Schiller, 1980). Similar results are found with threshold tests using solutions made with pure capsaicin (Rozin et al., 1981).
3. There is a significant but small decrease in capsaicin sensitivity in American chili likers, measured by their salivation to a series of increasing concentrations of capsaicin. It requires about 1.5 log(2) units more of capsaicin to achieve comparable salivation increases in strong likers as opposed to neutral/dislikers.

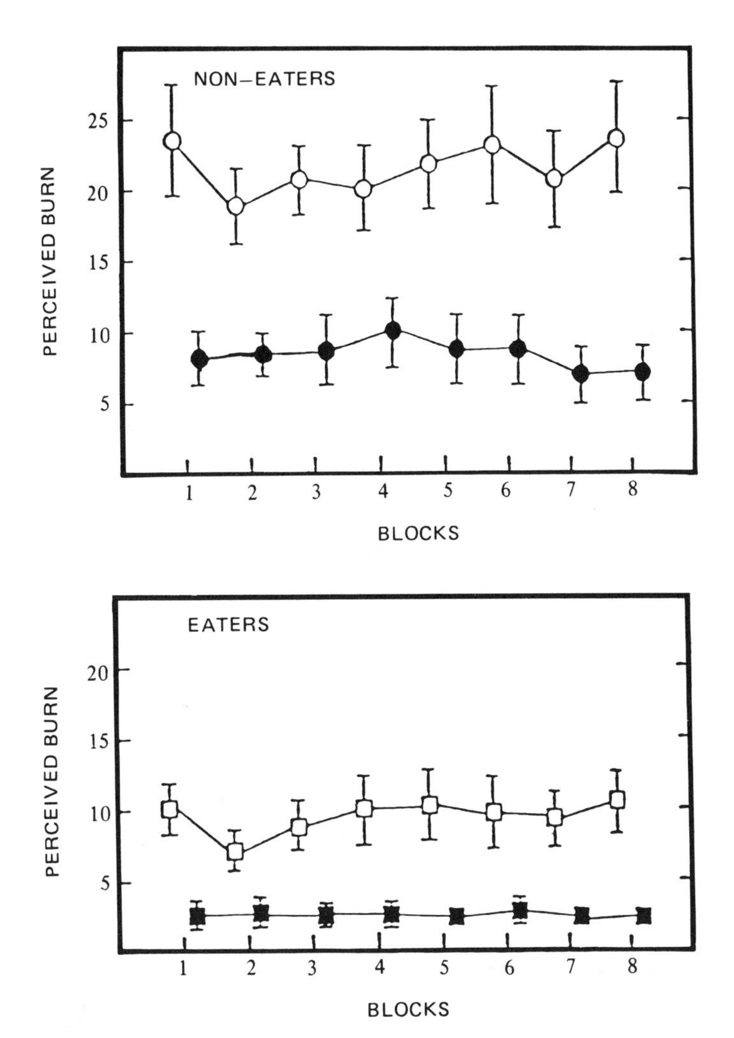


**FIGURE 3** Magnitude estimation (mean ratings, plus or minus 1 standard error of the mean), relative to a sound standard, of the intensity of burn of capsaicin solutions, by chili eaters and chili noneaters. Open symbols show the ratings just after expectoration of a capsaicin rinse solution, and closed circles show the rating after four intervening judgments on qualities of other solutions. This cycle or block (rinse, burn rating, four test solutions, burn rating) was repeated eight times (Lawless et al., 1985).

4. The correlation between the detection threshold and the level that produces a peak preference is positive (as desensitization would predict) but rather weak (0.29-0.39 in different measurements). For the sake of comparison, the correlation between peak preference and tolerance (the highest level that will be voluntarily accepted) is in the range of 0.69-0.83 (Rozin and Schiller, 1980).

These findings suggest that there is a small desensitization effect which may modulate responses to chili pepper but cannot account for the liking for chili pepper. Note that chili likers report that they *like* the burn; it isn't that they fail to sense it.

Desensitization on a major scale may occur under extreme conditions. Two adult Mexicans who ate extremely large amounts of chili pepper, including frequent ingestion of whole, fresh pepper, showed no aversion to our strongest stimulus, 262,000 SU, nor did they show any physiological signs (e.g., sweating) in response to such oral doses (Rozin and Schiller, 1980). These two individuals continued to enjoy highly piquant foods.

Desensitization is measured in terms of threshold. Suprathreshold response to capsaicin-induced burn differs much more markedly between likers and nonlikers than does threshold. This could be an amplification of a small threshold difference. However, it is more likely that it is generated at a different level in the system. Both Lawless et al. (1985) and Cowart (1987) report markedly higher ratings of capsaicin burn intensity in chili dislikers (see Cowart, this volume). Subjects in the Lawless et al. (1985) study had their magnitude estimations anchored with loudness intensity of auditory stimuli, so that the comparisons between the two groups had a common metric (Fig. 3).

## IV. THE ACQUISITION PROCESS

### A. The Natural History of Chili Preference Development

Studies on the development of alcohol and tobacco preferences suggest that the acquisition of preferences for innately unpalatable substances can be divided into two phases. First, initial exposures occur in the absence of a liking for the substance; curiosity and social pressure, particularly the desire to appear adult, often motivate the novice (Albrecht, 1973). Some ingestion may be "forced" by lack of availability of alternatives or by incorporation of the substance into obligatory ritual practices (Damon, 1973). Social pressure seems to be the dominant force in the first stage. For those person who enter the second stage, the sensory properties become pleasant in themselves. It is this transition that is of fundamental importance in the study of the acquisition of values and central interest in understanding the liking for chili pepper (Table 4).

**TABLE 4** Process of Exposure and Internalization

| | STAGE ONE (exposure) | STAGE TWO (internalization) |
|---|---|---|
| CHILI PEPPER | Exposure to increasing amounts, under mild social pressure → | Preference becomes internalized by development of a liking for the taste |
| | ↓ Consumption ceases; ↓ Consumption continues, but only under social pressure | |
| COFFEE OR TOBACCO | Exposure, under strong social pressure from peers, and motivation by desire to "be adult" → | Internalization by habitual use |
| | ↓ Consumption ceases; ↓ Consumption continues, but only under pressure | Internalization by anticipation of positive effects |
| | | Internalization by anticipation of avoidance of negative withdrawal effects (addiction) |
| | | Internalization by development of a liking for the taste ("affect") |

*Source*: From Rozin, 1982.

The development of preferences for chili pepper in various cultural contexts seems to follow this two-stage model, with a relatively early age of shift to the second stage, in comparison to alcohol, tobacco, or coffee (Table 4). Young children seem to be protected from exposure to chili (at least at moderate to high levels) (see Rozin and Schiller, 1980 for Mexico; Hauck et al., 1959 for Thailand; Bergsma, 1931 for east Africa). Depending on the culture, preferences based on liking for the flavor/burn seem to appear in the range of 4-11 years of age (e.g., Hacker and Miller, 1959: ages 10-11; Rozin and Schiller, 1980: ages 4-7).

American (college students; N = 57) subjects were asked: "How did you get to start eating chili?" (Rozin and Schiller, 1980). The most common responses among chili likers were that it was used at home (37%), that the parents put it on the food (29%), and that the first exposures were in restaurants or eating out (18%). In the response to the question: "How did you come to like chili?" the most common answer was that it was never disliked (43%). The next most common responses were development of a taste (23%), through exposure (23%), and enhancement of the flavor of food (11%). Although almost half of the subjects claimed never to have disliked chili pepper, interviews with parents in Mexico and the United States suggest that it is rare for children under 2 years of age to like chili pepper. Indeed, it is used on the mother's breast in a number of cultures to discourage nursing (Jelliffe, 1962). On the other hand, one parent reported a definite preference for piquant foods in a 1-1/2-year-old child, and the author has been present on two occasions when a young adult tried piquant food for the first time and liked it.

A group of 207 college students was asked on a questionnaire: "If you like chili pepper or other spices that produce a burn, indicate how many times you tried it before you liked it?" I do not pretend that people can respond to this question accurately, but their response is still likely to be informative. Most responses fell between 2 and 100 times, although 14.8% of subjects claimed to have liked chili pepper on the first tasting (Table 5).

Interviews with Mexican adults about their own early experiences with chili, interviews with Mexican mothers regarding the early exposure of children to chili, and observations of meal times in Mexican village homes all indicate uniformly that chili is introduced gradually into the young child's diet (Rozin and Schiller, 1980). Although infants may incidentally taste piquant adult foods (in which the chili pepper is cooked in with the food), no attempt is made to introduce hot foods to them. Gradually, from about 3 years on, small amounts of chili (salsa) are placed on tortillas and the accompaniments. No specific rewards are given to children for eating piquant foods, but they observe the avidity with which parents and older siblings consume it. All informants and observations lead to the conclusion that by 5 or 6 years of

**TABLE 5** Number of Times to Liking of Chili or Other Irritant Spices

| Number of times | N | % |
|---|---|---|
| 1 | 29 | 14.8 |
| 2-5 | 50 | 25.5 |
| 6-25 | 25 | 12.8 |
| 26-100 | 92 | 46.9 |

*Note*: N = 196 American undergraduate students who like chili pepper.
*Source*: From Rozin, unpublished.

age, children seem to like piquant food, and voluntarily add salsa to their food.

These observations are confirmed by direct preference measurements on subjects in a Oaxaca village, in the age range of 2 years to adulthood (Rozin and Schiller, 1980). A common snack, purchased in a few small stores in the Mexican village under study, was flavored powder in a small cellophane package. One type of snack consisted of four different types of fruit flavorings with appropriate coloring, each mixed with sugar and a sour powder (presumably citric acid). This snack was called *sal de dulce.* Another type of snack, with a distinctive deep red color that differed from the color of the fruit snacks, consisted of salt and ground dried chili pepper (*sal de piquante*). This snack was purchased and consumed frequently by residents, primarily but not entirely children. All of the children in the village elementary school and a selection of adults and younger children (down to age 2) were given a preference test in which five (four different sal de dulce and the sal de picante) were displayed, and the subjects were asked to pick the one they wanted.

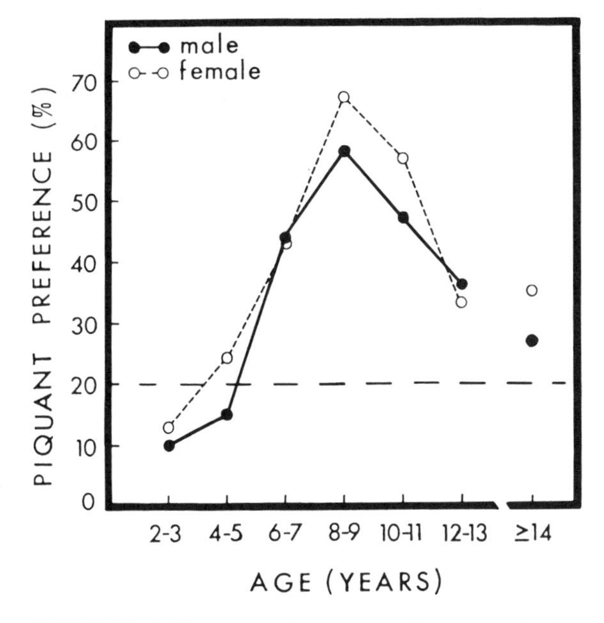


**FIGURE 4** Preference for a salty-piquant snack (sal de picante) vs. four sweet-sour snacks (sal de dulce) by Mexican villagers, age 2 to adult. The dashed line indicates the predicted level of piquant selection if the subjects chose randomly. The points represent data from a minimum of eight to a maximum of 32 subjects. (From Rozin and Schiller, 1980.)

They were then allowed to open and eat the selected item. Age and sex were recorded for each subject. All subjects were familiar with these items and were aware of the relation between color and taste. Altogether, 265 subjects were run (52% female). The results are totally consistent with verbal reports and observations: a preference for the piquant snacks over the sweet-sour snacks emerges by age 6-7. The preference remains high, peaking at age 8-9 (Fig. 4). Note that the less preferred sweet-sour items are highly palatable; the preference measure indicates relative palatability. The remarkable finding is that in the early school years, a chili-salt snack is preferable to a sweet-sour fruit snack. There is no obvious explanation for the drop from the peak after 8-9 years of age; the peak might reflect an enhanced preference because of the importance at that age of demonstrating an adult preference. There were no sex differences in preferences.

### B. Changes in Response that Occur During Preference Development

Chili likers differ from chili dislikers or chili-naive people in a number of ways with respect to their reaction to experiencing chili seasoning. I will summarize these differences, each of which seems to correspond to a more or less gradual developmental change. The occurrence of gradual developmental change is an inference from the observation of current differences in reactions to chili in different people, and the general observation that the changes tend to occur gradually. A number of authors who have discussed chili pepper refer to this gradual acquisition (Schweid, 1980; Heiser, 1985). Charles Heiser, a botanical expert on chili pepper, gradually came to like the burn of chili pepper, in the process of working with the peppers.

Rozin et al. (1982) evaluated preferences for chili pepper in 40 American subjects by offering them a series of crackers with increasing piquancy, from 0 to 8500 SU. Subjects tried each cracker and rated how much they liked the sensations produced every 10 sec for the first minute, and every 30 sec thereafter, using a scale that ran from −100 (the worst possible taste imaginable) through 0 (neutral) to +100 (the best possible taste imaginable). Subjects continued up the series of crackers unless they felt that the next cracker would be too unpleasant, in which case the series was terminated. Almost all of the subjects could be classified in one of three categories. Strong dislikers (N = 16) essentially disliked all chili-adulterated crackers. The blank cracker was at least as pleasant as any other cracker (Fig. 5A). The existence of a substantial group of strong dislikers indicates that there is not a very low level of irritation that is universally likable; rather the sensation is probably totally negative, at any level, to novices. Strong likers (N = 7) found all levels of chili enhancing, both while the food was still in the mouth (the first minute or so), and during the residual isolated burn, which may have lasted many

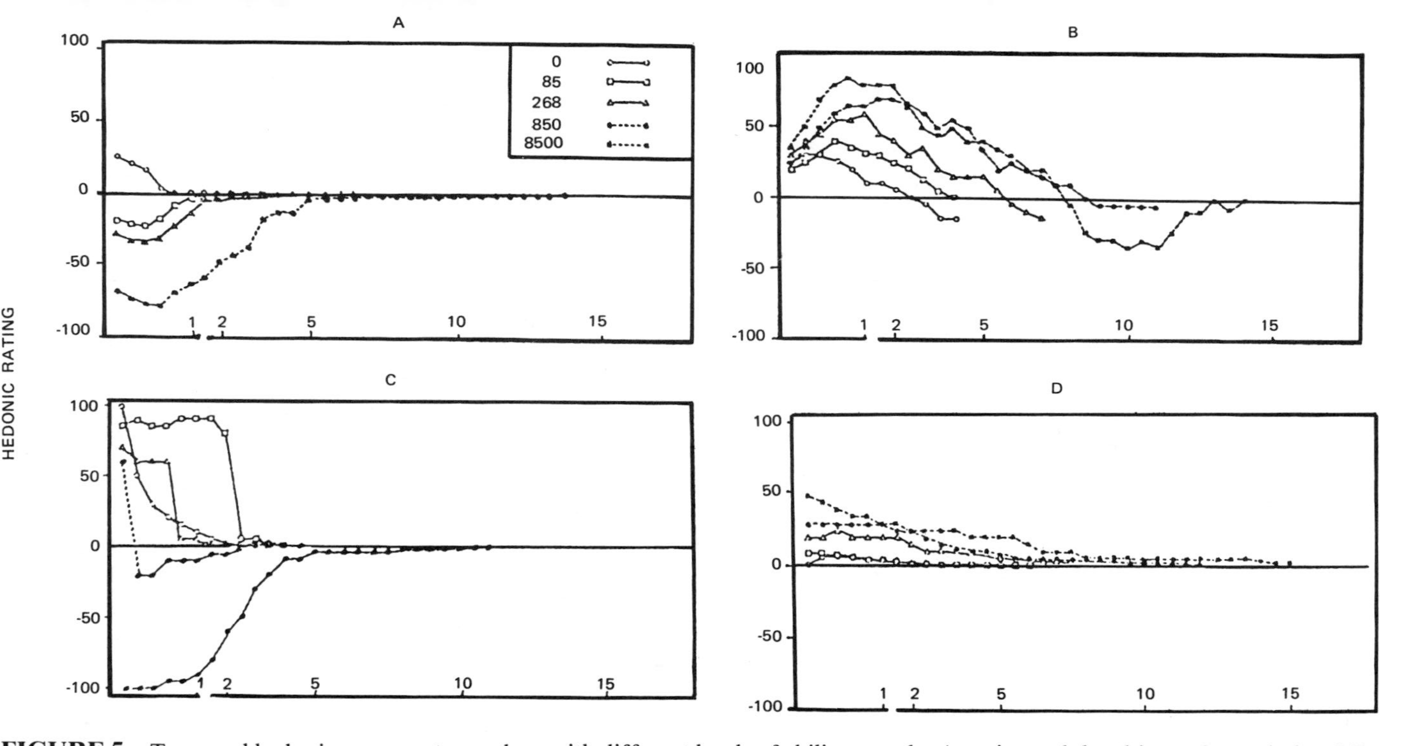


**FIGURE 5** Temporal hedonic response to crackers with different levels of chili pepper by American adult subjects. Open circles, 0 Scoville units (SU); open squares, 85 SU; open triangles, 268 SU; closed circles, 850 SU; closed squares, 8500 SU. Each panel shows the response to each of the stimuli for one representative subject. Panel A represents a typical disliker. Panel B represents a person who likes the burn with the flavor, but not the isolated burn. Panel C represents a subject who likes low levels of burn but not high levels. Panel D represents a typical liker. (From Rozin et al., 1982.)

minutes (Fig. 5D). Moderate likers (N = 14) showed a shift from like to dislike as the piquancy of the crackers increased. In some cases, the flavor enhancement of the burn was evident at all levels, but there was a negative response to the stronger residual, isolated burns (Fig. 5B). In other cases, the full sensation for the weaker crackers was positive, while the whole sequence of sensations for the stronger crackers was negative (Fig. 5C). It seems highly likely that most of the strong likers went through a moderate-like stage. The results of this study and other research (e.g., Rozin and Schiller, 1980), suggest the following developmental changes.

1. Over weeks to years, the most preferred level of piquancy rises. In early stages, only slight burns are preferred. However, this process does not continue indefinitely. People stabilize at some level and subsequently continue a peak preference at that level. Both Mexican and American subjects were presented with a series of corn snacks with increasing levels of piquancy (from 0 to 262,000 SU, with each stimulus twice the level of its predecessor). Subjects were asked to indicate whether they preferred each snack to its predecessor. They were quite consistent in this task in that once they ceased to prefer a snack to its predecessor, they almost invariably rated subsequent snacks as less palatable than their predecessors. The preferred level was set at that snack which was preferred to both its predecessor and follower. The sequence was continued until the subject declined to try the next stronger snack. The last consumed snack was designated the tolerance level. A similar series of measurements was made using a different, tortilla vehicle. For Mexican subjects between the ages of 4 and 15 years, the correlation between age and preference was 0.20 (n.s.) for the corn snacks and 0.52 ($p < 0.025$) for the tortilla crackers. The correlations between age and tolerance for these same subjects were 0.41 ($p < 0.05$) and 0.28 (n.s.), respectively. In contrast to these results, suggesting an increasing preference for stronger stimuli with age, corresponding correlations for Mexican adults (ages 18-56) were $-0.07$ for age-preference and 0.01 for age-tolerance (Rozin and Schiller, 1980). These results support the view that a gradual increase in preference subsequently stabilizes. In the natural setting, where chili is a normal part of the diet and is introduced well before the fifth year of life, maturity seems to be the point at which preferences stabilize.

2. Preference initially appears as enhancement of the flavor of other foods. As shown clearly in the first minute of the curves in Fig. 5B-D, chili enhances the flavor of a food, the more so as the preference increases (this is, of course, a major part of the definition of preference increase). The major justification given by Mexicans for eating chili is that it adds flavor to foods (see above).

3. There is an extension of the food contexts in which the burn is perceived as pleasant. Liking for the burn seems to begin in the context of a

**TABLE 6** Contexts for Chili Pepper for American College Students

| Context | Percent who eat chili in that context |
|---|---|
| Pizza | 54 |
| Meat/fish | 47 |
| Hoagies (submarine sandwiches) | 44 |
| Vegetables | 36 |
| Snacks/crackers | 14 |
| Beverages | 14 |
| Sweets | 4 |
| Dairy | 4 |
| Fruits | 4 |

*Source*: Rozin (unpublished).

specific set of foods and/or food/meal situations. With exposure, the range of contexts that are appropriate grows. This sequence is suggested by questionnaire results from American college student subjects, who were asked to indicate whether they ate chili pepper with a variety of foods, as listed in Table 6 (Rozin, unpublished). The basic pattern, similar to the pattern of use in Mexico, is for predominant use with main course items, and marginal use with sweets, beverages, dairy, and fruits. One may presume that those who use chili pepper with the infrequently used items also began with the more common items.

4. There is a reduction in the masking effect of chili on other flavors. As mentioned above, chili dislikers feel that it masks other food flavors, whereas likers do not hold to this view. Although this difference has not been demonstrated in laboratory sensory tests (Lawless et al., 1985), that is not to say that people are misreporting. Rather, it may be that the emergence of other flavors with experience with chili is more a cognitive than a sensory effect.

5. At later stages, there is a preference for the isolated burn of chili. Some chili likers like the burn that remains after all of the flavor of the food has disappeared (see particularly Fig. 5D). This seems to be a later development in the acquisition of preference, in the sense that it is less common than flavor enhancement. Another way to describe this characteristic is that the burn can be enjoyed outside of the immediate food context. However, we do not know that a pure isolated burn that did not arise from consuming a food would be pleasant. Such a burn might be unpleasant to almost everyone; our chili-liking subjects in studies in which they rinsed with capsaicin solutions (Rozin et al., 1981) generally found these solutions and the burn they produced unpleasant, even at low levels.

## C. Theories of Preference Development

We do not know the mechanisms involved in the development of a liking for the burn of chili pepper. Unfortunately, we know very little about the mechanisms behind any acquired likes, for foods or other entities (see reviews by Beauchamp, 1981; Booth, 1982; Rozin, 1984; Birch, 1987; Rozin and Vollmecke, 1986). All of the mechanisms that have been suggested for other foods may be relevant for the case of chili pepper. Proposed mechanisms for foods in general can be divided into three categories: mere exposure, general mechanisms (often Pavlovian) that may account for acquisition of all food preferences, and special mechanisms that require that the food in question be initially aversive.

### 1. Mere exposure

There is no doubt that exposure is almost a necessary condition for liking; it is unlikely that one will develop a liking for a food that one has never tried. Zajonc (1968) holds that mere exposure induces liking, i.e., exposure acts not to allow other mechanisms to operate, but itself increases liking. This is a "null" position, in the sense that it is hard to eliminate the possibility of something else operating during the exposure period. Exposure to foods without any notable consequence can lead to increased liking (e.g., Pliner, 1982). Surely in the natural setting there is a great deal of exposure to chili pepper that should be more than sufficient to induce liking. Thus, insofar as one considers mere exposure a theory of liking, chili pepper is a reasonable candidate.

### 2. General (primarily associative) mechanisms

There is evidence for three types of associative mechanisms that may contribute to the acquisition of likes for foods. One is Pavlovian pairing of a food/flavor with positive postingestional consequences (Booth et al., 1982). Such effects are not always robust and may be limited to satiety effects, since there is no evidence that other types of postingestional consequences (relief of gastric pain, general feeling of well-being, drop in fever, etc.) have any effect in inducing liking (Pliner et al., 1985). These data imply that explanations of the liking for chili pepper in terms of changes in temperature that it produces are unlikely. Insofar as satiety is a potent US in a Pavlovian paradigm, the normal exposure to chili pepper could enhance liking through this route: chili pepper is typically eaten with satiating foods, in a meal context.

A second Pavlovian mechanism is pairing of a food/flavor (CS) with an already positive flavor (US). For example, Zellner et al. (1983) showed that pairing a flavor with a sweet taste enhances the liking for that flavor in comparison to a different flavor presented an equal number of times but not paired with sweetness. It is probable that some of the liking for coffee is induced by early exposure to sweetened coffee. Chili pepper liking could be fostered by such pairing, since the chili is eaten with all of the basic main course foods, although it is not paired much with particularly palatable sweet dessert tastes. Furthermore, the salivation produced by chili pepper is said to enhance the flavor of food and may be a component of the pleasant oral US.

The most well-documented and almost certainly the most potent force in creating likes is social (see Birch, 1987; Rozin, 1988 for reviews). The perception that a food is enjoyed or valued by respected others seems to be the critical social event. This can be construed as a Pavlovian linkage between positive social expressions of others (US) and the food (CS), or a more cognitive framework can be applied. In either event, the experimental evidence, coming largely from the laboratory of Leann Birch, documents the importance of such factors in children. There is abundant evidence for the operation of social factors in the acquisition of chili preference. In those cultures where chili is a part of the flavor principle, children observe the enjoyment of chili pepper in parents and older siblings in every meal. Furthermore, Birch and her colleagues (1982) showed that explicit rewards given for eating a food retard the development of a preference for that food. In the Mexican home situation, rewards for eating chili are *not* given (Rozin and Schiller, 1980). Rather, there is mild social encouragement. The acquisition of liking for chili pepper among adults, as in the United States, often occurs in the context of encouragement from friends who eagerly consume it. Finally, as indicated above, the only clear cases of acquired preferences for chili in animals involve social mediation. In summary, social factors are a very likely influence in the development of a liking for chili pepper.

### 3. Mechanisms that presume an initially aversive response

Two mechanisms for liking chili pepper have been suggested that depend on an initial negative response. They are opponent-endorphin responses and benign masochism (enjoyment of constrained risks).

#### a. Opponent-endorphin responses

Typically, a liking for chili arises after a number of unpleasant, painful oral experiences with chili pepper. Social forces encourage people to continue to sample unpleasantly piquant foods, which would otherwise be avoided after the first taste. Opponent process theory (Solomon and Corbit, 1974; Solomon, 1980) provides a model for just such a series of events. Within the framework of hedonic homeostasis, it holds that departures from hedonic (or other) equilibria (known as the A process) are reduced by the generation of processes opposite to the process that initiates the departure. The theory holds that this opponent, or B, process is initially weak, sluggish, and short in duration. With repeated stimulation of the A process (e.g., by successive exposures to

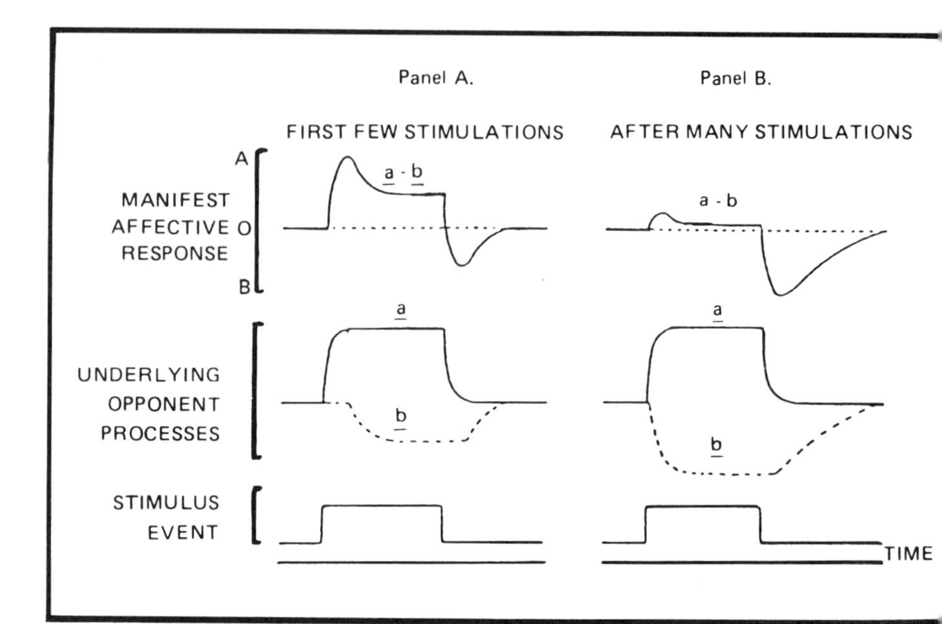


**FIGURE 6** Schematic representation of opponent process theory. Panel A represents the operation of the system for the first few stimulations. Panel B represents the operation after repeated stimulations. The manifest response is the summation of the two underlying processes. (From Solomon and Corbit, 1974.)

a stimulus such as nicotine or pain), the B process becomes stronger, more rapid in onset, and longer in duration (see Fig. 6). Ultimately, the B process dominates and accounts for withdrawal effects. According to the theory, the potentiation of the B process depends on reinstitution of the A process before the previous B process has dissipated (Fig. 6).

A variant of this theory holds that the opponent process is not innately linked to the A process but is a conditioned response whose adaptive value is to cancel the perturbation produced by a stimulus (Siegel, 1977; Schull, 1979). The conditioned opponent view does not require that stimulation occur with short intervals; it simply holds that the conditions that promote Pavlovian conditioning will promote the development of conditioned opponents. Either theory would account for a negative to positive hedonic shift by presuming that the opponent process became stronger than the A process. If, after many exposures, there is a purely positive response, with no initial indication of negativity, the conditioning theory would account for this by holding that the opponent process anticipated the arrival of the stimulus. Both of these theories have potential application to the acquisition of a liking for chili pepper; both types of processes may occur.

A likely physiological basis for chili liking within an opponent process frame is the secretion of brain endorphins. Substance P, known to be released in some neurons by topical stimulation by capsaicin, is a stimulant for the endorphin system. If the endorphins are presumed to be a part of an innate B process, this endorphin response could be accentuated by exposure according to the Solomon model. Alternatively, endorphin secretion could result from compensatory conditioning according to the Siegel model. In either case, the idea is that increasing levels of endorphin response to the same painful input could produce a positive feeling and convert the pain to pleasure. Indeed, one of the most successful areas for the application of both opponent process theories is heroin addiction, an opiate endorphin-related system. These theories fit well with Weil's (1976) description of the chili liker's experience as a rush, a form of "mouth surfing."

The gradual development of liking, the "high" that sometimes is described on eating chili, the conversion from pleasure to pain, and the substance P/endorphin link all suggest the reasonableness of these theories. However, there are some problems. First, a liking for chili pepper and other forms of pain seems to be almost uniquely human, yet the processes invoked are demonstrably common in animals. Second, chili dislikers do not typically show a positive hedonic effect as the negativity of a taste of chili wears off (Rozin et al., 1982); opponent effects are not often seen in the hedonic response to pepper stimulation. Third, Solomon's theory holds that the enhanced B process will return to normal levels with disuse. There is no evidence that chili preference abates after a period of weeks to years of nonstimulation. The conditioning view does not predict this dissipation. However, it does predict that when a person comes to like a particular food that is served with chili pepper, she should experience especially high pleasure if the same food is served without hot pepper, since the opponent pleasure response would be induced without induction of the A process. It also predicts that if this same food is presented many times without chili pepper, the next time it appears with chili pepper it will not be particularly pleasant (extinction of the Pavlovian positive hedonic CS). These two predictions of the Siegel compensatory conditioning approach have not been tested, but it seems unlikely that the predicted results would appear.

The direct test of opponent endorphin models would be either to measure increases in endorphins in chili likers in response to eating chili, or to show that blockage of the endorphin system blocks the liking for chili pepper. I have been involved in two attempts to do the latter. In both cases (two unpublished studies, one in collaboration with Schull, and the other with O. and C. Pomerleau), chili likers consumed and rated crackers of varying hotness under conditions of intravenous administration of either vehicle or vehicle plus naloxone, double blind. The results of both studies were less than definitive. There was a marginal statistically significant tendency for subjects under naloxone to rate piquancy as slightly less pleasant. These differences are in the predicted direction but small in comparison to the robustness of the chili-liking phenomenon. Given that there are multiple endorphin systems, blocked to different degrees by different blockers, and that dose of blockers seems to be a critical variable in this literature, it would seem that systematic studies with varying doses and different blockers would be necessary to test the endorphin hypothesis. This has yet to be done.

b. Benign masochism (enjoyment of constrained risks).

Liking chili pepper is like liking to ride a roller coaster. In both cases, the body senses danger and behavior normally follows which would terminate the stimulus situation. In both cases initial discomfort becomes pleasure after a number of exposures. Yet chili pepper still burns in the mouth, and the sympathetic system is still highly aroused as the roller coaster plunges toward earth. It is as if the mind realizes that these activities are actually safe, but the body does not. This body/mind disparity may be a source of feelings of mastery and pleasure, a case of body over mind. We have suggested that this form of "benign masochism" or enjoyment of "constrained risk" is a particularly human quality (Rozin and Schiller, 1980; Schweid, 1980). It appears abundantly in humans, in dangerous sports activities, amusement park rides, preferences for innately unpalatable substances, watching frightening or sad movies, or taking painfully hot baths. While such activities are a common part of human life, they are very rare in animals. Thus, one virtue of a benign masochism interpretation of a liking for chili pepper is that it accounts for the virtual absence of this phenomenon in animals.

There is a modest amount of evidence that suggests a role for benign masochism in the liking for chili pepper. It is not uncommon for people to like the body's defensive responses, nose and eye tearing, made in response to consuming hot peppers. Eleven of 15 Mexican adults who reported such effects from eating chili pepper claimed to like them, as did 11 of 32 Americans (Rozin and Schiller, 1980). The preferred level of chili pepper is often very close to the highest level that would be tolerated, suggesting that liking is related to pushing the limits of pain tolerance (Rozin and Schiller, 1980). Nine of 36 Mexican adults rated the most preferred level of piquancy in a series of crackers of increasing piquancy at the same level as the tolerance level, and 13 more of these subjects rated the preferred level just one log(2) level below tolerance (Rozin and Schiller, 1980). Furthermore, people who prefer higher levels of piquancy tend to show a smaller difference between the preference and tolerance levels; the correlations between preferred level and tolerance − preferred level is about -0.40 in a few studies of Mexican and Americans (Rozin and Schiller, 1980).

**TABLE 7** Correlations (Pearson) Between Liking for Chili Pepper and Liking for Risk-Related Activities

| Activity | Pearson r with chili pepper liking |
|---|---|
| Amusement park rides | 0.22 |
| Gambling | 0.19 |
| Nose running or eye tearing from eating chili pepper | 0.17 |
| Dangerous sports | 0.09 |
| Roller coaster rides | 0.03 |
| Pain from running/jogging | 0.03 |

*Note*: N = 150-250 American undergraduates. All ratings on a 9-point hedonic scale.
*Source*: From Rozin; unpublished.

Zuckerman (1979) developed a sensation-seeking scale which measures a personality variable that relates to what we have called benign masochism. One subscale of sensation seeking is called thrill seeking, and it seems to be particularly close to benign masochism. The existence of Zuckerman's personality scales indicates that there is some positive correlation between indices of thrill-seeking behaviors in different domains. My data support this weakly. Chili preference in Americans correlates 0.11 with a combination of three measures of benign masochism in other domains: liking sad movies, hot baths, and dangerous sports (Rozin and Schiller, 1980). In a different unpublished study of about 200 American undergraduate subjects, there were generally positive correlations between liking chili pepper and liking other risk-related activities (Table 7). The overall pattern is for a positive but very modest relation. One problem for the benign masochism view is that, although in traditional Mexican populations males are much more oriented to thrill seeking and "macho" behavior, there is not a significant sex difference in the liking for chili pepper (Rozin and Schiller, 1980).

More generally, there is a literature suggesting a weak positive relation between sensation seeking or measures like it and liking for strong or spicy foods (Child et al., 1969; Kish and Donnenwerth, 1972; Brown et al., 1974; Zuckerman, 1979; Logue and Smith, 1986). Logue and Smith (1986) included specific questions on liking for chili pepper and Mexican foods in their study on American subjects. They found a significant 0.16 correlation of liking for chili pepper with the sensation-seeking scale, although the liking for Mexican foods did not show a significant correlation. Surprisingly, on the thrill seeking subscale, neither the chili pepper nor Mexican food-liking scores showed a significant correlation. [See chapter 11 in this volume, for a different link between personality (private body consciousness) and chili pepper ingestion.]

### 4. Summary

There are multiple models for coming to like chili pepper. There are convincing arguments against each of them, which indicates either that all are incorrect, or that there is more than one route to liking. But since there is evidence *for* most of the models as well as against them, it is most likely that at least some of the proposed models *do* have a role in fostering liking, whether or not there are new models to come. Although the end point, liking the flavor and burn, is simple in the case of chili pepper, the routes to this state are surely multiple.

## V. CONCLUSIONS

Some may find this chapter irritating, in keeping with the sensation produced by its topic. Many basic issues have been raised, but few resolved. The widespread acceptance of chili pepper poses problems in many areas. It is a particular challenge for culinary historians, since chili pepper, on the face of it, would appear unlikely to be adopted. The strong liking for chili in light of its initial unpalatability is a challenge for the psychology of affect and for the study of the acquisition of culture. The multiple sensations produced by capsaicin challenge the notion of a single type of receptor for chemical irritation.

There are very few papers on the sensations produced by chili pepper, its cultural history or cultural context, or the acquisition of liking or the psychology of chili use. Because it is one of the most widely consumed substances in the world, the absence of literature is surprising. The study of chili pepper for its own sake, as an important part of human life, is more than justified. But chili pepper also offers us a tool to study some basic processes. Just as capsaicin has become a tool in the study of thermoregulation and neurotransmitters, the use of chili pepper offers relatively easy windows to the study of basic behavioral processes. One of these is the process of acquisition of liking in general. The reversal of liking that occurs with chili pepper occurs on a massive scale and represents one of the most dramatic changes in affective response that one can find. It should be possible to produce this under controlled conditions, as we (Rozin and Kennel, 1983) did with chimpanzees. This would allow the analysis and evaluation of different models for the acquisition of liking. It is a particularly good model system for studying the acquisition of liking for IUSs, including alcohol, tobacco, and coffee. Chili pepper has a special advantage; unlike many other widely used IUSs, chili is currently consumed for one reason: it tastes good. In this respect, it is a simplified system.

I hope that this chapter can stimulate more research on this remarkable food. Meanwhile, some will continue to like it hot. The questions are, why only some? and why anyone?

## ACKNOWLEDGMENTS

The preparation of this chapter and some of the research described herein was supported by National Institutes of Health Bio Medical Research Support Grant to the University of Pennsylvania (No. 2-507-RR-07083-18 SUB 17) and by the John D. and Catherine MacArthur Research Program on Determinants and Consequences of Health-Promoting and Health-Damaging Behavior. Thanks to Barry Green for constructive comments on the manuscript.

## REFERENCES

Albrecht, G. L. (1973). The alcoholism process: A social learning view. In: *Alcoholism: Progress in research and treatment* P. G. Bourne & R. Fox (Eds.). Academic Press, New York, pp. 11-42.

Anderson, R. K., Calvo, J., Serrano, G., and Payne, G. C. (1946). A study of the nutritional status and food habits of Otami Indians in the Mezquital Valley of Mexico. *Am. J. Pub. Health* 36: 883-903.

Andrews, J. (1984). *Peppers: The Domesticated Capsicums.* University of Texas Press, Austin.

Beauchamp, G. K. (1981). Ontogenesis of taste preference. In: *Food, Nuitrition and Evolution* D. Walcher and N. Kretchmer (Eds.). Masson, New York.

Bergsma, S. (1931). Gastric and duodenal ulcer in black people of Abyssinia. *Arch. Inter. Med.* 47: 144-148.

Birch, L. L. (1987). Children's food preferences: Developmental patterns and environmental influences. In: *Annals of Child Development*, Vol 4. G. Whitehurst and R. Vasta (Eds.). JAI, Greenwich, CT, pp. 171-208.

Birch, L. L., Birch D., Marlin, D. W., and Kramer, L. (1982). Effects of instrumental eating on children's food preferences. *Appetite* 3: 125-134.

Booth, D. A. (1982). Normal control of omnivore intake by taste and smell. In: *The Determination of Behavior by Chemical Stimuli.* J. Steiner and Ganchrow (Eds.). Information Retrieval, London, pp. 233-243.

Booth, D. A., Mather, P., and Fuller, J. (1982). Starch content of ordinary foods associatively conditions human appetite and satiation, indexed by intake and eating pleasantness of starch-paired flavors. *Appetite* 3: 163-184.

Brown, L. T., Ruder, V. G., Ruder, J. H. and Young, S. D. (1974). Stimulation seeking and the change-seeker index. *J. Consult. Clin. Psychol.* 42: 311.

Buck, S. H., and Burks, T. F. (1986). The neuropharmacology of capsaicin: Review of some recent observations. *Pharmacol. Rev.* 38: 179-226.

las Casas, Bernardo de (1909). Tears of the Indians. George Bell and Sons, London. (original work published 1552).

Child, I. L., Cooperman, M., and Wolowitz, H. M. (1969). Esthetic preference and other correlates of active versus passive food preferences. *J. Personal. Social Psychol.* 1: 75-84.

*Codex Mendoza* (1978). *Aztec manuscript.* Production Liber., Fribourg, Switzerland. Commentaries by K. Ross.

Columbus, C. (Colon, C.) (1493/1986). *Los cuatro viajes Testamento.* Alianza Editorial, Madrid. Edicion de Consuelo Varela.

Cowart, B. (1987). Oral chemical irritation: Does it reduce perceived taste intensity? *Chem. Senses* 12: 467-479.

Crosby, A. W., Jr. (1972). *The Columbian Exchange: Biological and Cultural Conse quences of 1492.* Greenwood Press, Westport, CT.

Damon, A. W., Jr. (1973). Smoking attitudes and practices in seven preliterate societies. In *Smoking Behavior: Motives and Incentives* W. L. Dunn (Ed.). V. H. Winston, Washington, D.C., pp. 219-230.

Diaz del Castillo, B. (1963). *The Conquest of New Spain.* Penguin, New York. (Trans. J. M. Cohen. Original work published in 1568: Historia verdadera de la conquista de la Neuva Espana).

Dib, B. (1989). After two weeks of habituation to capsaicinized foods, rats prefer this to plain food. *Appetite* (in press).

Dold, H., & Knapp, A. (1948). Uber die antibakterielle Wirkung von Gewurzen (Antibacterial effects of spices). *Zeitschrift fur Hygiene und Infektionskrankh* 128 (56): 696-706.

Dua-Sharma, S., and Sharma, K. N. (1980). Capsaicin and feeding responses in Macaca mulata: A longitudinal study. *Proceedings of the International Conference on Regulation of Food and Water Intake,* Warsaw, 1980

Galef, B. G., Jr. (1989). Enduring social enhancement of rats' preferences for the palatable and the piquant. *Appetite* (in press).

Govindarajan, V. S. (1985a). Capsicum: Production, technology, chemistry and quality, Part I. History, botany, cultivation and primary processing. *CRC Crit. Rev. Food Sci. Nutr.* 22: 109-176.

Govindarajan, V. S. (1985b). Capsicum: Production, technology, chemistry and quality, Part II. Processed products, standards, world production and trade. *CRC Crit. Rev. Food Sci. Nutr.* 23: 207-288.

Govindarajan, V. S. (1986). Capsicum: Production, technology, chemistry and quality, Part III. Chemistry of the color, aroma and pungency stimuli. *CRC Crit. Rev. Food Sci. Nutr.* 24: 245-355.

Govindarajan, V. S., Rajalakshmi, D., and Chand, N. (1987). Capsicum: Production, technology, chemistry and quality, Part IV. Evaluation of quality. *CRC Crit. Rev. Food Sci. Nutr.* 25: 185-282.

Hacker, D. B., and Miller, E. D. (1959). Food patterns of the Southwest. *Am. J. Clin. Nutr.* 7: 224-229.

Halasz, Z. (1963). *Hungarian paprika through the ages.* Corvina Press, Budapest.

Hauck, H. M., Hanks, J. R., and Sudsaneh, S. (1959). Food habits in a Siamese village. *J. Am. Diet. Assoc.* 35: 1143-1148.

Heiser, C. B. Jr. (1969). *Nightshades: The Paradoxical Plants.* W. H. Freeman, San Francisco.

Heiser, C. B. Jr. (1985). *Of Plants and People.* University of Oklahoma Press, Norman.

Hilker, D. M., Hee, J., Higashi, J., Ikehara, S., and Paulsen, E. (1967). Free choice consumption of spiced diets by rats. *J. Nutr.* 91: 129-131.

Jancsó, N. (1960). Role of the nerve terminals in the mechanism of inflammatory reactions. *Bull. Millard Filmore Hosp.* 7: 53-77.

Jancsó, N. (1968). Desensitization with capsaicin and related acylamides as a tool for studying the function of pain receptors. *Proceedings of the 3rd International Pharmacolocial Meeting, 1966: Pharmacology of Pain.* Pergamon Press, Oxford, pp. 33-55.

Jancsó-Gábor, A., and Szolcsányi, J. (1969). The mechanism of neurogenic inflammation. In: *Inflammation: Biochemistry and Drug Interaction.* A Bertelli and J. C. Houck (Eds.). Excerpta Medica, Amsterdam, pp. 210-217.

Jelliffe, D. B. (1962). Culture, social change, and infant feeding. Current trends in tropical regions. *Am. J. and Clin. Nutr.* 10: 19-45.

Kish, G. B., and Donnenwerth, G. V. (1972). Sex differences in the correlates of stimulus seeking. *J. Consult. Clin. Psychol.* 38: 42-49.

Koufax, S. (1966). *Koufax.* Viking Press, New York.

Krishnamurthy, V., De, S. S., and Subrahmanyan, V. (1948). Effect of supplementation with tamarind and chilli on the growth of young rats on a poor South-Indian rice diet. *Curr. Sci.* 17: 51-52.

Lee, S. O. (1963). Studies on the influence of diets and lipotropic substances upon the various organs and metabolic changes in rabbits on long-term feeding with red pepper. I. Histopathologic changes of liver and spleen. *Korean J. Intern. Med.* 6: 383-400.

Lawless, H., Rozin, P., and Shenker, J. (1985). Differences in perception of capsaicin between people who frequently consume and like chili pepper and others who rarely consume and dislike it. *Chem. Senses* 10: 579-589.

Logue, A. W., and Smith, M. E. (1986). Predictors of food preferences in adult humans. *Appetite* 7: 109-125.

Maga, J. A. (1975). Capsicum. *CRC Crit. Rev. Food Sci. Nutr.* 7: 177-199.

Markandaya, K. (1954). *Nectar in a Sieve.* John Day, New York.

Moore, F. W. (1970). Food habits in non-industrial societies. In: *Dimensions of Nutri tion* J. DuPont (Ed.). Colorado Associated Universities Press, Boulder, pp. 181-221.

Nopanitaya, W. (1974). Effects of capsaicin in combination with diets of varying protein content on the duodenal absorptive cells of the rat. *Am. J. Dig. Dis.* 19: 439-448.

*Nutrition Reviews* (1986). Metabolism and toxicity of capsaicin. *Nut. Rev.* 44: 20-22.

Parry, J. W. (1969a). *Spices, Vol. 1. The Story of Spices. The spices Described.* Chemical Publishing, New York.

Pickersgill, B. (1969). The domestication of chili peppers. In: *The Domestication and Exploitation of Plants and Animals.* P. J. Ucko and G. W. Dimbley (Eds.). Gerald Duckworth, London, pp. 443-449.

Pliner, P. (1982). The effects of mere exposure on liking for edible substances. *Appetite* 3: 283-290.

Pliner, P., Rozin, P., Cooper, M., and Woody, G. (1985). Role of specific postingestional effects and medicinal context in the acquisition of liking for tastes. *Appetite* 6: 243-252.

Porszasz, J., Gyorgy, L., and Porszasz-Gibiszer, K. (1985). Cardiovascular and respiratory effects of capsaicin. *Acta Physiol. Acad. Scient. Hung.* 8: 60-76.

Rosengarten, F. Jr. (1969). *The book of Spices.* Livingston, Wynnewood, PA.

Rozin, E. (1982). The structure of cuisine. In: *The Psychobiology of Human Food Selection.* L. M. Barker (Ed.). AVI, Westport, CT., pp. 189-203.

Rozin, E. (1983). *Ethnic Cuisine: The Flavor Principle Cookbook.* Stephen Greene, Brattleboro, VT.

Rozin, E., & Rozin, P. (1981). Culinary themes and variations. *Natural Hist.* 90: 6-14.

Rozin, P. (1978). The use of characteristic flavorings in human culinary practice. In: *Flavor: Its Chemical, Behavioral and Commercial Aspects.* C. M. Apt (Ed.). Westview Press, Boulder, CO, pp. 101-127.

Rozin, P. (1982). Human food selection: The interaction of biology, culture and individual experience. In: *The Psychobiology of Human Food Selection.* L. M. Barker (Ed.). AVI, Westport, CT, pp. 225-254.

Rozin, P. (1984). The acquisition of food habits and preferences. In: *Behavioral Health: A Handbook of Health Enhancement and Disease Prevention.* J. D. Matarazzo, S. M. Weiss, J. A. Herd, N. E. Miller, and S. M. Weiss (Eds.). John Wiley, New York, pp. 590-607.

Rozin, P. (1988). Social learning about food by humans. In: *Social Learning: A Comparative Approach.* T. Zentall and B. G. Galef, Jr. (Eds.). Erlbaum, Hillsdale, N.J., pp. 165-187.

Rozin, P., Ebert, L., and Schull, J. (1982). Some like it hot: A temporal analysis of hedonic responses to chili pepper. *Appetite* 3: 13-22.

Rozin, P., Gruss, L., and Berk, G. (1979). The reversal of innate aversions: Attempts to induce a preference for chili peppers in rats. *J. Compar. Physiol. Psychol.* 92: 1001-1014.

Rozin, P., and Kennel, K. (1983). Acquired preferences for piquant foods by chimpanzees. *Appetite* 4: 69-77.

Rozin, P., Mark, M., and Schiller, D. (1981). The role of desensitization to capsaicin in chili pepper ingestion and preference. *Chem. Senses* 6: 23-31.

Rozin, P., and Schiller, D. (1980). The nature and acquisition of a preference for chili pepper by humans. *Motiva. Emotion* 4: 77-101.

Rozin, P., and Shenker, J. (1989). Liking oral cold and hot irritant sensations: Specificity to type of irritant and locus of stimulation (in press).

Rozin, P., and Vollmecke, T. A. (1986). Food likes and dislikes. *Ann. Rev. Nutr.* 6: 433-456.

Sass, L. J. (1981). Religion, medicine, politics and spices. *Appetite* 2: 7-13.

Schull, J. (1979). A conditioned opponent theory of Pavlovian conditioning and habituation. In: *The Psychology of Learning and Motivation*, Vol. 13. G. Bower (Ed.). Academic Press, New York.

Schweid, R. (1980). *Hot peppers: Cajuns and Capsicum in New Iberia, Lousiana.* Madrona, Seattle.

Siegel, S. (1977). Learning and psychopharmacology. In: *Psychopharmacology in the Practice of Medicine.* M. E. Jarvik (Ed.). Appleton-Century-Crofts, New York, pp. 59-70.

Simoons, F. J. (1961). *Eat Not This Flesh.* University of Wisconsin Press, Madison.

Solanke, T. F. (1973). The effect of red pepper (Capsicum frutescens) on gastric acid secretion. *J. Surg. Res.* 15: 385-390.

Solomon, R. L. (1980). The opponent process theory of acquired motivation. *Am. Psychologist* 35: 691-712.

Solomon, R. L., and Corbit, J. (1974). An opponent-process theory of motivation: Temporal dynamics of affect. *Psychol. Rev.* 8: 119-145.

Stevens, D. A., and Lawless, H. T. (1987). Enhancement of responses to sequential presentation of oral chemical irritants. *Physiol. Behav.* 39: 63-65.

Szolcsányi, J. (1977). A pharmacological approach to elucidation of the role of different nerve fibers and receptor endings in mediation of pain. *J. Physiol.* (Paris) 73: 251-259.

Szolcsányi, J., and Bartho, L. (1980). Impaired defense mechanism to peptic ulcer in the capsaicin-desensitized rat. In: *Advances in Physiological Sciences, 29: Gastrointestinal Defense Mechanisms.* Gy. Mozsik, O. Hanninen, and T. Javor (Eds.). Pergamon Press, Oxford.

Szolcsányi, J., and Jancsó-Gábor, A. (1973). Capsaicin and other pungent agents as pharmacological tools in studies on thermoregulation. In: *The Pharmacology of Thermoregulation.* E. Schonbaum and P. Lomax (Eds.). Karger, Basel, pp. 395-409.

Szolcsányi, J., and Jancsó-Gábor, A. (1975). Sensory effects of capsaicin congeners. Part I. Relationship between chemical structure and pain-producing potency of pungent agents. *Arzneim. Forsch.* 25: 1877-1881.

Szolcsányi, J., and Jancsó-Gábor, A. (1976). Sensory effects of capsaicin congeners. Part II. Importance of chemical structure and pungency in desensitizing activity of capsaicin-type compounds. *Arzneim. Forsch.* 26: 33-37.

Time-Life (1970). *Foods of the World. Recipes: Pacific and Southeast Asian Cooking.* Time-Life, New York.

Todd, P. H. Jr., Bensinger, M. G., and Biftu, T. (1977). Determination of pungency due to capsicum by gas-liquid chromatography. *J. Food Sci.* 42: 660-665.

Tokey, L. (1932). Uber Kapsizismus. (Capsicum, toxic symptoms after excessive use of paprika: case). *Monatschrift Psychiat. Neurol.* 82: 346-355.

Weil, A. (1975). Hot! Hot!—I: Eating chilies. *J. Psychedelic Drugs* 8: 83-86.

Zajonc, R. B. (1968). Attitudinal effects of mere exposure. *J. Personal. Social Psychol.* 9(2): 1-27.

Zellner, D. A., Rozin, P., Aron, M., and Kulish, C. (1983). Conditioned enhancement of human's liking for flavors by pairing with sweetness. *Learn. Motivat.* 14: 338-350.

Zuckerman, M. (1979). *Sensation Seeking: Beyond the Optimal Level of Arousal.* Hillsdale, N.J.: Lawrence Erlbaum.

# Chapter 12 Discussion

**Dr. Finger:** I was wondering, regarding acquisition of a preference for chili, if there might be another stage. From the chemical properties of it, wouldn't you expect it to be secreted in the mother's milk? Based on our experience with our children, that happens.

**Dr. Rozin:** There are two issues. Capsaicin is very poorly absorbed. We don't known how poorly, but quite poorly. We know that some of it goes out because there is an old Hungarian saying, "paprika burns twice," which is essentially direct evidence for the fact that some of it doesn't get absorbed. It seems to get absorbed at rather low levels. It is amazing how low the pharmacological effects of eating a lot of chili are compared to administrating capsaicin systemically. So I don't know how much gets in mother's milk. No one has ever tasted the mother's milk of people who are heavy chili eaters to see if, like garlic, it gets through. Then there is the question of whether what is in mother's milk influences the preferences of children. And the answer to that is there is very little evidence that that's true. In fact, there is no evidence that it's true. There is very little evidence that foods eaten very early in life affect later preferences. If we know that the foods we eat in the first year or so do not have a big effect, then mother's milk is probably going to be a weak stimulus. But, it's possible, and it would certainly be worth doing; I don't think anyone has done it.

**Dr. Silver:** Given that capsaicin is not absorbed very well, do you think the desensitization—that repeated eaters don't feel the burn as much—is psychological or physiological?

**Dr. Rozin:** I don't know the answer to that. There is of course topical desensitization as Dr. Szolcsányi has shown, and that can last for quite a while.

And since frequent eaters usually have eaten it in the last few days—that's what we mean by a frequent eater—it could be that the effect is topical and not systemic. I don't know the answer. I don't know of any studies which have had chili eaters not eat it, for example, for a month. It's hard to do that. But if you could, and if their threshold goes back to normal, it would indicate that it is topical desensitization.

**Dr. Silver:** Are they less sensitive to capsaicin burn on other places?

**Dr. Rozin:** That's a good question. I don't know of any research or evidence on that point. Of course, capsaicin is not very volatile, so you don't get it in the nose.

**Dr. Green:** We have a little almost anecdotal data related to whether people who eat a lot of capsaicin are less sensitive to it somewhere else on the body. We looked at sensitivity to capsaicin on the forearm, and although we've tested only a small number of subjects, nothing obvious jumps out at us. We've had people who eat a lot of it be just as sensitive as people who don't.

The question I have relates to why people don't like capsaicin in the eye or some other area of the body. Perhaps the mouth is special because it's related to consumption and it's important that we like what we eat. That would not explain why we like chili pepper as much as we do, but it might explain why we like it in the mouth more than we do in some other areas.

**Dr. Rozin:** Let me just say this: the psychology of the whole mouth, as opposed to the chemical senses, has been virtually unstudied. This is a peculiar place. This is the entry into the body, *the entry* into the body, where almost everything that gets in comes in. And actually, most of my work now is about that. I don't work now on chili peppers. I'm working on why people put things into their mouths and the psychology of it. But you can turn that around and say the mouth should be the place where you are most sensitive to putting anything dangerous. People are much more resistant to putting anything of any questionable nature into their mouth than they are to touching it with any other part of their body, except the vagina in females. So this is a very sensitive place and in some sense you would think you would be most on guard because this is what gets things into your body. So there are two sides to this, I couldn't say on which side of that I would vote.

**Dr. Green:** I agree with that. But it seems to me that what happens is that the mouth is sensitive initially, but if nothing bad happens, then perhaps it becomes important to like whatever the substance is because it is a source of nutrition.

**Dr. Rozin:** I like that point. But, animals have the same nutritional needs as we do and they don't seem to get past this sensory barrier the way we do. I think it's a fascinating question. It's also true that some people do like burn on their skin. Ointments and linaments sell very well, and that's also painful in some sense. I don't know what that's about.

**Dr. Szolcsányi:** If I recall correctly, you have done experiments on the recognition threshold of likers and nonlikers. Is that correct?

**Dr. Rozin:** We did a study on preference thresholds.

**Dr. Szolcsányi:** But this could be an approach to finding differences between the desensitization effects of the nerve endings and the psychological effects. Recognition threshold measurements; you could ask subjects to recognize salt, bitter, and chili sensations. The task of the subject would be to decide which is which and then if it turns out that the recognition threshold of chili is shifted up, it should relate to real desensitization of receptors and not to psychological effects due to liking.

**Dr. Rozin:** Detection thresholds have been measured. Beverly Cowart presented some data I collected and she talked about other data. There is a small rise in the detection threshold in people who are regular eaters, but that could either be topical or systemic, I think. If topical effects can last for a week or two, we certainly didn't control for that.

**Dr. Szolcsányi:** If somebody could tell that this is not simple tap water but this is hot stuff, and this threshold differs from the normal population, then it could be related to the desensitizing effect, not preference.

**Dr. Rozin:** I think we did that, and the answer is yes. The problem is, as you saw from the data Beverly presented, people who are capsaicin eaters show lower magnitude estimates; this is data from Harry Lawless, myself, and Joel Shenker. They show a lower rating of the magnitude of the burn, but the desensitization couldn't account for that because they show it over an enormous range. The threshold difference is very small unless you have a very funny model of how to map thresholds onto other functions. These people like all levels of burns, from little burns to powerful burns, and their threshold difference is slight. So I don't think you can account for the preference difference by desensitization. But it might contribute, as Beverly suggested.

# 13
# Effects of Menthol on Nasal Sensation of Airflow

**Ronald Eccles**
University of Wales
College of Cardiff
Cardiff, Wales

## I. INTRODUCTION

When discussing nasal sensory receptors one immediately thinks of olfaction and olfactory receptors, but the nose as the entrance of the respiratory tract has other equally important sensory receptors which sample the chemical and physical properties of the inspired air. Stimulation of these sensory receptors can lead to respiratory reflexes which protect the lungs from inspiration of chemical irritants or water (Angell James, 1969). These protective nasal reflexes have been studied in detail but very little attention has been paid to the sensory receptors that give us our normal appreciation of breathing.

Our awareness of respiration and the regular respiratory rhythm comes primarily from stimulation of nasal sensory receptors rather than from lung stretch receptors or proprioceptors in respiratory muscles of the chest and diaphragm. Each inspiration draws in air through the nasal passages and this air stream stimulates sensory receptors in the nose and gives us the regular cool sensation of normal breathing.

Nasal congestion associated with infective or allergic rhinitis blocks the nasal passages and deprives us of the reassuring nasal sensation of breathing. A blocked nose is uncomfortable as continuous oral breathing leads to a dry mouth and the oral cavity is not as efficient as the nose in conditioning the inspired air. However, the oral cavity also possesses temperature receptors which are stimulated on inspiration and in this respect the oral cavity may take over some of the respiratory function of the nose.

Menthol-containing products are widely used for the treatment of nasal congestion associated with the common cold and remedies based on plant extracts containing menthol have been used for hundreds if not thousands of years. Despite the widespread use of menthol not only in common cold remedies but in toothpastes, shaving foams, and cosmetics, there has been relatively little scientific work on its pharmacological and physiological effects. There is now an awakening of interest in menthol and menthol analogs not only in the development of new commercial products but also right down to the basic level of scientific investigations on the effects of menthol on calcium conductance in sensory receptors. There are far more questions than answers to be put forward in this area of investigation, but this makes it all the more interesting and exciting for the scientist involved in basic research on menthol and for the company interested in developing new menthol-containing products.

This chapter looks at the sensation of nasal air flow as regards sensory receptors and nasal reflex activity, and also discusses the effects of menthol on the nose.

## II. SENSORY RECEPTORS AND NERVE PATHWAYS INVOLVED IN NASAL SENSATION OF AIR FLOW

The olfactory epithelium is a distinct area of the mucosal lining of the nasal cavity and the structure and properties of the olfactory receptors are well documented. When it comes to other nasal sensations, such as the cooling effect of inspired air, the itching sensation of nasal allergy, and nasal pain, it is generally assumed that these nasal sensations are served by branches of the trigeminal nerve.

Most of the sensory nerve fibers to the nasal vestibule and the respiratory mucosa of the nose and paranasal sinuses are supplied by the two branches of the trigeminal nerve: the ophthalmic and maxillary nerves as shown in Fig. 1 (Eccles, 1982; Cauna, 1982). The ethmoidal nerve is a branch of the ophthalmic nerve and the maxillary nerve supplies sensory branches to the sphenopalatine ganglion.

The ethmoidal branch of the trigeminal nerve supplies the anterior mucosa of the nose and the squamous epithelial lining of the nasal vestibule. This anterior area of the nose has an important sensory role as it is the first to make contact with the inspired air and it may be particularly important as far as detecting the cooling sensation of inspired air.

The maxillary division of the trigeminal nerve gives branches to the sphenopalatine ganglion which pass through the ganglion without synapse and

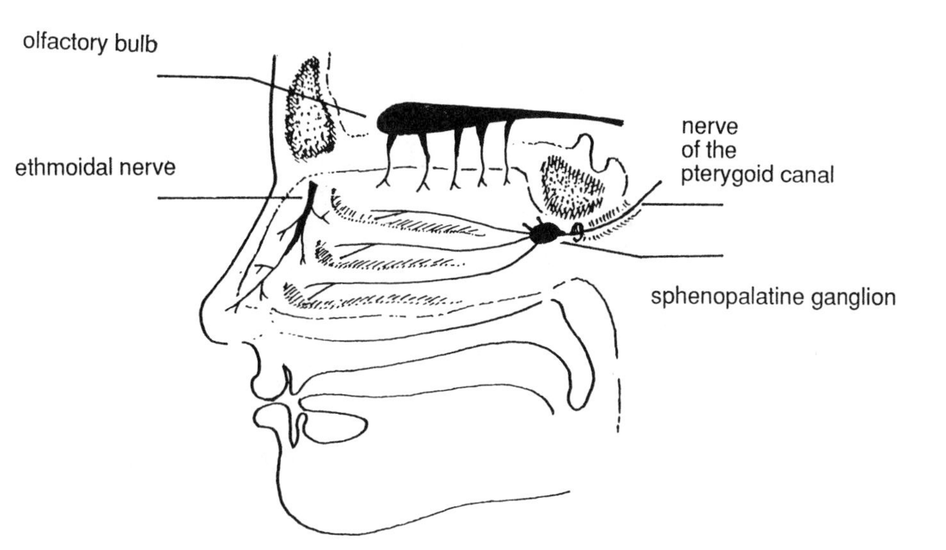


**FIGURE 1** Diagrammatic representation of the sensory innervation of the nose.

supply the middle and posterior parts of the nose and the paranasal sinuses. All of the respiratory mucosa of the nose is supplied by these nerves and they may have an important role in nasal symptoms such as itching, sneezing, pain, and the reflex nasal secretion associated with nasal inflammation.

There is some evidence from animal experiments and clinical observations that the nasal mucosa receives sensory fibers from the seventh and tenth cranial nerves and the upper thoracic spinal nerves, and these pathways may explain referred pain and headaches due to nasal disease but their role in other sensations is unknown (Eccles, 1982).

The nervus terminalis is well developed in animal species with a functional vomeronasal organ but this organ is vestigial in man and although the nervus terminalis serves the anterior portion of the nasal septum in man, its role in human nasal sensation is obscure.

As far as nasal sensation of air flow is concerned, the nose can be divided into four areas on the basis of the histological appearance of the tissues lining the nasal passage as shown in Fig. 2 (Mygind et al., 1982).

The anterior portion of the nose forms the nasal vestibule and is lined with skin, almost as though a portion of the facial skin had been turned inward to form the vestibule. At the entrance into the nares the skin is covered with short, stiff hairs, or vibrissae, and these are better developed in the male than the female, like a secondary sexual characteristic dependent on male sex hormones. The vibrissae are extremely sensitive to certain kinds of mechanical

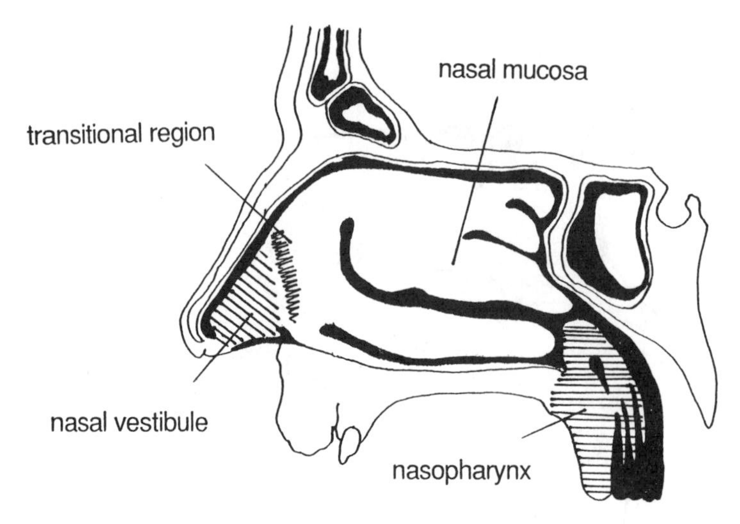


**FIGURE 2** Topographic distribution of epithelial types concerned with nasal sensation of air flow. The nasal vestibule is lined with skin which changes into a thin, stratified, squamous epithelium which is only moderately keratinized as one moves toward the nasal cavity. The transitional region develops microvilli and merges into a ciliated pseudostratified columnar epithelium which forms the respiratory nasal mucosa responsible for air conditioning. The ciliated respiratory nasal mucosa merges into the stratified squamous epithelium of the nasopharynx.

stimuli, as light deformation produces a sensation of acute itch or tickle and may induce violent sneezing. However, forceful inspiration and expiration which evenly bends most of the hairs produces no appreciable sensation (Cauna, 1982). The vibrissae do not appear to participate in the sensation of nasal air flow and their main role appears to be to form a tight protective screen or filter and to protect the respiratory tract from inspiration of large particles.

The nasal vestibule is lined with a stratified squamous epithelium which appears to have sensory properties similar to those of the facial skin. Cauna (1982) states that nasal thermoreceptors which detect nasal air flow are situated in the nasal vestibule. The nasal vestibule narrows at the entrance to the nasal cavity and this nasal valve area forms the smallest cross-sectional area of the whole respiratory tract.

The nasal valve area is strategically placed at the entrance of the nasal cavity and the inspired air velocity is maximal at this point. Any cooling or mechanical stimulus from the inspired air would be greatest at this point and the valve area may be of major importance as far as the sensation of nasal air flow is concerned.

Just past the nasal valve the skin gradually changes into the respiratory nasal mucosa. The transition zone between the vestibule and the respiratory area is characterized by the presence of high and narrow dermal papillae, each of which is filled with a long capillary loop (Alverdes, 1930). The functions of these interesting papillae are not known but they may have a role as sensitive mechano- or thermoreceptors.

The structure and distribution of sensory receptors in the nasal vestibule in man is assumed to be similar to facial skin but, as Cauna (1982) states, these receptor organs have not received any attention so far.

Experimental studies on the anesthetized cat have demonstrated that sensory receptors in the nasal vestibule are readily stimulated by gently blowing a jet of air into the nose (Davies and Eccles, 1985, 1987). These sensory receptors are innervated by the infraorbital branch of the trigeminal nerve and are probably mechanoreceptors rather than thermoreceptors. The effects of an air jet stimulus applied to the nasal vestibule on nasal EMG activity are shown in Fig. 3.

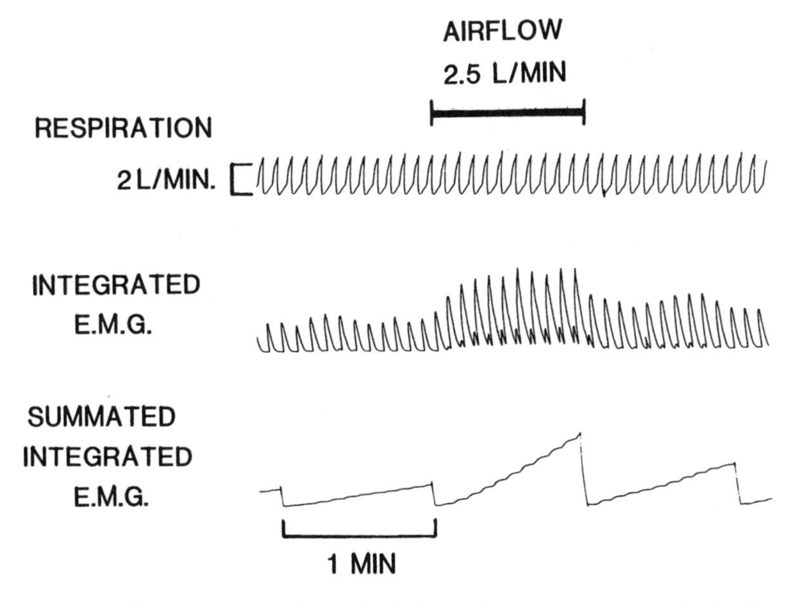


**FIGURE 3** The effects of a nasal air jet applied to the nasal vestibule of an anesthetized cat. Nasal electromyographic activity (EMG) was recorded from the dilator muscles of the nostril. The jet of air at 25 °C was directed into the nasal vestibule at a flow rate of 2.5 L/min, for a period of 1 min. Note the reflex increase in the nasal EMG activity caused by the air stimulus and the absence of any effect on respiration. (From Davies and Eccles, 1987.)

In contrast to the skin of the nasal vestibule, the nasal mucosa has a limited capacity for localization of stimuli and discrimination of sensory modalities (Cauna, 1982). The nasal mucosa apparently lacks thermoreceptors (Cauna and Hinderer, 1969) and, although irritation of the mucosa may initiate reflex sneezing and secretion, subjectively the mucosa only discriminates tactile and painful stimuli.

We are therefore left with the rather unsatisfactory conclusion that nasal air flow receptors in man are most likely found in the nasal vestibule region rather than the respiratory mucosal region, but that the evidence for this conclusion is very flimsy and that there is no supportive histological evidence about the types of sensory receptor found in these areas of the nose.

## III. NASAL AIR FLOW AND RESPIRATORY REFLEXES

The potent respiratory and cardiovascular reflexes initiated by nasal stimulation with irritant vapors or irrigation with water are well documented (Tomori and Widdicombe, 1969; Angell James, 1969; Allison, 1974). Stimulation of the nasal mucous membrane causes a reflex reduction in breathing or apnea in the expiratory position associated with bradycardia and hypertension or hypotension. These responses are well developed in diving animals and are often referred to as a diving reflex. In man the reflexes are protective as they prevent inspiration of water or irritants.

In contrast to stimulation of the nasal mucosa with irritant gases or water, there is little information available on the reflex effects of stimulating air flow receptors in the nose during normal respiration.

In recognition of our poor understanding of the role of nasal air flow receptors in the pathogenesis of sudden infant death and sleep apnea, there has been an increase of interest in nasal air flow as regards the regulation of respiration in health and disease. Nasal air flow during normal respiration has been shown to influence respiratory pattern and contribute to the smoothness of respiratory movements (Ramos, 1960). Experiments on man by McBride and Whitelaw (1981) and Douglas et al. (1983) led to the hypothesis that there are upper airway flow receptors which influence respiratory timing and respiratory drive. McBride and Whitelaw (1981) demonstrated that a stream of cold air circulated through the nose inhibited the involuntary inspiratory muscle activity that occurred during breath holding; and Douglas et al. (1983) showed that the pattern of breathing was different according to whether the nasal or oral airway was used and that this effect on breathing pattern was independent of airway resistance.

In studies on the anesthetized cat, Davies and Eccles (1985, 1987) demonstrated that a gentle flow of air directed into the nasal vestibule caused reflex changes in the electromyographic activity of nasal muscles without any of the more potent reflex effects on respiration or blood pressure. These experiments demonstrated that there are sensory receptors in the nasal vestibule which respond to air flow or gentle mechanical distortion. The reflex activation of nasal EMG activity caused by the stimulation of the nasal vestibule was abolished on section of the branches of the trigeminal nerve supplying the nose, clearly demonstrating the role of these nerves in the reflex.

Nasal air flow has also been shown to influence nasal EMG activity in man and this finding indicates that nasal air flow receptors may be important in the control of upper airway accessory respiratory muscles which stabilize the airway during inspiration (Eccles and Tolley, 1987).

If nasal air flow receptors have an important role to play in the regulation of respiration, then limitation or abolition of this sensory input would be expected to disturb respiration. There is now good evidence that this is the case as an inability to breathe through the nose has been recognized as a cause of disordered breathing during sleep (Lavie et al., 1983; White et al., 1985). In normal subjects anesthesia of the nasal passages or taping of the nostrils reduced or abolished the nasal sensory input created by air flow and in both these cases there was an increased incidence of periods of apnea and disturbed breathing pattern during sleep.

Nasal congestion associated with infective or allergic rhinitis frequently leads to nasal obstruction with partial or complete block of nasal air flow. In these common nasal disorders, and particularly with the common cold, this nasal obstruction could lead to discomfort during the day due to a decrease in the subjective sensation of nasal air flow and discomfort at night due to sleep disturbance.

Extracts of essential plant oils such as menthol are widely used for the treatment of nasal congestion and the case will be made that these products act on nasal air flow receptors and may reflexively alter breathing patterns and the activity of upper airway accessory respiratory muscles as well as giving a pronounced subjective relief from the congestion.

## IV. MENTHOL PRODUCTION AND CHEMISTRY

Menthol has probably been used for the treatment of nasal congestion and other respiratory problems since man first crushed the leaves of a menthol-containing plant and found that on breathing the volatile essential oils contained in the plant he obtained a cool, clear feeling of improved nasal air flow. It was not a very great step then to move to boiling these leaves and breathing in the relief-giving vapors. Extracting essential oils by steam distillation of commercially grown crops of *Mentha* species is still the major source of menthol. Isolation of menthol from peppermint oil competes with partial or total syntheses and the processes are elegantly described by Bauer and Garbe (1985).

Peppermint oil is obtained from steam distillation of the flowering herbs *Mentha arvensis* and *Mentha piperita* of which there are many subspecies and varieties.

*Mentha arvensis* is grown commercially in Brazil, Japan, Paraguay, and China. Brazil produced 3000 tons of oil and 3000 tons of (−)menthol in 1973. Peppermint oil from the flowering shrub *M. arvensis* may contain 70-80% free (−)menthol and this can be crystallized out. Since the crystalline product contains traces of peppermint oil, this menthol has a slightly herbal minty note. Pure (−)menthol is obtained by recrystallization from low-boiling-point solvents. The dementholized peppermint oil still contains 40-50% free menthol and this together with menthone can be further treated to extract or synthesize menthol.

Peppermint oil can also be made from *M. piperita*, which is grown widely in France, Italy, Great Britain, and North America. North America is the leading producer with 2000-3000 tons/year in a number of states, mainly Oregon and Washington, from two *M. piperita* varities "Mitcham" and "American." The product is an almost colorless pale to greenish yellow liquid with a characteristic peppermint odor. *Mentha piperita* oil often contains

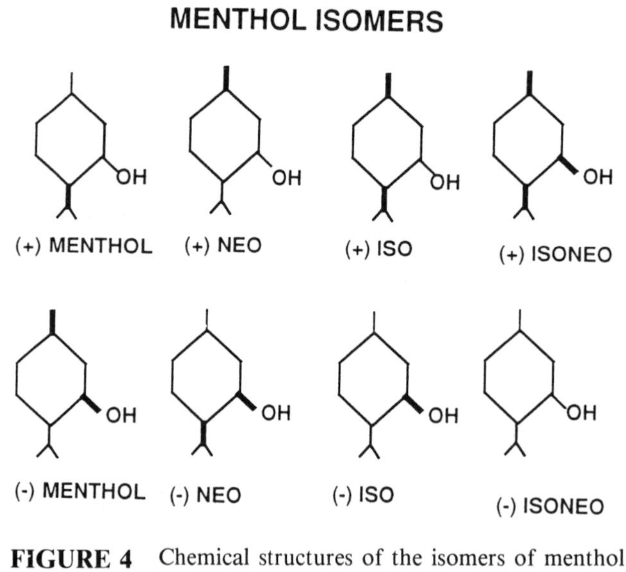

MENTHOL ISOMERS

OH OH OH OH

(+) MENTHOL (+) NEO (+) ISO (+) ISONEO

OH OH OH OH

(-) MENTHOL (-) NEO (-) ISO (-) ISONEO

**FIGURE 4** Chemical structures of the isomers of menthol. (−) Menthol is the isomer found most commonly in plant extracts and it is this isomer which is used in nasal treatments and other products.

up to 50% (−)menthol but because of its high price it is not used commercially for the production of menthol but rather is used primarily as a flavoring for toothpastes and chewing gum.

Menthol can also be extracted or synthesized from other essential oils such as (+)citronella oil, eucalyptus dives oil, and Indian turpentine oil.

Menthol carries three asymmetric carbon atoms in the cyclohexane ring and therefore occurs as four pairs of mirror image isomers as shown in Fig. 4 (Bauer and Garbe, 1985). (−)Menthol is the isomer which occurs commonly in plant extracts and it has the characteristic peppermint odor and at the same time exerts a cooling action. The other isomers of menthol have a similar but not identical odor but do not have the same cooling action as (−)menthol. The mirror image isomers each have identical physical properties apart from their specific effects on the rotation of light. Neomenthol, neoisomenthol, menthol, and isomenthol differ slightly in their boiling points with a range of 211.7-218.6°C. They also differ in their physical properties as (+)neomenthol is a colorless liquid at room temperature and isomenthol and menthol are white crystals.

## V. EFFECTS OF MENTHOL ON THE NOSE

Menthol-containing products are widely used for the treatment of nasal congestion in children and adults. Menthol may be presented as a vaporizing ointment and this is widely used on children, particularly before sleep. It may also be presented in a lozenge or in an inhaler or nasal spray.

Experiments on normal volunteer subjects have demonstrated that menthol has a marked subjective effect on nasal sensation of air flow without causing any objective change in nasal resistance to air flow.

In one of our first investigations we studied the effects of breathing menthol, camphor, or eucalyptus vapors on nasal resistance to air flow and nasal sensation (Burrow et al., 1983). This study demonstrated that the aromatic vapors had no effect on nasal resistance but greatly improved the sensation of nasal air flow and made the subjects feel that their nasal air flow had improved. In this same group of subjects 5 min exercise on a cycle ergometer caused a 70% decrease in nasal resistance to air flow but the subjects did not feel any improvement in nasal sensation of air flow. The marked decongestant effect of exercise demonstrated that the subjects were capable of a decongestant response and the result also indicated that the subjective and objective measurements of nasal air flow did not agree. After a 70% reduction in nasal resistance to air flow subjects did not perceive any improvement in nasal air flow, yet with no objective change in nasal resistance after breathing menthol subjects scored a marked improvement in nasal sensation of air flow.

In the above study (Burrow et al., 1983), nasal resistance to air flow was measured only on expiration. Since animal studies had demonstrated that menthol increased the activity of the inspiratory nasal muscles around the nostril in the anesthetized cat (Davies and Eccles, 1985), studies were performed on human subjects to investigate the effects of menthol vapor on both inspiratory and expiratory nasal resistance to air flow. Inhalation of a mixture of aromatic vapors, camphor, menthol, oil of pine needles, and methyl salicylate had no effect on inspiratory or expiratory nasal resistance to air flow (Eccles et al., 1987a).

The marked subjective effects of menthol on nasal sensation of air flow found in the above studies on normal subjects indicated that menthol was acting on sensory nerve endings in the nose. We proposed that menthol was sensitizing nasal sensory receptors which normally detected nasal air flow and thus creating the sensation of improved nasal air flow (Eccles et al., 1987b).

If menthol was exerting some nonspecific chemical action on nasal sensory receptors, then one would expect that all the isomers of menthol would have a similar sensitizing effect on nasal air flow. In order to test the specificity of the isomers we investigated the effects of (−)menthol, (+)neomenthol, and (+)isomenthol on nasal sensation and nasal resistance to air flow (Eccles et al., 1988a).

The menthol isomers were administered by inspiring from a wick inhaler containing the respective isomer at a concentration of 62.5 mg dissolved in 1 ml of liquid paraffin oil. Nasal resistance was measured by anterior rhinomanometry and nasal sensation of air flow was scored on a visual analog scale with the ends labelled "extremely clear" and "extremely blocked." The results of the experiments on 40 volunteer subjects are shown in Fig. 5. The act of sniffing air or a neutral substance such as vanilla created an increased awareness of nasal sensation of air flow and the subjective scores for menthol inhalation were therefore always compared with a control response (Eccles et al., 1987b).

The nasal sensation score after inhalation of (−)menthol was significantly greater than the control response whereas the responses for (+)neomenthol and (+)isomenthol were similar to the control. Inhalation of the aromatics or air control had no effect on inspiratory or expiratory nasal resistance.

These results indicate that (−)menthol has a specific pharmacological action on nasal sensory nerve endings as the closely related menthol isomers did not improve nasal sensation of air flow. This finding has been further strengthened by studying the effects of (−) and (+)menthol on nasal sensation. These isomers differ only in their optical rotation of light and are identical in their physical properties.

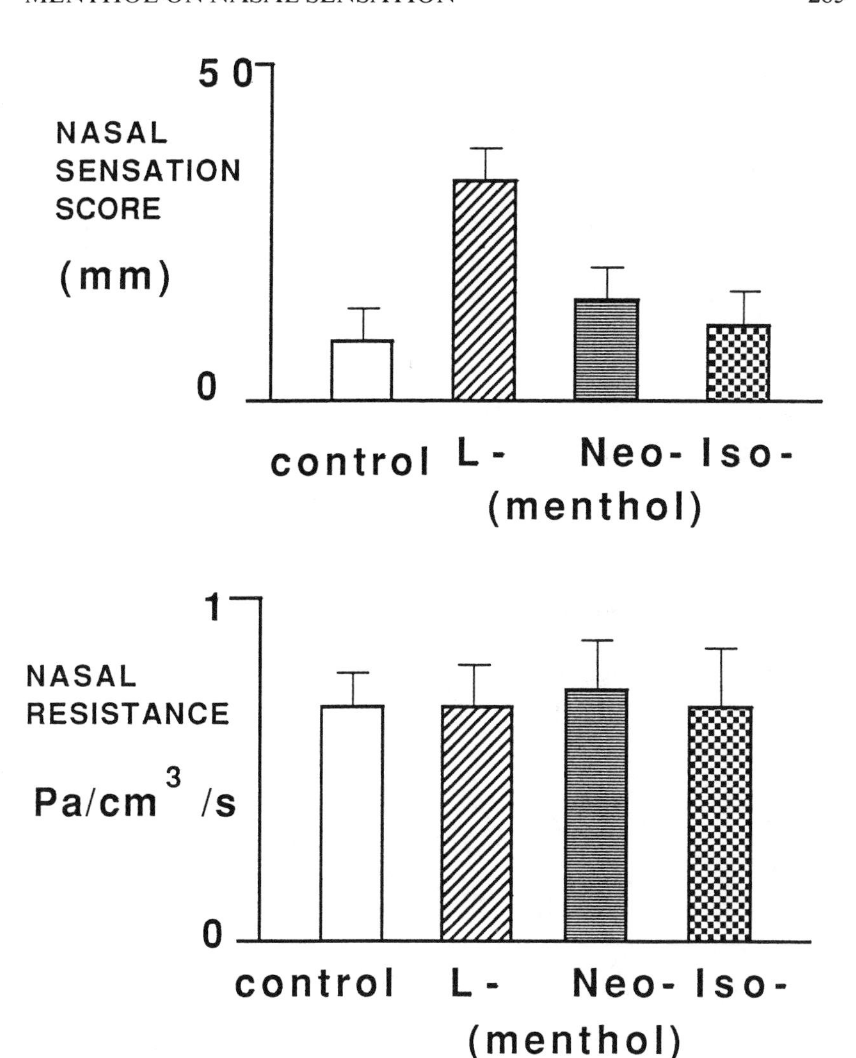


**FIGURE 5** Effects of inhalation of menthol isomers on nasal sensation and nasal resistance to air flow. The upper diagram shows the mean scores (± SEM, n = 40), registered on the visual analog scale. All the scores were toward the end of the scale marked "extremely clear" as regards nasal sensation of air flow and are measured in mm from the center of the scale. The lower diagram illustrates the mean inspiratory nasal resistance values (± SEM, n = 40). (From Eccles et al., 1988a.)

An investigation into the effects of (−) and (+)menthol on nasal sensation of air flow in 20 normal volunteers was carried out using wick inhalers containing 125 mg of the isomers dissolved in 1 ml of liquid paraffin oil (Eccles et al., 1988b). In this study vanilla was used as a neutral substance in a control inhaler and the results for nasal sensation are shown in Fig. 6.

Inhalation of (−)menthol caused a marked increase in the scores for nasal sensation of air flow with the subjects scoring close to the end of the visual analog scale marked "extremely clear." The scores after inhalation of (+)menthol were not significantly different from the control scores after inhalation of vanilla vapor.

These results clearly show that (−)menthol is the active isomer as regards actions on nasal sensory nerve endings. Menthol does not act by volatization and a direct cooling action; if this were the case, the results for (−) and (+)menthol would have been identical since the isomers have identical physical properties. The cool sensation of improved nasal air flow almost certainly results from a menthol-receptor interaction at or near sensory nerve endings. Nasal irritation and sneezing were never observed after inhalation of menthol and this is further evidence for a more specific action on sensory nerve endings in the nose.

The site of the nasal sensory receptors responsible for the sensation of nasal air flow is not known, although the above discussion indicates that the

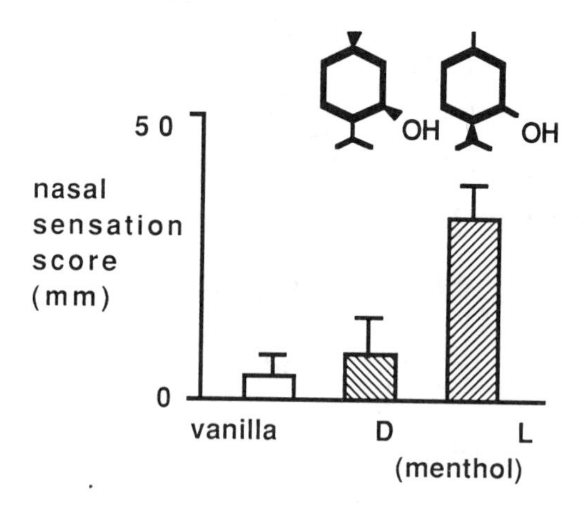


**FIGURE 6** Effects of inhalation of vanilla and the (−) and (+) isomers of menthol on nasal sensation of air flow. The histogram illustrates the mean scores in mm on a visual analog scale (± SEM, n = 20) where a score of 50 mm indicates maximum sensation of air flow. The figure also illustrates the molecular structure of the (+) and (−) isomers of menthol which are mirror images. (From Eccles et al., 1988b.)

skin-lined nasal vestibule may be an important area for further investigation. Studies by Jones et al. (1986) indicate that application of a local anesthetic spray to the nasal mucosa may paradoxically improve the sensation of nasal air flow. This finding is difficult to explain and it differs from our own studies in which application of a local anesthetic spray was shown to decrease the sensation of nasal air flow (Eccles et al., 1988c).

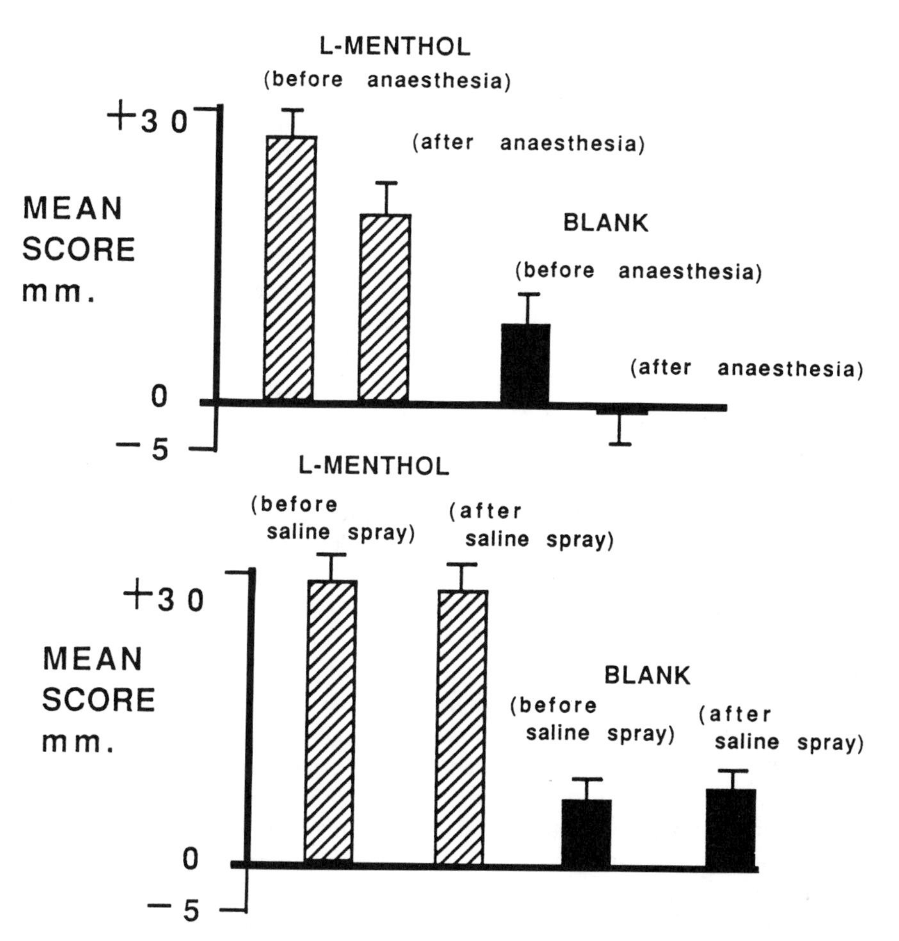


**FIGURE 7** Effects of nasal anesthesia on sensation of nasal air flow. The histograms illustrate the mean scores on the visual analog scale (± SEM). A score of +50 mm indicates maximum sensation of nasal air flow. The top figure shows the scores for inhalation of (−)menthol and a blank control before and after anesthesia of the nose with a lignocaine spray. The bottom figure shows the results after administration of a control saline spray. (From Eccles, Morris and Tolley, 1988c.)

In our own studies application of a local anesthetic spray to the nasal mucosa reduced the improvement in sensation of nasal air flow caused by inhalation of menthol vapor (Eccles et al., 1988c). The effects of the anesthetic spray and a control saline spray on the visual analog scores are shown in Fig. 7. The improved sensation of air flow due to inspiration through a blank wick inhaler was abolished after nasal anesthesia and the menthol response was significantly reduced but not nearly abolished.

These results indicate that sensory receptors in the transition region of the nasal vestibule may be responsible for some component of the sensation of nasal air flow as the local anesthetic spray would have anesthetized sensory receptors in the nasal mucosa but would not have affected sensory receptors in the skin lining the nasal vestibule.

## VI. MECHANISM OF ACTION OF MENTHOL

The sensitizing action of menthol on nasal sensory receptors responsible for the sensation of nasal air flow is probably due to a specific action of menthol on the calcium channels of nasal trigeminal nerve endings. Studies on cat lingual and nasal cold receptors at 25 °C have shown that menthol increases the activity from these receptors and that calcium application abolished this response (Schafer et al., 1986). The authors conclude that menthol may act by reducing the activation of calcium-stimulated outward current by an impeding effect on a calcium conductance, thereby inducing depolarization and a sensitization or stimulation of the sensory receptor. These results have been supported by studies on *Helix* neurons which also demonstrated a sensitizing effect of menthol on calcium-dependent inactivation (Swandulla et al., 1986).

Menthol also influences calcium conductance in gut smooth muscle and this may be the basis of its action in the symptomatic treatment of irritable bowel syndrome (Taylor et al., 1984, 1985).

The results of our own studies described above indicate that menthol exerts its effects by interaction with a pharmacological receptor which in some way influences calcium conductance. This receptor may be some component of the cell membrane controlling calcium conductance.

Further support for a drug receptor interaction is provided by Watson et al. (1978), who investigated the cooling sensation induced by various synthetic compounds related in structure to menthol and found that there was no association between minty smell and cooling. They also demonstrated that slight changes in the molecular structure of these compounds greatly influenced their cooling property and this would be expected if the cooling action was due to drug receptor type of interaction rather than some simple physical or chemical property.

The mechanism of action of menthol is at the very basic level of influenc ing calcium conductance. This basic mechanism of action can explain how menthol influences tissues as different as nasal sensory nerve endings and gut smooth muscle.

## VII. CONCLUSIONS

Menthol is widely used in the treatment of nasal congestion and our present state of knowledge indicates that menthol acts on sensory receptors in the nose supplied by the trigeminal nerve. These sensory receptors are most likely cold receptors but mechanoreceptors cannot be excluded from the sensation of nasal air flow. Menthol stimulates and sensitizes these receptors and creates the sensation of improved nasal air flow. In this respect menthol provides symptomatic relief for nasal congestion and at present there is only very limited evidence which indicates that menthol has any decongestant activity. However, animal and human studies have shown that menthol does influence the activity of upper airway accessory respiratory muscles around the nose and if this reflex action extends to other accessory respiratory muscles then menthol may provide relief by stabilizing the airway. This would be of particular advantage during sleep and could be important in sleeping infants.

Menthol acts on calcium channels in sensory nerve endings to influence calcium conductance and the studies on the effects of menthol isomers on nasal sensation indicate that this is a specific pharmacological interaction involving some kind of menthol receptor. Further studies on menthol and related chemicals with even more potent actions on calcium conductance in sensory nerve endings could provide key information which could be used in the development of new drugs for the treatment of inflammatory nasal disease.

## REFERENCES

Allison, D. J. (1974). Respiratory and cardiovascular reflexes arising from receptors in the nasal mucosa. M.D. thesis, University of London.

Angell James, E. (1969). Nasal reflexes. *Proc. Roy. Soc. Med.* 62: 1287-1293.

Bauer, K., and Garbe, D. (1985). *Common Fragrance and Flavor Materials: Preparation, Properties and Uses.* VCH, Weinheim.

Burrow, A., Eccles, R., and Jones, A. S. (1983). The effects of camphor, eucalyptus and menthol vapor on nasal resistance to airflow and nasal sensation. *Acta Otolaryngol* (Stockh.) 96, 157-161.

Cauna, N. (1982). Blood and nerve supply of the nasal lining. In: *The Nose: Upper Airway Physiology and the Atmospheric Environment.* D. F. Proctor and I. Andersen (Eds.). Elsevier, Amsterdam, pp. 45-69.

Davies, A. M., and Eccles, R. (1985). The effects of nasal airflow on the electromyographic activity of nasal muscles in the anaesthetized cat. *J. Physiol.* 358: 102P.

Davies, A. M., and Eccles, R. (1987). Electromyographic responses of a nasal muscle to stimulation of the nasal vestibule in the cat. *J. Physiol.* 391: 25-38.

Douglas, N. J., White, D. P., Weil, J. V., and Zwillich, C. W. (1983). Effect of breathing route on ventilation and ventilatory drive. *Resp. Physiol.* 51: 209-218.

Eccles, R. (1982). Neurological and pharmacological considerations. In: *The Nose: Upper Airway Physiology and the Atmospheric Environment.* D. F. Proctor and I. Andersen (Eds.). Elsevier, Amsterdam, pp. 191-214.

Eccles, R., Griffiths, C. G., Newton, C. G., and Tolley, N. S. (1988b). The effects of D and L isomers of menthol upon nasal sensation of airflow. *J. Laryngol. Otol.* 102: 506-508.

Eccles, R., Griffiths, D. H., Newton, C. G., and Tolley, N. S. (1988a). The effects of menthol isomers on nasal sensation of airflow. *Clin. Otolaryngol.* 13: 25-29.

Eccles, R., Lancashire, B., and Tolley, N. S. (1987a). The effect of aromatics on inspiratory and expiratory nasal resistance to airflow. *Clin. Otolaryngol.* 12: 11-14.

Eccles, R., Lancashire, B., and Tolley, N. S. (1987b). Experimental studies on nasal sensation of airflow. *Acta Otolaryngologica.* (Stockh.) 103: 303-306.

Eccles, R., Morris, S., and Tolley, N. S. (1988c). The effects of nasal anaesthesia upon nasal sensation of airflow. *Acta Otolaryngol.* (Stockh.) 106: 152-155.

Eccles, R., and Tolley, N. S. (1987). The effect of a nasal airflow stimulus upon human alae nasi e.m.g. activity. *J. Physiol.* 394: 78P.

Jones, A. S., Lancer, J. M., Shone, G., and Stevens, J. C. (1986). The effects of lignocaine on nasal resistance and sensation of airflow. *Acta Otolaryngol.* (Stockh.) 101: 328-330.

McBride, B., and Whitelaw, W. A. (1981). A physiological stimulus to upper airway receptors in humans. *J. Appl. Physiol.* 56: 500-505.

Mygind, N., Pedersen, M., and Nielsen, M. (1982). Morphology of the upper airway epithelium. In: *The Nose: Upper Airway Physiology and the Atmospheric Environment.* D. F. Proctor and I. Andersen (Eds.). Elsevier, Amsterdam, pp. 71-97.

Ramos, J. G. (1960). On the integration of respiratory movements III. The fifth nerve afferents. *Acta Physiologica Latino Americana* 10: 104-113.

Schafer, K., Braun, H. A., and Isenberg, C. (1986). Effect of menthol on cold receptor activity. Analysis of receptor activity. *J. Gen. Physiol.* 88: 757-776.

Swandulla, D., Schafer, K., and Lux, H. D. (1986). Calcium channel current inactivation is selectively modulated by menthol. *Neurosci. Lett.* 68: 23-28.

Taylor, B. A., Collins, P., Luscombe, D. K., and Duthie, H. L. (1984). The mechanism of the inhibitory action of menthol on gut smooth muscle. *Br. J. Surg.* 71(11): 902.

Taylor, B. A., Duthie, H. L., and Luscombe, D. K. (1985). Calcium antagonist activity of menthol on gastrointestinal smooth muscle. *Br. J. Clin. Pharmacol.* 20: 293-294.

Tomori, Z., and Widdicombe, J. G. (1969). Muscular, bronchomotor and cardiovascular reflexes elicited by mechanical stimulation of the respiratory tract. *J. Physiol.* 200: 25-49.

Watson, H. R., Hems, R., Rowsell, D. G., and Spring, D. J. (1978). New compounds with the menthol cooling effect. *J. Soc. Cosmet. Chem.* 29: 185-200.

White, D. P., Cadieux, R. J., Lombard, R. M., Bixler, E. O., Kales, A., and Zwillich, C. W. (1985). The effects of nasal anaesthesia on breathing during sleep. *Am. Rev. Resp. Dis.* 132: 972-975.

# Chapter 13 Discussion

**Dr. Finger:** I was wondering, does menthol have any effects if the person or animal is breathing humidified warm air, so that the incoming air is the same temperature as body temperature?

**Dr. Eccles:** If it's the same temperature as body temperature, I couldn't answer that. Certainly, in the cat experiments we looked at warming and cooling the air and it had no effect upon the reflex activity we were getting out. There are a whole series of experiments as to what determines sensation of nasal air flow. Certainly, there is a thermal effect, cooling; and if you warm the air to body temperature you would overcome that. There is probably also a mechanical effect, perhaps stimulating mechanoreceptors, and we may have some mixed activity, a mixed thermal-mechanoreceptor. We don't know what receptors are involved. But the experiments you are suggesting I have not performed.

**Dr. Finger:** Mere warming wouldn't necessarily do it, because you still have evaporative cooling. Did you try humidifying the air when you warmed it for the cat or was it simply . . .

**Dr. Eccles:** In some experiments, we tried humidifying it and that didn't have any effect, and we warmed it to 40 °C in the cat and there was no change. But I'm not saying that the receptors we are looking at in the cat are necessarily the same as those in the human. I think there could well be species differences. The animal experiment is providing us with a baseline, but we can't necessarily jump to the human being.

**Dr. Kobal:** At the last joint meeting of the British and German Physiological Societies, I saw a poster by Swandulla and his colleagues which indicated that menthol is specifically opening the calcium channels of cold receptors in birds. Do you think this is the fact that you studied?

**Dr. Eccles:** I've not gotten into the actual mechanisms, although I put it in the chapter because that is another area where one wants to look at how menthol is actually affecting the membranes. You said opening the calcium channels; if I remember correctly, it was actually inhibiting flux from the calcium channels, but it is in any case affecting calcium channels in nerve endings. It also affects calcium channels in gut smooth muscle. If you tend to think how menthol acts when you take peppermint oil for irritable bowel syndrome, well again, it appears that it is acting on calcium channels and smooth muscle. So it's having a very broad and basic action at the level of calcium conductance in smooth muscle and in sensory nerve endings. What I would really like to get at is, how is the menthol attaching to a calcium conductance channel? Is that the receptor? Why is it such a specific action for 1-menthol? It seems very odd that we've got this rather specific action of menthol. I can't see any biological basis for it at present.

**Dr. Silver:** That was a very nice presentation. From some of the preliminary results from my lab, we haven't seen any differences in ethmoid nerve responses to 1- and d-menthol in the rat. It could be species differences. There may be a hint of something at very high concentrations. What concentrations of menthol were you using in these experiments?

**Dr. Eccles:** In the cat we've gone from painting menthol onto the nasal vestibule at very high concentrations to very low concentrations for vapor, and in all of these instances we get potentiation. But one has to be careful because in very high concentrations, menthol can be a local anesthetic, and it can become very irritating, too. Dr. Rozin was talking about menthol as an irritant. I'm talking about a completely different action, probably at much lower levels, where we get a more specific effect of menthol. It is interesting that you were talking about the l and d activity because I know Barry Green looked at this in the mouth and didn't see such big differences. We checked our supplier because we had been trying to obtain pure d-menthol, but it is often contaminated with l, and the percent contamination really varies quite a lot. You have to check that you've got a pure sample, because a 5% contamination of a d with l can ruin your results, because the l is very potent. That's one factor; the other thing is it is sometimes difficult to relate nerve recording to subjective sensation and we may be looking at perhaps not one receptor but effects upon a group of receptors in different nerves all getting a central input which is eventually interpreted as the sensation of nasal air flow. That is much more difficult to interpret.

**Dr. Green:** I am intrigued by your results and also what appears to be a difference between our results. It is possible, as you suggest, that my d-menthol was not pure. Did you have subjects make judgments of coolness of l-and d-menthol in the nose to relate that to the sensation of air flow?

**Dr. Eccles:** No, we didn't score for coolness, we only scored on an analog scale, "extremely clear," "extremely blocked." In retrospect, we could have done that easily; it would only take a matter of a minute or so, perhaps 30 sec for each subject to do that. But I don't know how it would come out with l- and d- on coolness.

**Dr. Green:** I think Tom Finger's question is a very interesting one. It was going through my mind as well. Have you ever considered just having subjects rate the sensation of air flow with different temperature air streams?

**Dr. Eccles:** I think there is a whole series of experiments that should be done on temperature, flow, and humidity. One gets into a very interesting area of what is the optimal environment for us to be in for comfort.

**Dr. Cain:** There are a series of compounds produced by Wilkinson Sword; what about them? They don't have the odor of menthol but they have the specific cooling effect, and they would seem to help you isolate a cooling effect from other attributes of menthol.

**Dr. Eccles:** I have not done any experiments on those, but I have tried them myself. They are nonvolatile compounds and usually come as white powders. Although they don't have much of a smell, if you put your nose close, there is a very vague hint of a minty odor. If you sniff the powder rather like a snuff you get an odd feeling. First of all, you get an intense cooling effect in the anterior part of the nose and this tends to persist; you feel clear, and then it goes away and after a period of 10 min, you get a very pronounced cooling effect in the nasal pharynx. My interpretation is that something anterior is being affected by these compounds, and then it is transported across the mucosa, where there is no effect, and when it gets to the nasal flanks you get another effect and you feel cool. We've got the compounds but we've not gone into any major studies. They are very interesting compounds; they are more potent than menthol in some ways, and yet nonvolatile. If you take them orally it is interesting. You get a pronounced cooling effect in the mouth but nothing in the nose, which is different from a menthol lozenge. If you take a menthol lozenge you get the oral effect and the nasal effect, and to me that indicates that with a menthol lozenge the vapor gets up the back of the nose and gives you the nasal effect.

# 14
# Trigeminal Nerve Stimulation
## Practical Application for Industrial Workers and Consumers

**Yves Alarie**
University of Pittsburgh
Pittsburgh, Pennsylvania

## I. INTRODUCTION

Kratschmer (1870) was the first to report the variety of reflex reactions from stimulation of nasal trigeminal nerve endings by airborne chemicals. From his recordings of respiratory movements in rabbits during the application of irritants to the nasal mucosa, he described the effect as an expiratory tetanus (Alarie and Luo, 1986). With lower amounts of irritants it can be shown that the effect on respiration is quite characteristic. A bradypnea occurs due to a lengthening of the duration of expiration (Alarie, 1966). Thus concentration-response relationships can be obtained and the potency of airborne chemicals easily determined (alarie, 1966, 1973a,b, 1981; Alarie and Luo, 1986). A wide variety of airborne chemicals can stimulate nasal trigeminal nerve endings and are thus classified as "sensory irritants." Their effects in humans at concentrations below those needed to induce inhibition of respiration are described as "stinging or burning sensations" (Alarie, 1973a). Despite their wide variety of chemicals structures, sensory irritants seem to induce stimulation of trigeminal nerve endings by three basic mechanisms.

The first two mechanisms involve reactions with SH groups or with disulfide bonds in a receptor protein while the third one involves a purely physical interaction with the receptor protein (Alarie, 1973a,b; Neilsen and Alarie, 1982; Kristiansen et al., 1986; Kristiansen and Nielsen, 1988). While progress will certainly continue in elucidating the mechanisms of action of sensory irritants, the knowledge accumulated so far can be used in a variety of ways to protect workers as well as consumers from the toxic effects of these airborne chemicals. Some practical applications are given below.

## II. BIOASSAY

Using mice as experimental animals and body plethysmographs to record their respiration, it has been demonstrated that chemicals which are sensory irritants in humans will induce a characteristic bradypnea in those animals (Alarie, 1966, 1973a). Chemicals to be tested are vaporized or aerosolized and the respiratory frequency of the animals is recorded. The degree of respiratory depression is measured at different exposure concentrations of the chemicals to be tested. From a series of experiments at different exposure concentrations, a concentration-response relationship is developed. From least squares regression analysis of these data the concentration needed to cause 50% decrease in respiratory rate ($RD_{50}$) from normal is calculated. The $RD_{50}$ values are then taken to reflect the sensory irritating potency of the chemicals tested (Alarie, 1966).

## III. PRACTICAL APPLICATIONS

### A. Setting "Acceptable" Exposure Limits

Considering the large number of industrial chemicals with sensory irritating properties, an animal bioassay as given above can be of value if properly calibrated and validated to predict exposure levels below which industrial workers can be exposed with no complaint of sensory irritation. So far, using mice as experimental animals for the determination of the sensory irritating potency of 40 industrial chemicals, an excellent correlation has been found between their potency as determined in this bioassay and the threshold limit value (TLV) established as a "safe level of exposure" in industry (Alarie and Luo, 1986). Therefore for new chemicals being introduced or for old chemicals for which no TLV has yet been established, this bioassay can be used to rapidly determine the maximum exposure level to be permitted in industry. Once this level is established, further long-term toxicological studies can then be planned to confirm how appropriate it is.

### B. The Problem of Mixtures

Very often industrial workers are exposed to mixtures of airborne contaminants rather than to a single chemical. This problem has received very little attention (Holmberg and Lundberg, 1985; Nielsen and Bakbo, 1985).

### 1. Known mixtures

One way of investigating this aspect is to select appropriate models before testing known mixtures and to test for interactions. This was first done for formaldehyde and acrolein, two potent sensory irritants (Kane and Alarie, 1978). It was found that competitive agonism existed between acrolein and formaldehyde when present together. Thus from the potency of each chemical the level of reaction can be predicted for any mixture of the two. The same model can probably be used for mixtures of other potent sensory irritants which act by chemical interaction with the receptor protein, although only formaldehyde and acrolein mixtures have been tested so far.

For agents which act by a physical mechanism (Nielsen and Alarie, 1982) the same situation should exist. This was recently verified by Nielsen et al. (1988). For these agents their potency can also be predicted from physicochemical parameters (Nielsen and Alarie, 1982; Muller and Greff, 1984; Roberts, 1986). Therefore no further testing of competitive agonism is needed. As proposed by Nielsen et al. (1988), at the low concentrations permitted in industry, simple additivity can be used for these agents which includes a wide variety of common solvents.

### 2. Unknown or complex mixtures

Very often industrial workers are exposed to complex mixtures during an operation and little is known of the nature of the chemicals emitted. This often occurs during processing of polymeric materials at high temperature (Barrow et al., 1978). Also, complex mixtures are formed during photochemical oxidation in urban atmospheres (Kane and Alarie, 1978). In both cases such mixtures were evaluated for their sensory irritation potency. Knowing the potency of such mixtures can help in the determination of acceptable concentration to prevent sensory irritation in humans. Often consumer products will elicit complaints due to their irritating properties. Again, such mixtures have been evaluated in mice in attempts to reduce their offending properties (Nielsen and Bakbo, 1985).

### 3. Use of a standardized bioassay

In using this bioassay to predict acceptable levels of exposure in industry it should be pointed out that the calibration of this bioassay was done using male Swiss-Webster mice (ASTM, 1984; Alarie and Luo, 1986). Nielsen et al. (1984) and Kristiansen et al. (1986) showed that male Ssc:CF-1 mice have the same sensitivity as male Swiss-Webster mice. However, other strains of mice, as well as rats, have a different sensitivity (Alarie et al., 1980; Chang et al., 1981; Nikiforov and Ventrone et al., 1986) and therefore should not be used for this purpose without first comparing their sensitivity to male Swiss-Webster mice. It is also of practical importance to use mice with high sensitivity to avoid exposures at high concentrations, which can induce pulmonary irritation and/or systemic effects which interfere with recognition of the trigeminal reflex reaction induced by sensory irritants.

## IV. SPECIFICITY OF THE SENSORY IRRITANT RECEPTOR

As noted above, a wide variety of chemicals can stimulate the sensory irritant receptor and three different mechanisms may be involved (Alarie, 1973a; Nielsen and Alarie, 1982). For some chemicals, however, their interaction with the sensory irritant receptor must be viewed in the classical pharmacological content of agonist-receptor interaction rather than within the context of chemical reactivity or physical interaction. This is certainly true for capsaicin, which is a potent sensory irritant ($RD_{50}$ of 10.4 $\mu$g/liter) which acts quickly and at high concentration induces desensitization of the receptor (Alarie and Keller, 1973). Thus it acts at the sensory irritant receptor of the nasal mucosa in the same way as described by Jancso (1960) and Jansco et al. (1967) for other nerve endings. It is interesting to note that wide variations

**TABLE 1** Comparison of Potency (Measured as $RD_{50}$) of Sensory Irritants Related to Capsaicin

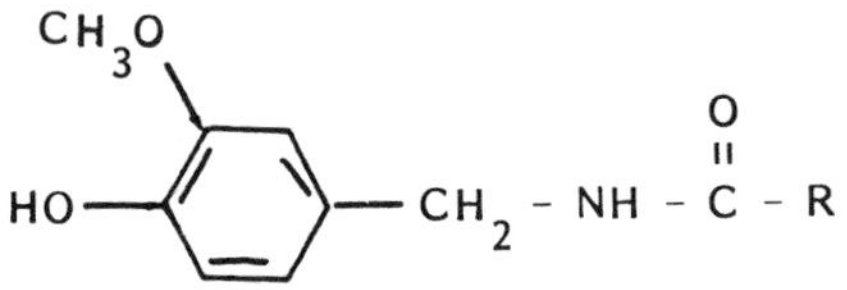

| R | $RD_{50}$ ($\mu$m/liter) | $RD_{50}$ ($\mu$g/liter) |
|---|---|---|
| $(CH_2)_8$ $CH=CH_2$ | 0.0212 | 6.8 |
| $(CH_2)_7$ $-CH=CH(CH_2)_7CH_3$ | 0.0244 | 10.2 |
| $(CH_2)_6$ $-S-CH(CH_3)_2$ | 0.0266 | 9.0 |
| $(CH_2)_7$ $-CH_3$ | 0.0279 | 8.2 |
| $(CH_2)_8$ $-CH=CHBr$ | 0.0280 | 11.2 |
| $(CH_2)_8$ $-CH$ | 0.0351 | 11.2 |
| $(CH_2)_7$ $-CH=CHCH_2CH=CH(CH_2)_4CH_3$ | 0.0655 | 26.2 |
| $(CH_2)_{10}$-Br | 0.0593 | 23.0 |
| $(CH_2)_{10}$-S-$CH(CH_3)_2$ | 0.0703 | 27.8 |
| $(CH_2)_5$ $-CH_3$ | 0.0856 | 22.8 |
| $(CH_2)_{10}$-$OCH_3$ | 0.104 | 36.0 |
| $OC_4H_9$ | 0.510 | 159.0 |
| $CF_3$ | 0.844 | 211.0 |

*Source*: Alarie, unpublished.

on the chain, as shown in Table 1, are associated with only a minor decrease in potency until short substitutions are made resulting in a large decrease in potency. The methoxy and hydroxyl groups on the benzene ring are essential for the activity since their removal results in inactive compounds.

Nicotine also stimulates the sensory irritant receptor ($RD_{50}$ of 35 μg/liter; Alarie and Wakisaka, unpublished) but unlike capsaicin desensitization follows quickly at all exposure concentrations.

Another chemical of interest is *cis*-4-cyclohexylmethylcyclohexylamine. The trans form is inactive. Again substitution of the cyclohexyl group by other groups, as given in Table 2, can change the potency. With substitution with an alkyl chain, the activity was lost abruptly at C-7.

Another series of chemicals of interest are diimines, which are known lachrymators in humans (Kliegman and Barnes, 1970). Their potency as sensory irritants in mice is given in Table 3. Thus the use of these chemicals may help in a better elucidation of the nature of the sensory irritant receptor of the trigeminal nerve endings in the nasal mucosa and cornea. With such

**TABLE 2** Comparison of Potency (Measured as $RD_{50}$) of Sensory Irritants Related to *cis*-4-Cyclohexylmethylcyclohexylamine

R
O
II
NH - C - $CH_3$

| R | $RD_{50}$ (μm/liter) | $RD_{50}$ (μg/liter) |
|---|---|---|
|  $CH_2$ | 0.188 | 44.7 |
| 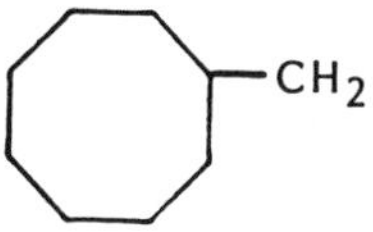 $CH_2$ | 0.0088 | 2.2 |
|  | 0.481 | 107 |
| n $C_4H_9$ | 0.276 | 58 |
| n $C_5H_{11}$ | 0.266 | 60 |
| n $C_6H_{13}$ | 0.033 | 7.8 |
| n $C_7H_{15}$ | Inactive | |
| n $C_8H_{17}$ | Inactive | |

*Source*: Alarie, unpublished.

**TABLE 3** Comparison of Potency (Measured as $RD_{50}$) of a Series of Diimines

| Diimines | $RD_{50}$ ($\mu$m/liter) | $RD_{50}$ ($\mu$g/liter) |
|---|---|---|
| $CH_3 - C(CH_3)(CH_3) - N = CH - CH = N - C(CH_3)(CH_3) - CH_3$ | 0.007 | 1.2 |
| N = CH − CH = N (cyclopentyl rings) | 0.014 | 2.5 |
| N = CH - CH = N (cyclopropyl rings) | 0.050 | 7.0 |
| $CH_3 - C(CH_3)(CH_3) - N = CH - CH = CH -$ (benzene ring) | 0.07 | 14 |
| N = CH − CH = N (cyclohexyl rings) | 0.12 | 26 |

*Source*: Alarie, unpublished.

chemicals, direct recordings of neural activity of the ethmoidal nerve (Silver et al., 1986) would permit a better differentiation of their characteristics since this approach is more sensitive than measuring the reflex induced change in respiratory frequency.

## REFERENCES

Alarie, Y. (1966). Irritating properties of airborne material to the upper respiratory tract. *Arch. Environ. Health* 13: 433-449.

Alarie, Y. (1973a). Sensory irritation by airborne chemicals. *CRC Crit. Rev. Toxicol.* 2: 299-366.

Alarie, Y. (1973b). Sensory irritation of the upper airways by airborne chemicals. *Toxicol. Appl. Pharmacol.* 24: 279-297.

Alarie, Y. (1981). Toxicological evaluation of airborne chemical irritants and allergens using respiratory reflex reactions. In *Proceedings of the Inhalation Toxicology and Technology Symposium*. BKJ Leong (Ed.). Ann Arbor Science, Ann Arbor, pp. 207-231.

Alarie, Y., and Keller, L. W. (1973). Sensory irritation by capsaicin. *Environ. Physiol. Biochem.* 3: 169-181.

Alarie, Y., and Luo, J. E. (1986). Sensory irritation by airborne chemicals: A basis to establish acceptable levels of exposure. In *Toxicology of the Nasal Passages.* C. Barrow (Ed.). Hemisphere, New York, pp. 91-100.

Alarie, Y., Kane, L., and Barrow, C. (1980). Sensory irritation: Use of an animal model to establish acceptable exposure to airborne chemical irritants. In: *Toxicology: Principles and Practice.* A. L. Reeves (Ed.). John Wiley and Sons, New York, pp. 48-92.

Alarie, Y., Ferguson, J. S., Stock, M. F., Weyel, D. A., and Schaper, M. (1987). Sensory and pulmonary irritation of methyl isocyanate in mice and pulmonary irritation and possible cyanidelike effects of methylisocyanante in guinea pigs. *Environ. Health Persp.* 72: 159-167.

ASTM (1984). Standard test method for estimating sensory irritancy of airborne chemicals. Designation E981. American Society for Testing and Materials, Vol. 11.04 pp. 681-696.

Chang, J. C. F., Steinhagen, W. H., and Barrow, C. S. (1981). Effect of single or repeated formaldehyde exposure on minute volume of B6C3F1 mice and F344 rats. *Toxicol. Appl. Pharmacol.* 61: 451-459.

Holmberg, B., and Lundberg, P. (1985). Exposure limits for mixtures. *Ann. Am. Conf. Ind. Hyg.* 12: 111-118.

Jancso, N. (1960). Role of the nerve terminals in the mechanism of inflammatory reactions. *Bull. Millard Fillmore Hosp.* 7: 53-77.

Jancso, N., Jancso-Gabor, A., and Szolcsanyi, J. (1967). Direct evidence for neurogenic inflammation and its prevention by denervation and by pretreatment with capsaicin. *B. J. Pharmacol.* 31: 138-151.

Kane, L., and Alarie (1978). Evaluation of sensory irritation from acrolein-formaldehyde mixtures. *Am. Ind. Hyg. Assoc. J.* 39: 270-274.

Kliegman, J. M., and Barnes, R. K. (1970). Glyoxal derivatives I. Conjugated aliphatic diimines from glyoxal and aliphatic primary amines. *Tetrahedron* 26: 2555-2560.

Kratschmer, F. (1870). Uber reflex von der nasenschleimhaut auf athmung und kreislauf. In *Sitzungsberichte der Kaiserlichen Akademie der Wissenschaften Mathematisch -Naturwissenschaftliche,* Vol 62, Wein, pp. 147-170.

Kristiansen, U., Hansen, L., Nielsen, G. D., and Holst, E. (1986). Sensory irritation and pulmonary irritation of cumene and n-propanol: Mechanisms of receptor activation and desensitization. *Acta Pharmacol. Toxicol.* 59: 60-72.

Kristiansen, U., and Nielsen, G. D. (1988). Activation of the sensory irritant receptor by C7-C11 n-alkanes. *Arch. Toxicol.* 61: 419-425.

Muller, J., and Greff, G. (1984). Recherche de relations entre toxicite de molecules d'interet industriel et properties physico-chimiques: test d'irritation des voies aeriennes superieures applique a quatre familles chimiques. *Food Chem. Toxicol.* 22: 661-664.

Nielsen, G. D., and Alarie, Y. (1982). Sensory irritation, pulmonary irritation, pulmoary irritation and respiratory stimulation by airborne benzene and alkylbenzene: Predictions of safe industrial exposure levels and correlation with their thermodynamic properties. *Toxicol. Appl. Pharmacol.* 65: 459-477.

Nielsen, G. D., and Bakbo, J. C. (1985). Exposure limits for irritants. *Ann. Am. Conf. Ind. Hyg.* 12: 119-133.

Nielsen, G. D., Kristiansen, H., Hansen, L., and Alarie, Y. (1988). Irritation of the upper airways from mixtures of cumene and n-propanol. Mechanisms and their consequences for setting industrial exposure limits. *Arch. Toxicol.* 62: 209-215.

Nielsen, G. D., Bakbo, J. C., and Holst, E. (1984). Sensory irritation and pulmonary irritation by airborne allyl acetate, allyl alcohol and allyl ether compared to acrolein. *Acta Pharmacol. Toxicol.* 54: 292-298.

Nikiforov, A. L., and Ventrone, R. (1986). Measurements of respiratory parameters in rats to evaluate the sensory irritation potential of aerosols. In: *Proceedings of Second International Aerosol Conference.* Pergamon Press, Oxford, pp. 323-326.

Roberts, D. W. (1986). QSAR for upper-respiratory tract irritation. *Chem. Biol. Interact.* 57: 325-345.

Silver, W. L., Mason, J. R., Adams, M. A., and Smeraski, C. A. (1986). Nasal trigeminal chemoreception: Responses to n-aliphatic alcohols. *Brain Res.* 376: 221-229.

# Chapter 14 Discussion

**Dr. Mason:** Is there any way to calculate the concentration of the material actually inspired by animals? You may know the concentration passing through the aquarium, but that concentration may not be the same as the amount breathed. Is there a way to tag compounds so that they could be identified in postmortem examinations of test animals?

**Dr. Alarie:** When you say the material inspired, we don't have any general rule, it depends on what kind of material. If they are solvents, you treat them as general anesthetic agents, which simply means you will reach an equilibrium with the concentration that is in the air. The time to reach that equilibrium is fully predictable based on their solubility coefficient. If you are talking of reactive chemicals—something like acrolein, for example—you can assume that 90-95% is retained. So if you assume that virtually all of it is retained, you are not going to make a very large error.

**Dr. Finger:** How does menthol fit into the picture that you just drew of these three main classes?

**Dr. Alarie:** I have not tested menthol but I know other people have. I have no idea of its potency, but one would predict that menthol would act by a physical mechanism. I don't see any possible chemical reactivity of menthol; maybe a small one but not a very large one.

**Dr. Finger:** You don't think it might be acting on an entirely different population of receptors?

**Dr. Alarie:** Not as far as its irritating properties are concerned. It may act on other populations of receptors as well as stimulating irritant receptors.

**Dr. Szolcsanyi:** What about capsaicin? You have done experiments with capsaicin, so how does it fit into your model?

**Dr. Alarie:** I would look at capsaicin just like I would look at nicotine; that is, it fits the receptor in a very specific way, not be a chemical reactivity and certainly not by a physical mechanism. We tested about 25 different congeners of capsaicin and we know quite well the structural requirement that is needed for its potency. You have done similar work on other sensory systems and your work and our work correlate very well.

**Dr. Cain:** With respect to the physical absorption process and your finding of a correlation between lipid solubility, vapor pressure, and so on, with irritation effects: can you account for differences from one series to another?

**Dr. Alarie:** Yes. You can account for differences from one series to another if you start correcting for such things as molecular volume, polarizability of the molecule, and other physical characteristics of these molecules. At this time there are two groups, one is in France and one is in England, who have done such correction. They find that the correlation coefficient gets better and better and better until you reach something like 0.95. So that is probably as high as we are going to go.

**Dr. Mozell:** Are you saying that the acceptability or nonacceptability of an irritant is independent of the first cranial nerve input?

**Dr. Alarie:** Well, we have irritants which have no olfactory qualities. The ones that were tested in the army volunteers were mostly of this kind. If you were to talk about something like ammonia, which has both odorant and irritant properties, or something like sulfur dioxide, which also has both, I am not sure. My own feeling from having exposed myself to many of these is that I don't like them because they sting rather than because they smell bad. It seems to me that there is a difference in how I accept ammonia and how I accept hydrogen sulfide.

**Dr. Mozell:** You don't accept either one.

**Dr. Alarie:** No, I don't accept either one, but I reject hydrogen sulfide because of its nauseating quality. Ammonia odors do not really bother me that much. I tend to wait until it gets high enough that it is beginning to sting to say now wait a minute, I don't want to be here, I don't want to work here.

**Dr. Mozell:** But there are a class of odors that I always wondered about, and maybe you can help me, that lead to a sort of suffocation, like honeysuckle on the vine. You take a whiff and you stop smelling it not because it stings but because you just can't bring any more of it into your lungs. That's pretty irritating too.

**Dr. Alarie:** I'm not familiar with that sensation or experience. You are the first to tell me that you would absolutely stop breathing because it smells bad. I've heard in the literature of people who possibly died of an irritating chemical such as sulfur dioxide or ammonia because they went into a laryngeal spasm. They couldn't breath; they literally suffocated. But I have never heard that that phenomenon could occur because of an odor.

**Dr. Mozell:** That's why I'm asking. We put a lot of emphasis of irritation on other than the first cranial nerve input, and I often wondered if we are overlooking the first cranial nerve input.

**Dr. Alarie:** I don't know. My personal feeling is it is the irritation that matters.

**Dr. Silver:** I was interested in your comments about mixtures. If I understood you correctly, you took two mixtures from the physical absorption class and put them together. Are they addictive?

**Dr. Alarie:** They are strictly additive and this works for those who are chemically reactive also. That is, if you take acrolein, for example, and mix it with another chemical of that same class, you can fully predict from the concentration of the two what the mixture will do. They act as competitive agonists in the classical sense of the pharmacologist.

**Dr. Silver:** And what happens if you take two from different groups and put them together?

**Dr. Alarie:** If you were to mix, say, ethanol vapor with acrolein, you have a chance that ethanol will react with acrolein because ethanol is a nucleophile. What you can do, and we have actually done this, is to place the animal in the chamber and bring in the acrolein so that you have a steady response with the acrolein, and then you add ethanol, and this can make the acrolein response disappear because ethanol is now reacting with acrolein. You can do the same with sulfur dioxide and acrolein because sulfur dioxide will form bisulfite when it hits the water and the bisulfite addition to the acrolein is a well-known reaction. It occurs very quickly, and you form an adduct and indeed you can play the same trick—you can bring the animal to a nice steady level with acrolein and then you add sulfur dioxide and you will see the response disappear. When you stop both of them, you will see the reaction reappear although there is nothing in the exposure chamber. You can play all kinds of tricks depending on the kind of compound you put in the mixture.

# 15

# Effectiveness of Six Potential Irritants on Consumption by Red-Winged Blackbirds *(Angelaius phoeniceus)* and Starlings *(Sturnus vulgaris)*

**J. Russell Mason**
U.S. Department of Agriculture
Animal and Plant Health Inspection Service
Science and Technology
and Monell Chemical Senses Center
Philadelphia, Pennsylvania

**David L. Otis**
U.S. Department of Agriculture
Science and Technology
Denver Wildlife Research Center
Denver, Colorado

The morphological organization of the peripheral trigeminal system in birds is not very different from that in mammals, but there appear to be broad functional (e.g., behavioral) discrepancies. The present experiments were designed to explore these behavioral discrepancies by testing avian responsiveness to a variety of potent mammalian irritants. In experiment 1 red-winged blackbirds (*Agelaius phoeniceus*) and starlings (*Sturnus vulgaris*) were assigned to 12 groups. Interspecific group pairs were presented with allyl isothiocyanate, ammonia, gingerol, mercaptobenzoic acid, piperine, or zingerone at five concentrations [0.0, 0.001, 0.01, 0.1, 1.0% (w/w)] in randomized, 2-hr, one-choice tests. The design of experiment 2 was similar to that of experiment 1, except that only allyl isothiocyanate, mercaptobenzoic acid, and piperine served as stimuli, and a broader concentration range of each stimulus [0.0-10.0% (w/w)] was used. In experiment 3, allyl isothiocyanate, mercaptobenzoic acid, and piperine again served as stimuli and concentrations again ranged from 0.0 to 10.0% (w/w), but different group pairs were tested with different concentrations of individual irritants in a double changeover design. The results of experiments 1 and 2 showed that allyl isothiocyanate, mercaptobenzoic acid, and piperine reduced consumption at high (≥1.0% w/w) concentrations. Ammonia, gingerol, and zingerone were ineffective. In contrast, the results of experiment 3 suggested that even the lowest concentration [0.001% (w/w)] of allyl isothiocyanate, mercaptobenzoic acid, and piperine were aversive. As such, while different experimental designs were consistent in the feeding responses elicited, apparent sensitivity was influenced by habituation and/or sensitization. We propose that changeover designs (experiment 3) are preferable to simple randomized stimulus presentations (experiments 1 and 2). We speculate that irritants may have practical uses as bird control compounds and as adjuvants to registered repellents in specialized settings.

## I. INTRODUCTION

In higher vertebrates, an important component of the common chemical sense (Parker, 1912) is the trigeminal system. This system consists of free nerve endings in the exposed surfaces of the eye, mouth, and nose. While the morphological organization of the peripheral trigeminal system in birds is not very different from that in mammals (Dubbledam and Veenman, 1978), there appear to be broad functional discrepancies (Kare and Mason, 1986; Mason et al. 1988). For example, the avian trigeminal system is responsive to odorants (e.g., Walker et al., 1979; Mason and Silver, 1983) but apparently does not mediate avoidance of strong mammalian irritants. Thus, pigeons (*Columba livia*) and gray partridges (*Perdix perdix*) are indifferent to ammonia (Soudek, 1929), and parrots (*Amazona* spp., Mason and Reidinger, 1983), pigeons (Szolcsanyi et al., 1986), and red-winged blackbirds (Mason and Maruniak, 1983) are insensitive to capsaicin, the pungent principle in *Capsicum* peppers.

The present experiments were designed to assess the responses of two passerine species (red-winged blackbirds [*Agelaius phoeniceus*], European starlings [*Sturnus vulgaris*]) to several potential irritants. In experiments 1 and 2, interspecific group pairs were presented with different chemicals (one chemical per pair) at a variety of concentrations. In experiment 3, interspecific group pairs were presented with all chemicals, but at only one of six concentrations. Two different paradigms were employed because pilot experiments had suggested that the responses of passerines to chemical irritants might be context-specific (i.e., "sensitivity" might depend on the method of stimulus presentation).

The irritants used were allyl isothiocyanate, ammonia, gingerol, mercaptobenzoic acid, piperine, and zingerone. Allyl isothiocyanate is the pungent principle in black mustard (*Brassica niger*), and recent evidence indicates that it is aversive to pigeons (Szolcsanyi et al., 1986). Ammonia is a prototypical trigeminal stimulus and reports suggest that passerines in feed lots avoid livestock diets containing high levels of nonprotein nitrogen (e.g., ammonia). Mercaptobenzoic acid was tested because of its commercial use in South Africa as an avian repellent (J. Thorpe, DuPont Co., pers. commun.). Gingerol and zingerone are the pungent components of ginger (*Zingiber officinale*) and piperine is the pungent component of black pepper (*Piper nigrum*).

## II. MATERIALS AND METHODS

### A. Experiment 1

Thirty-six experimentally naive male red-winged blackbirds (*Agelaius phoeniceus*) and 36 starlings (*Sturnus vulgaris*) served. All birds were individually caged (dimensions: 61 × 36 × 41 cm) under a 6:18 light-dark cycle (Mason and Reidinger, 1982). Water was always available, and before adaptation to experimental conditions the birds were permitted free access to Purina Flight Bird Conditioner (PFBC) and crushed shell grit.

Allyl isothiocyanate (Flavor Innovations, South Plainfield, NJ), piperine (ICI Biomedicals, Inc.), gingerol (PPF International, East Hanover, NJ), ammonia (Sigma), mercaptobenzoic acid (Sigma), and zingerone (Pfaltz and Bauer) were dissolved in 10 ml of ethanol, and then these solutions were mixed with PFBC to produce concentrations of 0.001, 0.01, 0.1, and 1.0% (w/w).

Five days before the experiment, all birds were adapted to a food deprivation regime. Deprivation involved removing all food and grit from the cages at dark onset. At light onset of the following day, 20 g of PFBC was placed in each cage. After 2 hr, consumption was assessed, and the birds were permitted free access to food and water until lights out.

On day 6, birds were assigned to 12 groups (six groups per species) on the basis of consumption. Briefly, the bird with the highest consumption was assigned to group 1, that with the second highest to group 2, and so on. This assured that the groups were balanced with respect to mean consumption.

The 12 groups were randomly assigned to six interspecific group pairs. Each pair was presented with five concentrations [0.0, 0.001, 0.01, 0.1, 1.0% (w/w)] of a different randomly selected irritant in a series of 2 hr no-choice tests. Over the next 15 days, presentations of each concentration were randomized such that each was presented three times. At the end of each 2 hr test, consumption was assessed, and birds were offered plain PFBC until lights out when food was removed.

After all tests had been given, mean consumption by each bird in each group for each irritant concentration was calculated. These means were assessed in a three-factor analysis of variance (ANOVA; irritants, species, concentrations), with repeated measures over concentrations. Tukey Honestly Significant Difference (HSD) tests were used to isolate significant differences among means.

### B. Experiment 2

Eighteen experimentally naive male red-winged blackbirds and 18 starlings served. These birds were caged, maintained, and adapted to a food deprivation regime as previously described. Each species was assigned to three groups on the basis of consumption, and the groups were randomly assigned to three interspecific group pairs.

On the basis of experiment 1 results, only allyl isothiocyanate, mercaptobenzoic acid, and piperine served as stimuli. Experiment 2 essentially served as a replication of experiment 1, except that a broader range of stimulus concentrations was used. Different group pairs were randomly assigned to a single irritant, presented at six concentrations [0.0, 0.001, 0.01, 0.1, 1.0, 10.0% (w/w)] in a randomized series of 2 hr, no-choice tests. Mean consumption by each bird in each group for each irritant concentration was calculated and these means were assessed in a three-factor ANOVA (irritant, species, concentration), with repeated measures over concentration. Tukey HSD tests were used to isolate significant differences among means.

### C. Experiment 3

Thirty-six experimentally naive male red-wings and 36 starlings were assigned to 12 groups ($n = 3$/group for each species) on the basis of consumption, as previously described. The groups were assigned to six interspecific pairs, and the pairs were randomly assigned to one of six concentration levels [0.0, 0.001, 0.01, 0.1, 1.0, 10.0% (w/w)]. At each of these levels, group pairs were given 2 hr, no-choice tests among allyl isothiocyanate, mercaptobenzoic acid, and piperine according to a double-changeover design (Heisterberg and Otis, 1983; Table 1).

Unlike experiments 1 and 2 in which stimulus presentations were randomized, the double-changeover design balanced presentations so that all irritants preceded and followed one another an equal number of times. As such, the design permitted estimation of residual and direct treatment effects, although the effective number of replicates for the former was less than the number for the latter (Federer, 1955).

An ANOVA procedure tailored for a three-treatment, double-changeover design (Federer, 1955) was used to assess the results. Tukey HSD post-hoc tests were used to isolate significant differences among means.

**TABLE 1** Double-Changeover Design for Stimulus Presentations

| | Square 1[b] bird | | | Square 2 bird | | |
|---|---|---|---|---|---|---|
| Presentation[a] | 1 | 2 | 3 | 1 | 2 | 3 |
| 1 | A[c] | B | C | A | B | C |
| 2 | B | C | A | C | A | B |
| 3 | C | A | B | B | C | A |

[a]Repeated for each concentration level.
[b]The two groups of three birds each are randomly assigned to the two Latin squares.
[c]A = piperine; B = mercaptobenzoic acid; C = allyl isothiocyanate.

## III. RESULTS

### A. Experiment 1

There were significant differences among irritants ($F(5,60) = 16.95$, $p < 0.00001$), between species ($F(1,60) = 813.82$, $p < 0.00001$), and among concentrations ($F(4,240) = 73.98$, $p < 0.00001$). Also, there were interactions between irritants and species ($F(5,60) = 8.29$, $p < 0.00001$), irritants and concentrations ($F(20,240) = 73.98$, $p < 0.00001$), and species and concentrations ($F(4,240) = 20.34$, $p < 0.00001$). Finally, the three-way interaction among irritants, species, and concentrations was significant ($F(20,240) = 9.20$, $p < 0.000001$).

Tukey tests ($p < 0.01$) indicated that overall, (a) allyl isothiocyanate and piperine reduced consumption relative to the other irritants, (b) red-wings exhibited lower consumption than starlings, and (c) consumption was lowest at the highest irritant concentration.

Post-hoc examination ($p < 0.01$) of the irritant by species and species-by-concentrations interactions suggested that allyl isothiocyanate, mercaptobenzoic acid, and piperine produced greater drops in consumption by starlings than by red-wings. In addition, while both 0.1 and 1.0% concentrations of irritant reduced consumption by starlings, red-wings were affected only by the latter.

Post-hoc examination ($p < 0.01$) of the irritant by concentration interaction indicated that only the highest concentration of allyl isothiocyanate and mercaptobenzoic acid reduced consumption. Piperine appeared to be more effective and significantly reduced consumption at both 0.1 and 1.0%.

Finally, post-hoc examination ($p < 0.01$) of the interaction among irritants, species, and concentrations reinforced inferences drawn on the basis of the main effects and lower order interactions (Fig. 1). While red-wings and starlings exhibited the same pattern of consumption across concentrations,

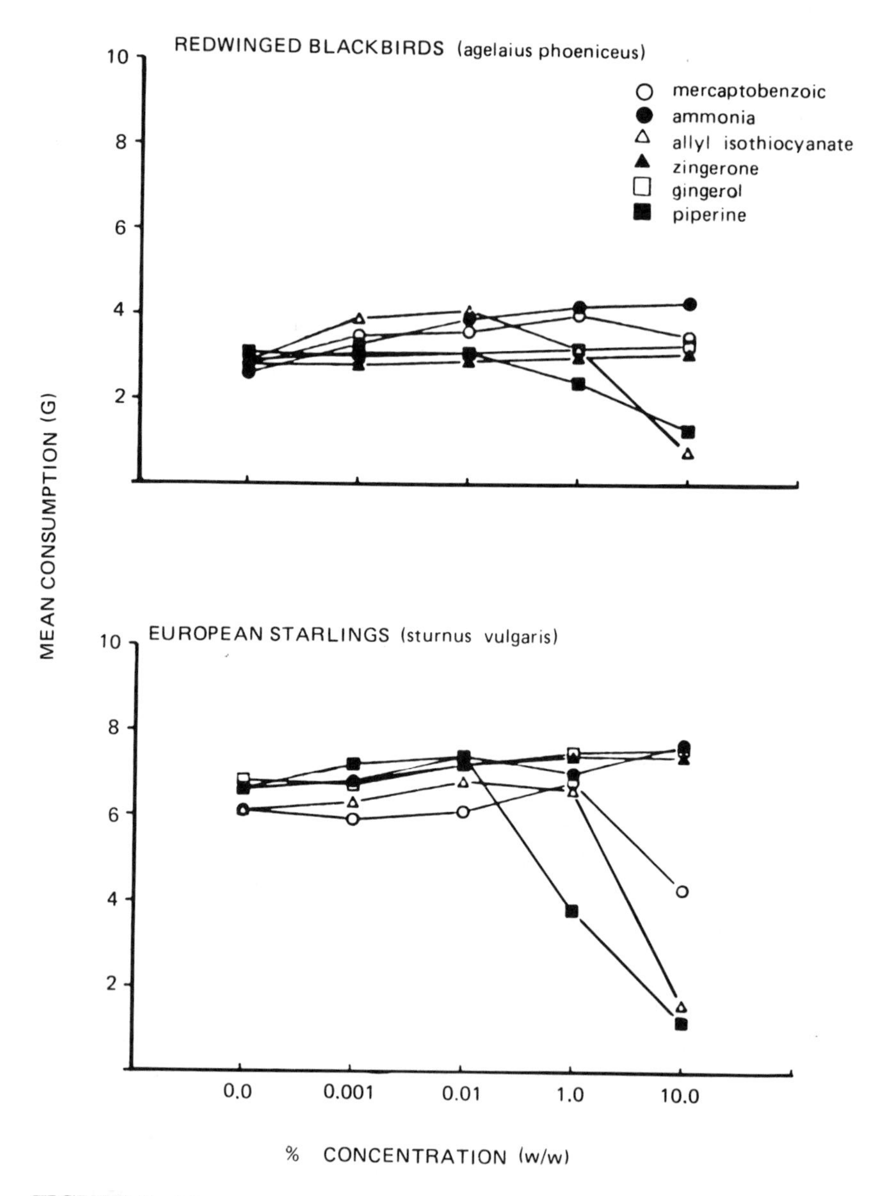


**FIGURE 1** Mean consumption of irritant adulterated PFBC in experiment 1 by red-winged blackbirds (top panel) and starlings (bottom panel).

starlings showed higher consumption than red-wings. Among irritants, only allyl isothiocyanate and piperine caused significant reductions in consumption for both species (Fig. 1), and this effect was observed only at concentrations equal to or exceeding 1.0%. Mercaptobenzoic acid significantly reduced consumption by starlings at a concentration of 1%, but feeding by red-wings was not similarly affected. Ammonia, gingerol, and zingerone had no effects on consumption by either species.

## B. Experiment 2

There were significant differences among irritants ($F(2,30) = 6.11, p < 0.006$), between species ($F(1,30) = 179.14, p < 0.00001$), and among concentrations ($F(5,150) = 351.79, p < 0.00001$). In addition, there were significant interactions between irritants and concentrations ($F(10,150) = 28.23, p < 0.00001$) and species and concentrations ($F(5,150) = 71.37, p < 0.00001$). Finally, the three-way interaction among irritants, species, and concentrations was significant ($F(10,150) = 7.00, p < 0.00001$).

Tukey tests ($p < 0.01$) revealed that overall (a) allyl isothiocyanate and piperine were more aversive then mercaptobenzoic acid, (b) starlings showed higher consumption than red-wings, and (c) consumption was lowest at the highest [1.0 and 10.0% (w/w)] irritant concentrations.

Examination of the irritant-by-concentration interaction revealed that the lowest consumption of all irritants occurred at the 10% concentration. While there was no difference between consumption at 10.0 and 1.0% for allyl isothiocyanate or piperine, consumption of mercaptobenzoic acid increased significantly from the former to the latter. Consumption of all irritants increased when the concentration was reduced to 0.1%, although piperine consumption was significantly lower than consumption of allyl isothiocyanate or mercaptobenzoic acid. There were no differences in consumption at concentrations lower than 0.1%.

Examination of the species by concentration interaction suggested that as in experiment 1, starlings were more sensitive than red-wings to food adulterated with the higher irritant concentrations. For the three highest concentrations (0.1-10.0%), consumption by starlings was less than that by red-wings.

Examination of the three-way interaction among irritants, species, and concentrations reinforced inferences drawn on the basis of the main effects and lower order interactions. All three irritants were repellant at 10% (w/w), but only mercaptobenzoic acid was more aversive at 10% than at 1.0% (Fig. 2). Between species, red-wings showed lower consumption than starlings at concentrations between 0.0 and 0.1% (w/w). At 1.0-10.0%, there were no differences between species.

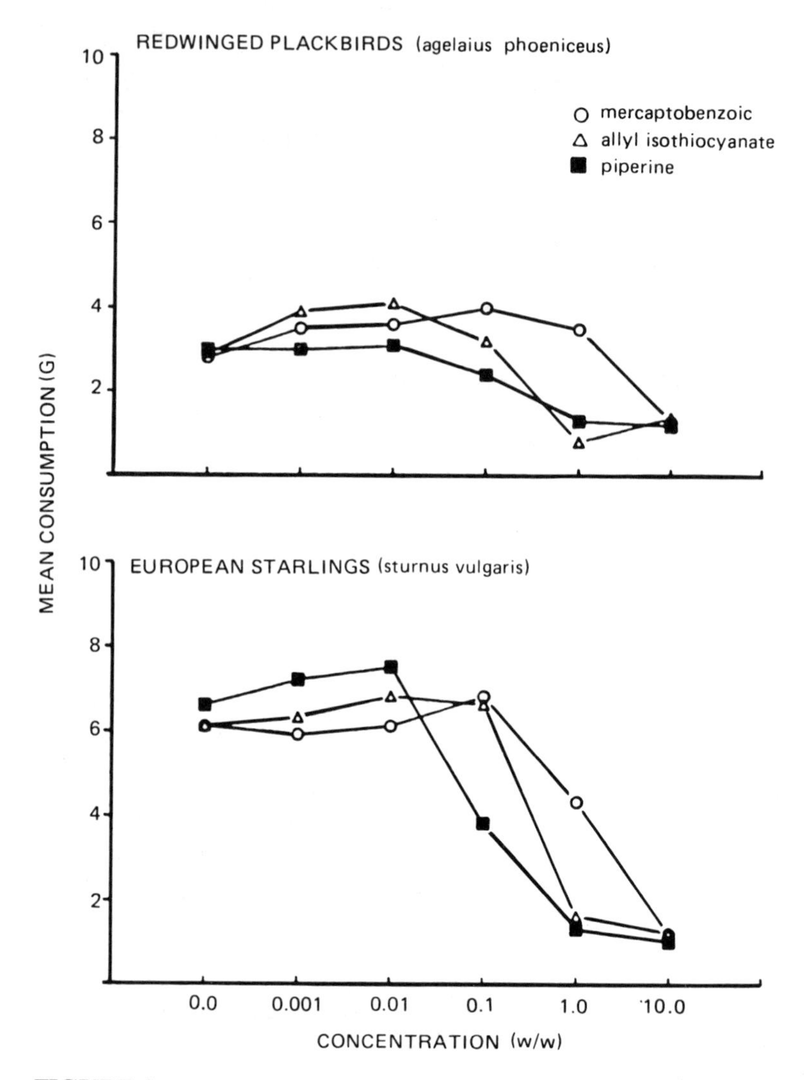


**FIGURE 2** Mean consumption of irritant adulterated PFBC in experiment 2 by red-winged blackbirds (top panel) and starlings (bottom panel).

## C. Experiment 3

The results of ANOVAs for each of the species/concentration combinations are summarized in Table 2. In general, factors associated with groups of birds, individuals within groups, and time periods were not significant, indicating that individual birds tended to be consistent in their response to treatment regimes as well as unaffected by repeated testing over a 3-day period.

**TABLE 2** Summary of ANOVA Results in Experiment 3

Red-winged blackbirds

| | df | F value concentrations 0 | $10^{-3}$ | $10^{-2}$ | $10^{-1}$ | 1 | 10 |
|---|---|---|---|---|---|---|---|
| Group | 1,4 | 0.001 | 1.4 | 0.015 | 0.5 | 0.1 | 0.5 |
| Period | 2,4 | **7.3** | 0.3 | 0.6 | 0.3 | 2.5 | 0.9 |
| Bird (group) | 4,4 | **9.9** | 1.0 | 1.8 | 0.9 | 0.2 | 0.5 |
| Period × group | 2,4 | 2.9 | 1.9 | 2.8 | 0.2 | 2.0 | 0.5 |
| Direct (ig. residual) | 2,4 | 0.6 | **43.0** | **27.0** | **18.0** | **12.8** | 0.6 |
| Residual (elim. direct) | 2,4 | 2.1 | 2.9 | 0.7 | 0.6 | 0.02 | 0.4 |
| Direct (elim. residual) | 2,4 | 1.0 | **37.2** | **19.5** | **14.2** | **10.4** | 0.2 |
| Residual (ig. direct) | 2,4 | 1.7 | **8.6** | **8.1** | 4.4 | 2.4 | 0.8 |

Starlings

| | df | F value concentrations 0 | $10^{-3}$ | $10^{-2}$ | $10^{-1}$ | 1 | 10 |
|---|---|---|---|---|---|---|---|
| Group | 1,4 | **30.1** | **7.5** | 1.6 | 0.2 | 1.6 | 6.8 |
| Period | 2,4 | 2.7 | 3.3 | 1.7 | 1.2 | 0.3 | 0.5 |
| Bird (group) | 4,4 | **30.1** | 1.1 | 0.9 | 0.9 | 0.2 | 2.0 |
| Period × group | 2,4 | 0.3 | 0.3 | 0.03 | 1.1 | 0.05 | 1.8 |
| Direct (ig. residual) | 2,4 | 1.3 | **181.8** | **22.1** | **38.9** | 2.2 | 0.8 |
| Residual (elim. direct) | 2,4 | 0.5 | **7.6** | 0.8 | **11.3** | 0.3 | 5.3 |
| Direct (elim. residual) | 2,4 | 0.8 | **175.1** | **18.5** | **36.6** | 2.4 | 0.5 |
| Residual (ig. direct) | 2,4 | 1.0 | **14.3** | 4.4 | **13.7** | 0.1 | 5.6 |

**Significant F values in bold type:**

| df | 0.05 | 0.01 |
|---|---|---|
| 1,4 | 7.7 | 21.2 |
| 2,4 | 6.94 | 18.0 |
| 4,4 | 6.39 | 16.0 |

Because residual effects are somewhat confounded with individuals in this design, overall treatment effects can be broken out in two ways: (a) direct ignoring residual effects and residual eliminating direct effects, or (b) direct eliminating residual effects and residual ignoring direct effects. For practical purposes, emphasis is placed on the two terms that eliminate (adjust) direct effects for the residual effects, and vice versa. For both species, direct treatment effects, adjusted for residual, were revealed at the lowest concentration (0.001%) and continued through the 0.1% level. At the two highest concentrations, direct treatment effects dissipated, with the exception of red-wings at the 1.0% level. Examination of Figs. 3 and 4 show that these differences were due to the greater effectiveness of allyl isothiocyanate in reducing consumption. There is no evidence that mercaptobenzoic acid and piperine differed in their repellant activity. At higher concentrations, all three compounds were equally effective in reducing consumption.

Results from all 12 groups were combined and analyzed by computing a three-factor ANOVA (species, concentrations, treatments) on the adjusted treatment means from the individual groups. An overall experimental error mean square was calculated by averaging mean squares over experiments. The analysis produced strong evidence for overall differences among treatments ($F(2,48) = 23.06$, $p < 0.001$) and concentrations ($F(5,48) = 89.54$, $p < 0.001$). Tukey HSD tests ($p < 0.01$) indicated that piperine and mercaptobenzoic acid were not different, but allyl isothiocyanate was significantly better than both. Consumption at all concentration levels was significantly reduced from control levels. Further, any levels more than one order of magnitude apart were significantly different. There was a significant species by

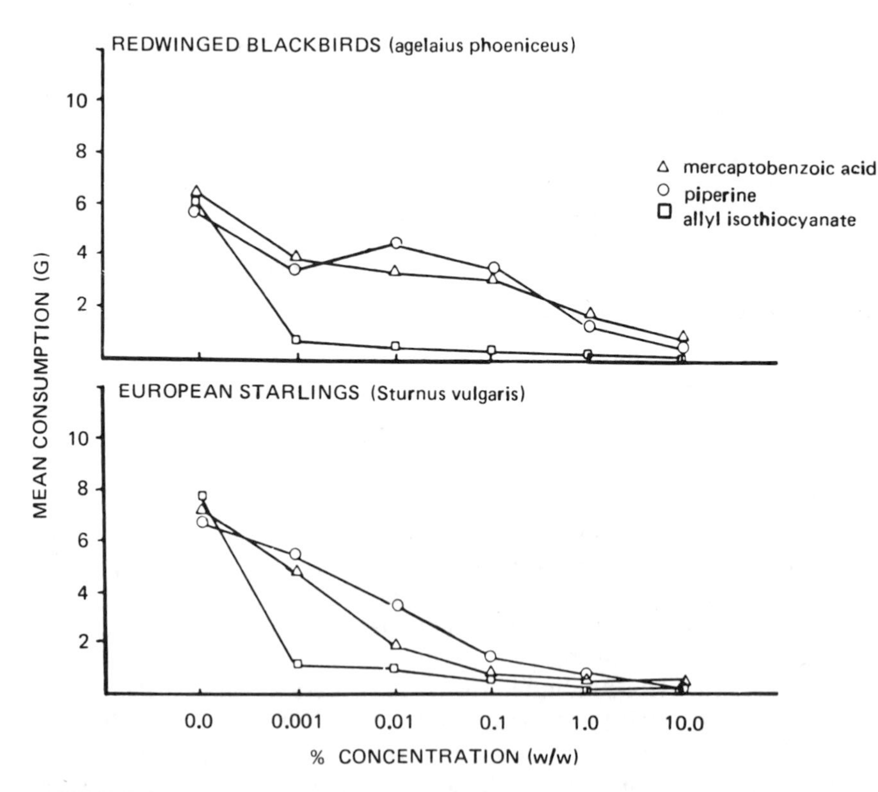


**FIGURE 3** Mean consumption of irritant adulterated PFBC in experiment 3 by red-winged blackbirds (top panel) and starlings (bottom panel).

concentration interaction ($F(5,48) = 4.11$, $p < 0.005$), due to the relative flatness of the red-wing dose-response curve for medium concentration levels of piperine and mercaptobenzoic acid (Fig. 3). The same phenomenon was responsible for evidence of a concentration-by-treatment interaction ($F(10,48) = 4.24$, $p < 0.001$).

The general lack of significant residual effects exhibited in the individual experiments is indicated in Table 2. Evidence for residual effects appeared for only two of the starling groups and for none of the red-wing groups. The average absolute adjustment in treatment means for residual effects was only 0.087 g/period (s.d. 0.068).

## IV. DISCUSSION

The primary goal of the present experiments was to assess the repellancy of six potential irritants. In addition, we aimed to evaluate the sensitivity of two experimental techniques for investigating repellancy. Allyl isothiocyanate, piperine, and mercaptobenzoic acid were offensive to both red-wings and starlings: ammonia, gingerol, and zingerone were apparently innocuous. For the latter substances, the lack of repellant effects lends support to the notion that there are functional differences between the mammalian and avian trigeminal systems. The factors which account for this discrepancy remain unclear, but avian trigeminal receptors may lack the appropriate structures to react with at least some irritants (Szolcsanyi et al., 1986).

For allyl isothiocyanate, piperine, and mercaptobenzoic acid, observed repellancy was dependent on the manner of presentation. When birds experienced all compounds at only one concentration (experiment 3), even the lowest level of each chemical (0.001%) was aversive. Conversely, when birds were offered all concentrations of only one compound (experiments 1 and 2), only the highest concentrations ($\geqslant 1.0\%$) were effective, and starlings appeared more sensitive than red-wings. These discrepancies are not readily explained in biological or behavioral terms. However, they do highlight differences in the efficiency of the changeover design (experiment 3), relative to standard repeated concentrations designs (experiments 1 and 2). In experiment 3, the error mean square for testing among species, concentrations, treatments, and higher order interactions was 0.32, and the standard error of a treatment mean was 0.23. In experiment 2, the error mean squares for testing treatment and concentration means were 1.29 and 0.29, respectively. Assuming 12 birds per group as in experiment 3 (for the purpose of comparison), then the standard errors of treatment and concentration means in experiment 2 were 0.33 and 0.15, respectively. Given these standard errors, then experiment 2 possessed only 70% (0.23/0.33) of the sensitivity of experiment 3 for detecting differences among treatment means. Conversely, the repeated concentrations design was about 50% more sensitive (0.23/0.15) than the changeover design in detecting differences among concentration means. Ignoring the possibility that carryover effects may have confounded the results of experiment 2, then a reasonable conclusion is that the changeover design sacrificed some power in detecting concentration differences for increased power in detecting differences among treatments (irritants).

Residual effects apparently did not play a significant role in experiment 3. Thus alternative changeover designs that do not assume the presence of such effects should be considered (cf. Federer, 1955, p. 441) in experiments aimed at isolating differences among irritants. Such designs are probably more powerful than the changeover design used in experiment 3 because more degrees of freedom are available for estimating experimental error.

## V. MANAGEMENT IMPLICATIONS

Allyl isothiocyanate, piperine, and mercaptobenzoic acid were offensive to birds in the present experiments. Of these, allyl isothiocyanate was the most effective. We speculate that any of these materials might serve as bird repellants in particular situations. For example, allyl isothiocyanate, perhaps incorporated into a paste, might be used as a fumigant to disperse birds from enclosed structures. In addition, starling and blackbird depredations at swine and cattle feed lots (Feare, 1975, 1980) can be reduced through the use of feeds that are unpalatable to birds but acceptable to livestock (Glahn and Mason, unpubl. data). Piperine is already added to swine feeds to improve "mouth feel" at the rate of 0.05-0.15% (w/w) (Cerny and Bazucha, 1983). The present experiments suggest that piperine may also have value as a bird repellant feed additive. In either case, the fact that these substances possess fungicidal and insecticidal properties (Matsurbara and Tanimura, 1966; Su, 1977; Madhyastha and Bhat, 1984; Das et al., 1985; Das and Choudhury, 1984; Nakatani et al., 1986) suggests that they may also act as preservatives of the baits into which they are incorporated. Finally, in experiment 1, gingerol was not offensive at any concentration. While this may (in part) reflect the insensitivity of the paradigm used in experiment 1, there is no evidence that low concentrations of gingerol are detected by birds. For mammals, gingerol (and piperine) enhances the uptake of substances from the gut (Atal et al., 1985). Conceivably, gingerol could produce the same effect in the avian gut without reducing overall food intake. As such, gingerol might be used to enhance the effectiveness of poultry feed additives and medications, and for pest birds, to improve the potency of ingested toxicants.

## REFERENCES

Atal, C. K., Dubey, R. K., and Singh, J. )1985). Biochemical basis of enhanced drug bioavailability by piperine: Evidence that piperine is a potent inhibitor of drug metabolism. *J. Pharmacol. Exp. Ther.* 232: 258-262.

Atal, C. K., Zutshi, U., and Rao, P. G. (1981). Scientific evidence on the role of ayurvedic herbals on bioavailability of drugs. *J. Ethnopharmacol.* 4: 229-232.

Cerny, A., and Bazucha, F. (1983). Additive for improving the mouth feel of pig feeds. Czech Patent CS-201823B.

Das, B. P., and Choudhury, B. (1984). Search for potential larvicides from carbizole. Trans. Bose. Res. Inst. (Calcutta) 47: 91-95.

Das, B. P., Choudhury, B., and Jamal, M. Y. (1985). Toxicity of some N-substituted carbizoles along with compounds containing methylenedroxyphenyl ring on mosquito larvae. *Orient. J. Chem.* 1: 99-102.

Dubbledam, J. L., and Veenman, C. L. (1978). Studies on the somatotopy of the trigeminal system in the mallard, *Anas platyrhyncous* L.: The Ganglion Trigeminale. *Netherlands J. Zool.* 28: 150-160.

Feare, C. J. (1975). Cost of starling damage to an intensive husbandry unit. *Proc. Br. Insecticide Fungicide Conf.* 8: 253-259.

Feare, C. J. (1980). The economics of starling damage. *Econ. Damage* 2: 39-54.

Federer, W. T. (1955). *Experimental Design.* Macmillan, New York, pp. 444-452.

Heisterberg, J. F., and Otis, D. L. (1984). A changeover test design to compare the relative efficacies of bird repellent seed corn treatments. In *Vertebrate Pest Control and Management Materials.* D. E. Kaukienen (Ed.). American Society for Testing and Materials, Philadelphia, pp. 98-109.

Kare, M. R., and Mason, J. R. (1986). The chemical senses in birds. In *Avian Physiology.* P. D. Sturkie (Ed.). Springer-Verlag, New York, pp. 59-73.

Madhyastha, M. S., and Bhat, R. V. (1984). *Aspergillus parasiticus* growth and aflatoxin production on black and white pepper and the inhibitory action of their chemical constituents. *Appl. Environ. Microbiol.* 48: 376-379.

Mason, J. R., Adams, M. A., and Clark, L. (1989). Anthranilate repellency to starlings: chemical correlates and sensory perception. *J. Wildl. Manage.* 53: 55-64.

Mason, J. R., and Maruniak, J. A. (1983). Behavioral and physiological effects of capsaicin in red-winged blackbirds. *Pharmacol. Biochem. Behav.* 19: 857-862.

Mason, J. R., and Reidinger, R. F. (1982). Observational learning of food aversions in red-winged blackbirds (*Agelaius phoeniceus*). *Auk* 99: 548-554.

Mason, J. R., and Reidinger, R. F. (1983). Exploitable characteristics of neophobia and food aversions for improvements in rodent and bird control. In *Test Methods for Vertebrate Pest Control and Management Materials.* D. E. Keukienen (Ed.). American Society for Testing and Materials, Philadelphia, pp. 20-39.

Mason, J. R., and Silver, W. L. (1983). Trigeminally mediated odor aversions in starlings. *Brain Res.* 269: 196-199.

Matsubara, H., and Tanimura, R. (1966). Utilization of constituents of peppers as an insecticide and as a synergist for pyrethrins or allethrin. *Bochu Ksgaku* 31: 162-167.

Nakatani, N., Inatani, R., Haruko, O., and Nishioka, A. (1986). Chemical constituents of peppers (*piper* spp.) and application to food preservation: naturally occurring antioxidative compounds. *Environ. Health Persp.* 67: 135-142.

Parker, G. H. (1912). *Smell, Taste and Allied Senses in Vertebrates.* Lippincott, Philadelphia.

Soudek, J. (1929). The sense of smell in birds. *Proc. Int. Congr. Zool.* 755.

Szolcsanyi, J., Sann, H., and Pierau, F-K. (1986). Nociception in pigeons is not impaired by capsaicin. *Pain* 27: 247-260.

Su, H. C. F. (1977). Insecticidal properties of black pepper to rice weevils and cowpea weevils. *J. Econ. Entomol.* 70: 18-21.

Walker, J. C., Tucker, D., and Smith, J. C. (1979). Odor sensitivity mediated by trigeminal nerve in the pigeon. *Chem. Senses Flav.* 4: 107.

# Chapter 15 Discussion

**Dr. Rozin:** I may have missed something along the way, but are all the tests you showed us consumption tests?

**Dr. Mason:** Yes. All the laboratory tests involve consumption.

**Dr. Rozin:** Is it consistent with what you found that there is no real nasal irritation sensitivity at all in these animals? That is to say, that all of the results could be due to the oral cavity.

**Dr. Mason:** With the exception of the first two slides where we plugged the nose and the responding is abolished, in that particular experiment I would argue that it's a nasal capsule effect. It could be oral acceptance when we cut olfactory nerves and trigeminal innervation of the nasal capsule but not the oral cavity. We do find these selective effects.

Birds are very much like amphibians; unlike mammals they have long olfactory nerves that can be sectioned. You can do the same thing with trigeminal innervation of the nasal capsule. Interestingly enough, Wayne Silver and I tried for some time to get trigeminal nerve recordings from starlings and I think, Wayne, you only got occasional units that responded to odor?

**Dr. Silver:** Yes.

# General Discussion

**Dr. Mason:** We have arrived at the general discussion. Neither Barry nor I want to impose any particular framework on the discussion, so please address any topic that you think needs to be discussed. Who would like to start?

**Dr. Silver:** I think this is a question that's been floating around since we started here, and one that Barry Green and I have discussed on at least a couple of occasions. This seems like a good place to pursue it. How should we define the common chemical sense? Is that a valuable term? What does it mean? What is the common chemical sense and what isn't the common chemical sense? I'm just free-associating. I'll pose the question and let you think about it first and then come back with what I think. Barry?

**Dr. Green:** Thank you for putting me on the spot. I don't have a definition of the common chemical sense. I think I saw the term common chemical sense as a problem because I came into the study of chemical irritation from the skin senses. I was studying touch and temperature and then began studying chemical sensitivity, so I naturally viewed the common chemical sense as a skin sense. I recognized its relationship to the pain sense. My first conversations with people here at Monell about the common chemical sense gave me the impression that most of the people studying it were coming from olfaction and taste and looking at it strictly as a specific chemical sense. Not much attention was given to its other possible functions, or the fact that it does reside in the skin and is, in a way, a multimodal sense. Historically, it seems to me it has either been viewed as a chemical sense in a very simple-minded way, or as a subset of the pain sense. I guess we are closer to the latter right now. But I'm not particularly comfortable with that either. There are a lot of issues that complicate the picture. We've heard a lot about the role of polymodal nociceptors, and we know they play a very crucial role in the sensitivity of the skin to chemicals. Undoubtedly that is true in the oral cavity as well, and it seems to be true in the nasal mucosa; but are there specific chemoreceptors we have not yet found that aren't part of the pain system or are sensitive to chemical irritants but not sensitive to things like heat and so forth? The electrophysiologists may have missed those because they typically look for polymodal nociceptors with heat or mechanical stimulation and later test to see if the units they find are also sensitive to chemicals. So we have to wonder if we really have at this time a good survey of the neurophysiology of the common chemical sense. And there's also the whole issue of quality and how it relates to the underlying physiology, and how that may be different in the nose vs. the mouth vs. the rest of the body. But I think I've said enough for the moment.

**Dr. Szolcsanyi:** Actually, I have not followed the historical aspects of the name. I know that Keele has used the term common chemical sense quite often, but I am not too happy with it either. I know that the term was related to the sensitivity to pain evoked by chemical agents, but I am not completely clear why they are "common." I have an idea it was just to distinguish that sense from the specific chemical senses of taste and smell. But I think it would be good to find a better term. I'm not sure whether we should look to the neurophysiological aspects for a definition because it is a descriptive sensation term. It is true that in the skin the polymodal nociceptors are the candidates for this. But the receptor types in the nose and mouth are not described exactly, and it may happen that they are not sensitive to thermal or mechanical stimuli and so they may not be strictly polymodal in that sense; they might have special characteristics. I have introduced a term for effects related to peptides and efferent effector functions, and that is "capsaicin-sensitive" receptors. This term has become popular because it is a pharmacological classification just like nicotinic receptor and atropinic receptor. I don't want to go so far as to say the capsaicin-sensitive receptor type is the final answer, but it would be clearer, at least for me, than the common chemical sense. Maybe we should call it the irritation sense, or something like that.

**Dr. Finger:** I think part of the historical perspective on this goes back to Parker's quote that Morley Kare read at the beginning of the symposium, since he was one of the first people to talk about this as a unified common chemical sense. I think that the distinction he was trying to draw was that it was common in terms of being common to cutaneous nerves, trigeminal nerves; that it wasn't carried by a specific group of cranial nerves in contrast to olfaction and taste. But there is a problem with that when you look at aquatic animals. Fish like searobins have a spinal chemosensitivity that is clearly different than what we have been talking about today as a chemical irritant sense; they are actually using a spinal chemosense in feeding. It doesn't make sense any longer to refer to that as a common chemical sense. So I think that Parker's use of the term, even though he coined it and made it popular, was not very effective either. And yet referring to trigeminal nasal oral chemosensitivity as an irritation sense is also a little difficult when it comes to consideration, for instance, of a response to menthol, where the evidence would lead me to believe it is being mediated by entirely different populations of nerve fibers. I'm not sure I would call the typical response to menthol, at least at low levels, an irritation response. So I'm not offering any solutions, but only further elaborating the problem, I suspect.

**Dr. Eccles:** I agree with Tom with respect to menthol. Certainly, in the experiments we've performed on menthol we've never believed we were acting upon the same type nerve endings as capsaicin does. We never see the typical sort of irritant response. In fact, in all the years we've been working with menthol, we've never induced a sneeze or lacrimation or nasal secretion and we've used some very high doses of menthol. I think one would have to reach extremely high doses before you would get any sort of burning sensation, which could be related to stimulation of the same sort of irritant receptors. So I think we are looking at at least two groups, the trigeminal group that serves irritation and protective reflexes, and capsaicin will act upon this group, and another group which is more related to thermal receptors or mechanoreception or a mixture of these. Menthol is a chemical that would act upon these nerve endings, but is it part of a common chemical sense? I would say no. It's part of thermoreception and not related to the capsaicin story. On the capsaicin story, I think that is interesting in itself because I've been trying to think—what came first?—the receptor on the sensory nerve ending for capsaicin, or capsaicin itself. It is a very difficult problem. Presumably the sensory nerve endings were around in the animals long before capsaicin was there to stimulate them. And there is some advantage in the plant kingdom for capsaicin to be present, either for protective means or for seed distribution or perhaps other areas which we don't understand at present.

**Dr. Mason:** Wayne Silver and I published some data 2 years ago, where we desensitized animals with capsaicin and they continued to respond to menthol. So, on that basis, one might argue that there are two quite separate systems. Just as an aside, with regard to capsaicin and ecological protective function, rodents are often seed predators and birds are often seed dispersal agents. It is interesting that rodents are sensitive to capsaicin and birds are not.

**Dr. Green:** There really is no question—I think perhaps Dr. Kobal brought it up earlier—that menthol stimulates cold fibers. And Dr. Szolcsanyi showed there was no effect of capsaicin on cold fibers, so there really is a very clear distinction. But something that is of interest to me, and to Dr. Szolcsanyi as well, is whether warm fibers are affected by capsaicin. I have contradictory psychophysical evidence of my own, and I know there is evidence that if it is injected into the hypothalamus, warm sensitive units there are affected by capsaicin. But the question is, are those the same type of units that you find in the skin? They might not be, and hence capsaicin may not affect warm fibers in the skin the way it does in the hypothalamus.

**Dr. Rozin:** I have a problem with the term irritation because, first of all, it has a very negative connotation, and since people seem to like these chemicals a lot, it is a little misleading. And in addition, I think a particular area of interest that is related to this is that we don't know much about very low levels of irritation, just above threshold, which people describe as tingling, and not particularly unpleasant. It seems quite a bit different from the burn that people describe at higher levels. I'm not even sure that those levels are aversive and there might be something very interesting in there about both the acquisition of a liking for capsaicin and its relation to the physiology. For example, I don't know if at threshold levels, where you just feel a tingling, there is any physiological, reflexive result. So I think there is an argument for not using irritation as a way to describe this system.

**Dr. Kobal:** I think that the term common chemical sense is not necessary. It is just the somatosensory system in the oral cavity and in the nose, and the difference is only that we have a different epithelium in the mouth than in the other areas of the body. Also, why don't we take the normal word for irritation that others do, which is the term "noxious," which indicates that damage may occur during stimulation with these chemicals. I really can't associate the word irritant to sensory physiology. What is irritated and what does that mean in a physiological way? There are receptors that are stimulated, so then you have a stimulus, or there is damage, you can use the term noxious stimuli. I think this is enough to describe everything that we have talked about these last 2 days.

**Dr. Rozin:** Capsaicin is not particularly noxious. Its beauty is that it tricks the body into thinking it's noxious. That is, it's the body's response to capsaicin that causes the irritation and inflammation. In fact, it is not a harmful agent by itself. So in capsaicin-desensitized animals, which can bathe in capsaicin, it is not necessarily going to harm itself. And since that is perhaps the prime stimulus for the system, to call the system noxious (I'm not speaking for the American Capsaicin Association, I assure you), I think is incorrect. In some sense capsaicin is a mimic of a noxious stimulus, which makes the system think that something bad is going to happen. The system mobilizes itself and produces an inflammatory response, as it were, to nothing; nothing that is a threat to it. So I wonder about that word as well.

**Dr. Alarie:** In industry, the word irritation and the words sensory irritation are very well accepted and the description of formaldehyde and sulfur dioxide and ammonia is that these are sensory irritants. In industry in general, they were arrived at by people who basically came from physiology. The textbook of noxious gas and so on came from Henderson from Yale, so there is nothing wrong with using the word irritation. If you look in the Webster dictionary, it gives you two definitions: one is for a stimulation of nervous entities and the other one is for inflammatory reaction. So when we describe the first action of formaldehyde, we say it is a sensory irritant, and in the second action of formaldehyde, if you have enough of it, you induce inflammation. I would disagree that capsaicin is not a toxic chemical. You can induce some rather strange things with capsaicin including destruction of thermoregulation. I think that that is "highly toxic, highly undesirable."

**Dr. Rozin:** That's for systemic capsaicin.

**Dr. Alaria:** No, no, if you inhale capsaicin in an aerosol, a sufficient amount of it will be systemically absorbed, and so the trigeminal nerve endings are there to warn you that there is something in your environment that you shouldn't tolerate at too high a concentration, that you should get away from. The first thing you feel is, of course, the stinging or burning, but if you go to higher concentrations you will have toxicity. We made a best guess and we said that 10 times above the $RD_{50}$ you will induce toxicity and for 11 chemicals that were looked at, indeed there was very significant toxicity in the nasal mucosa. So I personally feel very comfortable calling these materials sensory irritants.

**Dr. Walker:** One question I've had—it has occurred to me in talking about quality that people may have two things in mind. One is very perceptual and hard to get at; that is, does it feel like a different kind of sensation? Another, it seems to me, is the way that most of us in olfaction and trigeminal sensitivity would typically describe it. That is, a qualitative difference exists if you can discriminate it from something else, in the absence of an apparent intensity cue. Do you see what I mean? One way is to simply define it operationally by saying that non-olfactory quality discrimination occurs only if an anosmic subject can do it, providing that you have removed apparent intensity as a cue. Like hue discrimination when brightness is not a cue. But Harry and a few other people seem to be saying, which I can understand, that it wouldn't be a quality discrimination unless you could convince yourself that you felt different kinds of sensations that were mediated by the trigeminal nerve. And I think that unless that distinction is carefully looked at, you can run into problems later with someone saying, I've shown quality discrimination in the trigeminal system, and someone coming up and saying well, not really. It was unfair because you used the heat part of the sensation to do it, and someone else says no, I used the noxious sensation or something else. The second comment is that maybe you ought to have two terms. One is for what happens at low levels, before you get these violent reflexive kinds of changes, and it makes sense to call that irritation. The other thing is what happens at the lower levels where, for example, Cain's work with unilateral trigeminectomized humans would suggest that lots of things are happening before people are really calling the sensation they perceive truly irritating. There's lots of stuff going on there and maybe we ought to define it, if it is really nonolfactory, yet no one is calling it irritating. You have to take care of that concentration range.

**Dr. Alarie:** If you do recording from the ethmoid nerve, you will see impulse traffic long before you begin to see an effect on respiration. And so obviously something is happening; you are stimulating the sensory system, so it is still sensory irritation. You've demonstrated that you can stimulate these nerve endings by either directly recording from the fiber which carries those impulses or by measuring a reflex reaction which occurs after these impulses are integrated. Obviously, these reflex reactions are going to occur after a lot of traffic has been integrated. They don't happen with just a few fibers firing. I don't think it would be good for us to go into apnea with only a few molecules of sensory irritants. I think it would be very embarrassing for us, we would not be living too well if we were to do that.

**Dr. Lawless:** I think we should decide whether or not we are going to define this system, if it is a system, on the basis of its neurophysiology or on the basis of its subjective qualities, if there are such sensations. If we do the latter, I think it is prudent to be aware that there is a committee in ASTM to develop vocabularies for such things. They have, for example, adapted the term "chemical heat" for the sensations produced by capsaicin. They are now reserving pungency for the nasal pharyngeal sensations induced by horseradish, for example. It would be nice if we could all get our ducks in a row and begin to use the same type of terms, not one set for ASTM's food people in industry and another set for us more academically oriented individuals.

**Dr. Alarie:** If I may make a comment on the ASTM (I am a member of ASTM), it takes them forever to decide anything. So I think you may have to wait a very long time to get those definitions.

**Dr. Cain:** If you look up the definition of hearing, or the definition of vision, you'll find that things are defined there neuroanatomically or neurophysiologically. That is, if you look up the definition of vision it will say the operation that arises from the stimulation of the optic nerve or from the auditory nerve, so it defines it more morphologically/physiologically. But then there are those other modalities which we can't define quite so simply that way, so we duck to another mode. Like the position sense, where in that case we can't specify the specific nerves but we do believe that there is something that is a kind of unified experience that we call a sense of proprioception. The warmth sense is another example. Insofar as you are worrying about whether the common chemical sense should just be part of somesthesis, I don't think you have a great deal of trouble accepting the warmth sense as a useful term even though you may not be able to specify a nerve physiologically that underlies the warmth sense. You may have some fibers that respond to warmth and other things as well, and then it would be very hard for you to specify exactly what the physiological basis for the warmth sense is. Just a final point: the more that you try to get around using a term like "common," or whatever, "chemical sense," you will just keep failing. Common is perhaps the most unpleasant part of that definition. We don't know what common means, but the fact that it is a chemical sense is really indisputable, even though it may in fact sort of derive from somesthesis. The question is, should there be some adjective ahead of it and what should that adjective mean? Although I agree that one might define a substance as a sensory irritant without any ambiguity, I don't think calling this the "irritation sense" is going to stick because of things like cigarettes. When you smoke a cigarette you have a lot of common chemical effect, but it's not irritating, and the smoker will be able to distinguish that from irritation very readily. And on political grounds ASTM groups and so on will not want to say that such-and-such a cleaning product seems to have greater efficacy when it has some irritation in it. I just think it will be harder to get them to swallow that.

**Dr. Green:** Your argument that there is a lack of complete specificity in the skin senses is reasonable: even in the tactile sense there are mechanoreceptors that respond to cooling, for example. But it is generally believed that when you cool a mechanoreceptor and it fires, the CNS interprets that as pressure, not coolness, and there is a well-studied illusion, the Thaler illusion, that bears that out. What you have in what we are now calling the common chemical sense is the worst case of the skin senses insofar as specificity is concerned. These fibers really do seem to be part of the pain sense and yet they are multimodal in the broadest sense of the term.

**Dr. Cain:** But not all "irritations" are painful.

**Dr. Green:** That's right. It was brought up earlier in the meeting that you can have activity in nociceptors without inducing a painful sensation, and that presumably is true via mechanical stimulation and thermal stimulation as well as chemical stimulation. So that particular problem extends to the classically defined pain sense as well as to the common chemical sense.

**Dr. Eccles:** One point about referring to it as a chemical sense is if you consider what happens with inert dust taken in the eye or nose, it is going to bring about sneezing, hypersecretion, and irritation similar to inspiration of an irritant chemical. Now presumably if you are inspiring an inert dust, it can't in any way be considered as a chemical sense; it must be stimulating some sort of mechanoreceptor. But the irritation you get from dust in the eye or the nose must in some way be dependent upon the stimulating trigeminal nerve, and we can't consider this part of the chemical sense. Yet it is having a very similar reflex activity to a chemical stimulus.

**Dr. Alarie:** We have used dusts that are acutely inert, like silica. We have used very high concentrations and we did not induce this type of reflex reaction in mice. We have used a very, very high concentration of a liquid aerosol which has no chemical reactivity, e.g., polyethylene glycol, which we have used as a solvent to put ingredients into aerosol. Capsaicin is a good example. You have a fog of polyethylene glycol to which the animals are exposed and there is absolutely no reaction at all. You have no indication that there is a reflex reaction. So I don't think that we can then say that a chemically inert substance would stimulate the trigeminal reflexes or give you a trigeminal reaction. I have exposed myself to this polyethylene glycol aerosol and I feel absolutely nothing. I have done a few experiments recently with a new nebulizer which puts out very small particles of distilled water, or you can put anything in it, and if you get the stream in your nose you feel no irritation whatsoever. When it hits your throat you start coughing. So I have not seen "inert" or "acutely inert" things give me a sensation of trigeminal stimulation in one way or another, or induce this sensation in mice.

**Dr. Eccles:** But what about when you get dust in your eyes?

**Dr. Alarie:** No, now that is a different story. You start scratching the epithelial cells and they start leaking all kinds of things, and you've done real damage to those epithelial cells. I feel under those conditions it would be unfair to call that a physical stimulus. It's like if you scratch your skin or sandblast your skin, you would feel it.

**Dr. Eccles:** So you would relate such a response to actual damage.

**Dr. Alarie:** Yes, I think when you have dust in your eye and you feel pain, there has been some scratching of epithelial cells that has occurred.

**Dr. Rozin:** I want to change the subject slightly but I think it is related to the same issue. I think both Alarie and Mason mentioned that we don't know quite why this sense is there in nature. I'm really puzzled by this as we talk about it, because almost all of the primary stimulants of this system that we discussed are man-made products or they are products that are very well localized, particularly in plants, and in a way they are not in fact harmful if maintained at low levels. I want to contrast this with the bitter taste. In the bitter system, we have a "subsense" which produces avoidance as the irritant system does, and it correlates much better than the irritant system does with natural toxins. Now we don't know which came first; it is probably a coevolution problem because neither would work without the other. But we have a situation in bitter which I think is in some sense similar. We have chemicals in the environment and we have animals that avoid them quite widely phylogenetically. We have a much clearer notion of why the animals are avoiding them because things like caffeine and nicotine, all the aldehydes, are toxic substances at levels that could be ingested by animals. Now what I am wondering about, to follow up on both of you, is what is the original stimulus that got this system going, got it to be there? Because if we had a little idea of that it might help us functionally understand what this system is about. So I just wanted to raise that as a related issue—where did this system come from?

**Dr. Mason:** First off, for mammals bitter works nicely, but for birds, bitter is as enigmatic as capsaicin. For example, denatonium saccharide and denatonium benzoate are tremendous bitter stimuli for rats. However, birds will drink copious amounts of both. Birds respond to bitters like sucrose octa-acetate electrophysiologically, but many don't reject it behaviorally. Overall, then, for irritants or bitters, the "species exceptions" make speculations about ecological rationale generally unbelievable. Perhaps bitters and compounds such as capsaicin are targeted for particular kinds of animals, engaged in specific behaviors, i.e., the compounds are designed to shape the behavior of species. As an example, there are indications that anthocyanins in flowering plants, the purple in angiosperm fruits, are noxious to birds, and yet birds eat these fruits. Why are anthocyanins there? One possibility is to control when the fruits are eaten. Late winter berries, which birds eat, are heavily laced with anthocyanins. Perhaps the anthocyanins make the berries a less preferred food, so that they are eaten only when other forage is scarce, and incidentally, when seed dispersal is unlikely to be disrupted by rodents that are inactive. Or take grapes. Grapes have thick skins, perhaps to protect the seeds until they develop; anthocyanins in the skins assure that birds feeding on grapes will suck the seeds out. Thick skins therefore do not have negative effects on fecundity. Perhaps all "irritants" are simply the plant's strategy for controlling useful or harmful predators.

**Dr. Rozin:** But why are irritants around the seed but not in the seed?

**Dr. Mason:** Well, I don't know; maybe the reason is like the one given for the lack of pain receptors inside the skull. I suppose if the predator gets that far it really doesn't matter.

**Dr. Green:** When we talk about what the function is of the—I guess we can still call it the common chemical sense at this point—and how it developed, we may be talking about, in humans anyway, essentially two different systems. Perhaps in the nose it evolved for different reasons than it evolved in the skin, and perhaps the mouth is somewhere in between. In the skin—I am not an expert on this at all, perhaps someone else can talk about it more—the common chemical sense or the nerve fibers that are associated with it are involved in the inflammatory response. That the chemoreceptors that are sensitive to endogenous chemicals are also sensitive to exogenous chemicals may in a way be accidental; the function as detectors of environmental chemicals may have evolved from the other, primary functions having to do with local inflammatory processes and the hyperalgesia associated with cutaneous injury.

**Dr. Kobal:** I think again, one can very nicely reduce this by saying that the nociceptive system warns you against noxious stimuli coming from the outside, and this influences the respiration. I think nothing needs more to be proven than this: that the system warns the body of inspiring noxious stimuli. The other thing is, with the olfactory system we don't have a capability for localization as we do with the trigeminal system. The data I showed yesterday demonstrated that in the olfactory system you cannot discriminate between left and right stimulation, while in the trigeminal system it is very easy to do so. So the common chemical sense might have the function of orienting to the chemical world. This is much more important for animals than for us, but for us it might also be important that the moment you detect a noxious stimulus you know which nostril is stimulated the most.

**Dr. Eccles:** While Barry Green was talking, my mind was working over some other functions possibly related to these trigeminal nerve endings associated with inflammation. If you think of the upper respiratory tract of the nasal cavity, that is the prime area for the entrance of viruses into the body. If you are talking about the common cold and even things like the measles, this is how they get in. The symptoms associated with inflammation there would be hypersecretion, sneezing, very similar to the sort of thing you would get with chemical irritation. The common cold you associate with lacrimation, nasal congestion, sneezing, there is good evidence that this is mediated by histamine 1 receptors on trigeminal nerve endings. Histamine gets released as part of the inflammatory response. Are we perhaps looking at these nerve endings being stimulated now by plant substances and other substances? As Barry said, they are already sensitive to internally released chemical mediators, kinins, histamines, prostaglandins, and now we are externally applying chemical agents which also act upon them. Maybe it is a very basic mechanism related to reflex control and inflammation in the nasal cavity.

**Dr. Szolcanyi:** I have already shown you a picture in my talk that showed that sensation depends on the frequency of response, and there is a lot of other data in the literature both with respect to chemical agents and mechanical and noxious heat stimuli in humans from single-unit recordings. And in every case there is at first the firing and, later, at a given frequency, pain sensations begin to be felt. We also did experiments in the rabbit ear using ultraviolet irritation without touching the receptive field, and 30 mins later spontaneous firing of the nociceptors began and there was no pain response at all. So it is quite possible that nociceptors, and it is agreed that they are nociceptors, could fire without a definite pain sensation or even a sensation of irritation.

Back again to the name. What can we use? It could be a problem to define what an irritation is, but I think we can certainly call it a chemical sense, or a chemical sensation. In a given experiment we can call it a spice chemical sensation or irritant chemical sensation without overemphasizing the idea of a single system. I am saying this although I think all of these effects that have been called common chemical sensations could be desensitized by capsaicin.

**Dr. Silver:** But what about menthol?

**Dr. Szolcsanyi:** Cooling is often called a common chemical sense, but as early as the 1930s Hensel and Zotterman showed that menthol sensitizes cold receptors. So I don't think the sensitivity to menthol should be included in the common chemical sense because cold receptors have never been included in the common chemical sense.

**Dr. Silver:** My impression from Parker's work in 1912, where he originally used the term, is that he was simply saying that if a chemical didn't stimulate gustatory receptors or olfactory receptors, it acted through the common chemical sense. This would include menthol and capsaicin and every other compound. I think, though, that today we still are faced with the problem of how to define irritants. What are the natural irritants? Parker's original definition was for fish and it was originally done by squirting acidic acid or hydrochloric acid on the fish's flank. Well, I would imagine that most fish do not have hydrochloric acid squirted on their flanks naturally. It's tough enough to think of what a natural stimulus would be for an air breather; I don't know what it would be for a fish. Anyway, I kind of like the idea that Ron Eccles and Barry Green both suggested, if I understood them correctly, that these are really endogenous chemical detectors, not exogenous chemical detectors.

**Dr. Finger:** Today we have so far criticized the word common and criticized the word chemical and now I'd like to criticize the word sense. I'd like to suggest that in fact we are not looking at a single common chemical sense, that maybe there is an oral chemosense and a nasal chemosense, including for the oral chemosense both gustation and trigeminal chemosensitivity, and for the nasal chemosense both olfaction and trigeminal chemosensitivity. I'd like to draw the analogy to the 8th nerve sensory systems, where we have both an auditory sense and a vestibular sense and yet the actual transducing elements, the end organs, are very similar; they are both hair cell receptors using very similar physiological mechanisms. Nature, in its conservative way, had modified the gross surroundings of these hair cell mechanoreceptors to produce two very different sensory perceptions, and a good deal has to do with how our brain interprets the information coming into it. If we draw that kind of analogy, I would argue that maybe we should be considering "nasal chemical senses" and "oral chemical senses" rather than a common chemical sense as a unified thing. Nature had a perfectly good polymodal nociceptor and just put it to a different use, which was to detect chemicals trying to gain entrance into our bodies.

**Dr. Rozin:** I like that point a lot and I agree with it. I just want to point out the next step that you won't like. Now you want to call olfactory input that is associated with things in the mouth which we call taste, colloquially, as part of oral chemoreception, because that is how the system handles it; that is, you perceive it in your mouth and at some point in the brain it's handled very differently from odors you are sniffing. I happen to like that distinction because functionally organisms are dealing with what goes in their mouth and its sensory properties, which includes the olfactory input from the mouth as well as things that are out in the world that are inhaled through the nose and reach the olfactory sense that way. What this does is put us in a nice conflict, which is an interesting one, between functional and anatomical definitions of senses. This goes back to what Bill Cain said before.

**Dr. Green:** I just want to say, don't forget the skin. And how about corneal sensitivity, which is also trigeminal but isn't really directly related to the nose or directly related to the mouth? So I like Tom Finger's definition too, but I think the chemical sensitivity of the skin has to be added to it.

**Dr. Finger:** Let me just respond briefly to this. Again, it is a question, I think, of the polymodal nociceptor population just being put to a different epithelial use. I don't think our body treats corneal chemoreception anywhere near the way it treats nasal or oral chemoreception. With respect to why an anatomist might object to oral chemoreception including things through the nose, if you look at the evolution of the system, it evolved from fishes, where there is no connection between the oral and nasal cavities. So I think it makes perfectly good sense from that standpoint, too, to separate oral and nasal chemoreception.

**Dr. Green:** Before we wind this up, I want to pose one more general question: that is, where do we go next? I realize that's a big question. Does anyone want to take a bold step forward and say what they think the important issues are that we should tackle next?

**Dr. Cain:** I think the one that you mentioned, the question of responding to endogenous chemicals, is very important. And to what degree are there mediators functioning in the system? If we look at Alarie's classification, it says in two of three cases we have actual chemical changes in the tissue. In olfaction and in taste, we don't have chemical change; we have an absorption process. In these other cases, we have breaking of sulfide bonds and tying up of nucleophilic groups. Those are actual chemical changes that are occurring, and presumably this opens the possibility of some chemical mediator coming along and actually doing the stimulation. Those things like alcohols you made reference to, it's not yet clear except that you generally need very high concentrations in order to get stimulation of the modality. I like the idea that capsaicin, for instance, may be something that happens sort of accidentally to stimulate receptors that are normally stimulated by endogenous materials. I wonder for instance whether anyone has done neural recordings by pretreating the preparation with antihistamine first to see if you get an attenuation of the response. You can play all those pharmacological games with the tissue, provided you don't upset things too much, and find out whether you can in fact show that there are intermediates in the stimulation process. I think that the second issue which is really important is one that came up in connection with Harry Lawless's talk. We have to find out something about what makes materials bind in certain places, i.e., we have to understand the biophysics of the question, particularly in the contact domain more than in the inhaled domain.

**Dr. Silver:** Actually, we are just starting to do some of those pharmacological experiments. I think it's clear that menthol is affecting cold receptors. And we've thrown around polymodal nociceptors, but I don't think there is much evidence, very good concrete evidence, that what we are talking about is actually stimulating polymodal nociceptors in the nose or the mouth. I think that would be one thing to do. I think another thing, as Dr. Szolcsanyi said the other day, would be to characterize those fibers in the nose and mouth that are being stimulated by these compounds.

**Dr. Szolcsanyi:** As far as capsaicin is concerned, there is a lot of evidence that it directly stimulates the nerve endings. One piece of evidence is that after high doses of capsaicin are given there are ultrastructure changes in the receptors and the result is that histamine or other endogenous substances can no longer be produced. Another is that you can produce depolarization of the cell by nanogram doses of capsaicin in tissue culture. You cannot inhibit the stimulatory effect of capsaicin by an antihistamine or that sort of thing.

**Dr. Cain:** Do you think capsaicin is stimulating receptors that are normally activated by endogenous substances?

**Dr. Szolcsanyi:** Yes, of course. Bradykinin, for example, activates the same receptor. After capsaicin desensitization, the response to bradykinin is reduced.

**Dr. Mozell:** I think there is a whole series of experiments that we keep skipping. Basically, what kinds of discriminations can the common chemical sense, or whatever you call it, really make? We have a number of patients now that are documented as being anosmic; therefore any discriminations they can make of vaporous materials must be done by the other system and I would like to know what this other system can do. Is it just pain, or is it pain and something else? Several people have noted in this discussion over the last several days that it is the psychophysicist that tells us, the electrophysiologist, what to look for. And yet we haven't really been very well defined as to what we should look for. I think one of the ways we can do that is those people who have access to anosmic patients should do a very systematic study to determine what sorts of discriminations people left with the common chemical sense can make. I can't believe that it is only pain and cool. It seems to me that they might be able to discriminate different types of cool. I'm not even convinced yet, although you've said it several times, that the coolness of menthol is indeed due to the stimulation of cool receptors. Is there solid data on that?

**Dr. Szolcsanyi:** Yes, Zotterman's data.

**Dr. Mozell:** Okay. Anyway, I think one of the main things that ought to be done is to find a better definition of what sorts of discriminations this system that we've been discussing can really make. I would like to know, for instance, if at very low concentrations, before you get pain sensations, if discriminations can still be made without the olfactory system playing a role. We've looked at a few Kalman syndrome patients, but we are not a psychophysics lab and I'd like to see some really good psychophysics labs look at this sort of thing.

**Dr. Walker:** I think you're right. I think it would be very interesting to look at trigeminectomized patients and I'd be very curious to see if the presence or absence of that system plays a role even at "nonirritating" concentrations. My prediction would be that it would.

**Dr. Silver:** I just wanted to say that there are at least two animal studies, one by Walker and one by Mason and myself, that suggest that at least for some fairly benign compounds the trigeminal system cannot tell the difference between those compounds matched for equal intensity. So it looks like they can't tell quality differences between compounds but they can tell quantity differences. But again, it is just for a few compounds. And I think you are right; rather than have Russ sit in a room for a year with salamanders trying to answer that question, I think you can ask people and get an answer much more quickly.

**Dr. Green:** Chuck Wysocki and I intend to do that if we can find enough patients who have had the olfactory nerve transected. I believe Jim Walker is going to do the same sort of experiments.

**Dr. Walker:** I may be repeating myself, but I think it is important to be careful in these kinds of studies. Suppose we found that anosmic subjects could, at equal apparent intensities, make a discrimination. I still think we have to be careful about how we talk about that, and whether or not we want to call that an odor quality perception or an odor quality discrimination. I think it is extremely unlikely. . . with the different kinds of receptors that are probably there that you couldn't somehow pick a combination of chemicals such that you would get an apparent quality discrimination. I just think you have to be careful about this, about saying, ''OK, that is an odor quality'' and then say the trigeminal system can mediate a true odor quality discrimination.

**Dr. Cain:** Well, I agree with what you say. Max particularly knows about how deposition in the nasal cavity varies from one material to another, and I think that could be a key to discriminations that might be made between compounds. Some things are going to be sharper in their onset because of the way they deposit themselves in the mucosa; others are going to be more blunted and perhaps the waning part of the sensation could give you the ability to make discriminations. Some of these types of differences could be based on very subtle, temporal cues, like those used in discriminating sounds on the basis of their rise time or attack.

**Dr. Green:** I think the question of what constitutes a perceptual quality, and therefore a quality discrimination, is a difficult one. When you start thinking about the common chemical sense as a skin sense, you can run into complications when you try to define what a quality is. Because on the skin, changing the location can change the quality of sensation a stimulus produces. So if two chemical stimuli affect different areas, can they be considered qualitatively different on that basis alone? I think that it is reasonable to say they can. So you really can get into a conundrum in trying to define what you mean by quality and quality difference.

**Dr. Cain:** Yes, you have to settle on an operational definition of qualitative difference.

**Dr. Mozell:** That is extremely important because a pain in the throat and a pain in the tip of the nose and a pain half way back in the nose are all discriminable, just as touch on the forehand and touch up the arm is. It may be that some of these odorants, if you want to call them odorants, some of these vaporous chemicals, will stimulate different places and we can discriminate different places, as we can on the skin, in the nose. Here is a simple possible way to discriminate different odors. Instead of saying everything is painful or everything is irritating, it may be irritating in different places, and maybe those different places are what you have to look at. Maybe people can make discriminations based on location, but they don't realize it.

**Dr. Green:** I know we could go on much longer, but I'm afraid we're out of time. Perhaps we can meet again in the future with answers to many of these questions. On behalf of the organizers, I thank you all for coming.

# Index


A

Acetaldehyde:
  cerebral-evoked potentials to chemical stimuli, 124
    evoked potential mapping, 129
Acetic acid:
  single-fiber responses, 23
  trigeminal electrophysiologic threshold
    chemically-provoked ethmoid responses, 25, 26, 28, 29
Acetylsalicylic acid, cerebral-evoked potentials to chemical stimuli, 131, 132
Acetylcholine:
  chemically-induced burning, 149
  SP-induced transepithelial potential, 162
    interaction between trigeminal afferents and olfactory mucosa in amphibia, 62
Acinar cells, 61, 62, 64
  secretory activity in olfactory mucosa, 62, 65
Acrolein, 31, 299, 305, 307
A-delta fibers, 40, 171
*Agelaius phoeniceus* (Red-winged blackbirds), 309-320
[*Agelaius phoeniceus*]
  response of trigeminal nerve to irritant consumption, 309-320
Airborne irritants, 43, 45, 236
  inurement to, 53
  sensory irritants, 297-302
    trigeminal chemoreception, 297-302
  trigeminal input, 85, 297
Airjet stimulus, trigeminal nerve innervation, 279, 282
Alcohols, *see also* Ethanol, n-butyl alcohol
  trigeminal nerve innervation, 25-29, 31, 32
Alkylbenzenes, trigeminal nerve innervation, 31
Allosteric inhibition, 53
Allyl isothiocyanate, 309
α-Terpineol, trigeminal electrophysiological threshold, chemically-provoked ethmoid responses, 25, 26, 28, 29
Amino acids, sensory and pain transmission, 20
Ammonia, 11, 89, 90, 120, 306, 307, 309
  anosmic odorant detection, 97

[Ammonia]
cerebral-evoked potential to chemical stimuli, 124
evoked potential mapping, 129
chemical somesthesis, 47, 48
detection threshold, 85
dependence on trigeminal and olfactory nerve input, 85, 86
irritation of nasal mucosa, 55
laryngectomy
induced hyposmia, 81
loss of nasal air flow, 81
odorant detection threshold, 75-86
comparison with Kallman syndrome patients, 86
time course of adaptation, 50
Amyl acetate, 22, 101, 106, 108, 110
changes in respiration, 113
odorant concentration, 113
trigeminal receptor stimulation, 115
comparison of olfactory and trigeminal systems, 97
response to odorants, 106, 108, 109, 111, 112
trigeminal electrophysiologic threshold, chemically-provoked ethmoid responses, 25, 26, 28, 29
trigeminal nerve sensitization to, 100
single-fiber responses, 23
Amyl butyrate, 137
interaction between olfaction and common chemical sense, 53-55
reduction of odor by irritation, 53-55
Anethol:
chemosensory-evoked potentials, 126
sensory response differentiation, 124-134
Anise, stimulation of olfactory nerve, 96
Anosmics, 97, 101, 102, 104, 105, 108, 111, 113, 124
CSSEP, 134
duration-dependent irritation potential, 55
[Anosmics]
evoked potentials in, 138, 139
laryngectomy-induced, 71, 72
nonolfactory quality discrimination, 329
odorant recognition, 95-121
receptor stimulation, 29, 31
response to chemical stimuli, 29, 139
Antidromic stimulation, 8, 20, 64, 65
inducement of mucosal ciliary movement, 20
efferent functions of sensory endings, 20
NV ob, 64
Antisera, substance P and CGRP, 2, 6
Apnea, 55, 330
stimulation of nasal mucous membrane, 280
Asafetida, stimulation of olfactory nerve, 96
Ascorbic acid, chemically-induced burning, 149
Atropine, 64
blocking effect of, 69
SP-evoked excitatory responses, 63
Autonomic neurons, preganglionic:
activation by peptidergic sensory fibers, 9
spinal reflex system, 9
Avoidance behavior, free nerve ending mediation, 21
Axonal conduction, 142, 152, 158, 161
blockade of, 143
antinociceptive effect of capsaicin, 143
Axonal reflex:
neurogenic inflammation, 9
peptidergic fibers
distribution, 8
innervation, 10
physiological alterations
in mucosa, 14
in peripheral tissue, 9
sensory-effector functions of sensory endings, 20

**B**

Basal lamina, 5, 34
  bioactive peptides
    release of, 11
    site of release, 12
  collateralization of peptidergic fibers, 9
Benign masochism, mechanisms for preference development of chili pepper, 262-264
Benzaldehyde:
  anosmic odorant detection, 97
  chemosensory-evoked potentials, 126
    sensory response differentiation, 124-136
  nasal stimulation via physical interaction, 43, 44
Benzyl acetate:
  trigeminal electrophysiologic threshold
    chemically-provoked ethmoid responses, 25, 26, 28, 29
Benzyl amine, 22
Bioactive peptide release sites
  alterations in tissue physiology, 9
  basal cells, 12
    modification of proliferative rate of, 12
  blood cells, 12, 13
  ciliated epithelial cells, 11
    acceleration of mucociliary activity, 11
  intraepithelial secretory cells, 12
    sustentacular cells, 12
  nasal glands
    alterations of secretory activity, 12
    olfactory modulation, 12
  olfactory bulb
    modulation of neuropeptide activity, 11
    presence of CGRP-immunoreactive fibers, 13
  paracellular shunts, 11
    substance P increase of anion permeability, 11
[Bioactive peptide release sites]
  parasympathetic ganglia, 13
    modulation by peptidergic fibers, 13
Bitter taste system, contrasted with irritants, 333
Blast injection technique, 97
Blood pressure, changes in, irritation-induced, 55
Bombesin
  collateralization of peptidergic fibers, 12
    neurotrophic activity, 12
Bowman glands, 61-64
Bradykinin, 338
  activation of pain fibers, 3
  antagonists, 3
  chemically-induced burning, 149
  neurotrophic activity, 12
  production, 3
  reception, 3
Bradypnea, 297, 298
Brain mapping, 127, 128, 129, 130
  by multi-channel recordings, 133
Butanol, trigeminal electrophysiologic threshold, chemically-provoked ethmoid responses, 25, 26, 28, 29
Butyl acetate, 100, 101
  trigeminal electrophysiologic threshold, chemically-provoked ethmoid responses, 25, 26, 28, 29
Butyric acid, trigeminal electrophysiologic threshold, chemically-provoked ethmoid responses, 25, 26, 28, 29

**C**

Caffeine, 227
C afferents, *see also* C fibers, nociceptors
Calcitonin gene-related peptide, *see* CGRP
Camphor, 243
  effect on nasal resistance to air flow, 284
  single-fiber responses, 23

*C. annuum*, 233, 243
Capsaicin, 11, 40, 197, 201, 213-215, 300, 301, 305, 306, 310, 333, 338
absorption of in lactile fluid, 271
afferent impulse elicitation, 159-162
analgesic effects of, 158
cell death, 158
antibacterial effect, 238, 239
axonal effects, 143, 158, 161
blocking effect of, 69
chemical heat, 330
chili peppers, 232-235, 238-240, 245
location, 233
cholinergic effects, 235
common chemical sense, 326-338, 335, 336
CSSEP generation, 133
nociceptive nature, 133
depletion of substance P, 35, 64
desensitization and irritation, nasal, 35, 141-162
effects on trigeminal, nodosal dorsal root ganglia, 155-159
nociceptive and antinociceptive effects, 141-150
sensory-efferent function of capsaicin-sensitive sensory neurons, 159-161
single unit recordings from cutaneous nerves, 150-155
temperature-dependence, 173-176
differences in sensitivity to, 248
dorsal root ganglion, effects on, 155-159
reduction of outward potassium currents, 156, 157
role of mitochondria, 156
voltage clamp studies, 155-159
ecological protection function, 327
irritant qualities, 198, 199
irritation, oral, 171-198, 300, 301, 305, 306
human preference for burning, 241, 246, 247
[Capsaicin]
inhibitory effect on cold nerve fibers, 173-176
masking power of burn at sensory level, 248, 257
mechanical effect, vibration, 179-186
thermal effects, 35, 172-176, 181-183, 193-195
nociceptive effects of, 141
neurotoxicity, 143, 144, 156, 158
noxious mechanical stimulation, 35, 328, 329
noxious heat threshold, 143, 144, 153
pharmacology of, 234-236
private body consciousness tests, 219-221
red peppers, 148-150
ruthenium red, 156
suppression of transepithelial slow potentials, 64
toxicity of, 235
types, 233, 241
as a monolithic stimulus, 241
warm/cold fiber stimulation, 328
Capsaicin oleoresin, 237, 240
Capsaicin-sensitive afferents:
in the adult animal (CSA), 154, 155, 161
in the newborn animal (CSB), 154, 155, 161
*Capsicum*, 35, 198, 232, 310
Carbon dioxide, 40, 41, 130
cerebral-evoked potentials to chemical stimuli, 124, 126, 137, 139
evoked potential mapping, 129
chemical somesthesis, 46, 47
interaction between olfaction and common chemical sense, 53
odor reduction via irritation, 53-55
irritant role, 59
change in pH, 59
pain chemical stimuli, 123, 127, 131
nociceptor excitation, 131, 132

[Carbon dioxide]
sensory response differentiation, 123-134
monomodal stimulation of trigeminal nerve, 123, 124
Cardiac acceleration, pigeon:
classical conditioning, response to odorants, 102
Carvone, 243
CCS, *see* Common chemical sense
Chemesthesis
defined, vi
C fibers, 40, 158
conductance block of, 146
perineural application of, capsaicin, 146
neonatal treatment with capsaicin of cutaneous nerves, 153-155
site of capsaicin action, 235
unmyelinated, 2
CGRP (calcitonin gene-related peptide):
colocalization with substance P, 2
release into olfactory bulb, 13, 19, 20
activation of trigeminal sensory fibers, 13
similarity with substance P, 2
CGRP antisera, 6
CGRP-fibers:
collateralization of single fibers, 12
role in growth modification of basal cells, 12
peptidergic fiber tracing, glomerulus, 8
presence in olfactory nerve, 8
trigeminal nerve activation, 8
CGRP-immunoreactive fibers, 8, 13
nasal cavity, 8
nonsensory nasal epithelia, 7
similarity to nasal transepithelial fibers, 7
olfactory bulb, 13
innervation of nasal cavity, 13
reactivity to substance P antisera, 6
role in vomeronasal pump, 8
CGRP-SP fibers:
distribution within nasal cavity, 5, 6
amphibia, 5
rodents, 6
immunoreactivity to substance P and CGRP antisera, 2
intranasal distribution, 6
immunoreactivity to substance P antisera, 6
location in lamina propria, 6
nonhomogeneous distribution of intraepithelial fibers, 6
transepithelial peptidergic fibers in vomeronasal organs, 6-8
lamina propria, 6
modulation of olfactory activity, 12
Chemical somesthesis, 46-53, 131
summation, 46, 47
spatial, 46, 47
temporal, 47, 48
Chemonociception, 142
inhibition by capsaicin, 144, 145
Chemosomatosensory-evoked potentials, *see* CSSEP
Chemosensory-evoked potentials, 124-134
sensory response differentiation, 124-134
Chicken soup, sodium chloride concentration, 223, 224
source identification of personality variables, 217-227
Chili peppers, 46, 227
adaptive/evolutionary explanation of preference for, 237-240
digestive aid, 239, 259
medicinal uses, 234, 240, 246
vitamin supplementation, 233, 238, 239
characteristics, 233
current motivation for explanation of preference for, 237-250
component in skin linaments, 247
desensitization to, 248, 250
flavor enhancer, 240, 241, 246, 248, 259

[Chili peppers]
history of preference development for, 250-264
changes in response, 254-257
social mediation, 258, 259
two stages of preference acquisition, 250-254
mechanisms of preference development, 258-264
benign masochism, 262-264
opponent-endorphin responses, 259-261
preferences in animals, 236, 237, 259
uses in war, 232, 234
world production and consumption, 232, 233, 235
Chloroacetophenone, 31
Chlorobenzylidene malonitrile, single-fiber response, 23
Chloroform, irritation of nasal mucosa, 55
results, 55
Cholinergic receptors, 31
Christopher Columbus, chili peppers, 231
Cineole, single-fiber response, 23
Citric acid, 253
tastant sensation inhibition, 205
private body consciousness tests, 219, 220
Classical skin senses, oral irritation, 171-190
Coffee:
odorant stimulation of olfactory nerve, 97
two stages of preference acquisition for, 250, 251
Cold fibers, 328, 335
Collateralization, peptidergic, 8-14
Colocalization, 2
Common chemical sense (CCS), 3, 8, 9, 52, 55, 190, 310
carbon dioxide as an irritant, 59
chemical somesthesis, 46-53
classical skin sense, 171, 190
[Common chemical sense]
defining, 325-340
nasal chemosense, 336
oral chemosense, 336
function of pain, 171, 172
interaction with olfaction, 53-55
irritant sense, 46
mediation by free nerve endings, 21
avoidance behavior, 21
noxious chemicals, 21
oral irritation, 171-190
researcher's view, 56
restriction by horny epidermis, 22
toxicologist's view, 56
transmission of sensory information to CNS, 8, 9
likely route of transmission, 8, 9
Cornea, 40, 41, 114, 207
Cornified skin, 40
Cortical sensory-evoked potentials, brain response to chemical stimulation of trigeminal nerve, 123, 124, 127, 131
Cross-sensitization, NaCl irritation, oral, 187-190
CSSEP (Chemosomatosensory-evoked potentials), 124, 130, 131, 133, 134
brain responses to trigeminal chemoreception, 123-134
cerebral-evoked potentials to chemical stimuli, 124
in anosmics, 124
dependence on stimulus intensity, 131
determination of analgesic action, 131, 132
acetylsalicylic acid, 131, 132
distinction from olfactory-evoked potentials, 123-134
brain mapping, 133
establishment of cortical generator, 124, 127
sensory response differentiation, 124-134

[CSSEP]
generation of, 133
topographical distribution, 133
discovering a pure odorant, 133
C-terminal peptide, 62
interaction between trigeminal afferents and olfactory mucosa, amphibia, 62
Cutaneous nerves, single unit recordings of, 150-155
Cyclohexanone, 30
chemically-provoked ethmoid responses, 25, 26, 28, 29
single-fiber response, 23
Cyclopentanone, single-fiber responses, 23

D

*d*-Carvone, chemically-provoked ethmoid responses, 25, 26, 28, 29
Deionized water, 173, 177
Desensitization, 35, 149, 151, 248, 250, 271, 300
C fibers, 235
capsaicin pharmacology, 234
chemonociceptive stimuli, 141-162
intraepithelial fibers, 142, 143
red peppers, 148-150
species differences, 148
subepithelial fibers, 142, 143
defining, 141
noxious heat threshold, 143, 144
reflex latencies, 143, 144
oral irritants, 171, 190-195, 207-209, 215, 248
Dichorhinic mixtures, 59
interaction between olfaction and common chemical sense, 54, 55
odor reduction via irritation, 54, 55
Dienophiles, 31
Dihydrocapsaicin, 241, 245
Diamines, 301
Diphenylcyanoarsine, 31
Diphenyl ether, single-fiber response, 23
Diving reflex, 280
Dorsal root ganglia (DRG), nodosal:
capsaicin effects on sensory neurons, 155, 159
antidromic stimulation, 160
depolarization and conductance increase, 155-159
effectiveness of efferent response, 160
increase in cation permeability, 158
ultrastructural results, 157-159
DRG, *see* Dorsal root ganglia
*d*-Tubocurarine, stimulation of amphibian mucosa, 62

E

Edema, capsaicin-induced, 64, 65
association of sensitive and secretory function in trigeminal nerve, 64, 65
Endoplasmic reticulum, effects of capsaicin, 156, 158
EOG, *see* Odor-evoked electroolfactograms
Ethanol, 307
chemically-provoked ethmoid responses, 25, 26, 28, 29
distribution of irritants in oral cavity, 214
irritation, oral, 171-190
chemical effects, 187-190
mechanical effects, 179-186
thermal effects, 176, 190, 193, 194
Ether, single-fiber responses, 23
Ethmoid nerve, 23-25, 30, 40, 104, 302, 330
fiber characteristics, 35
innervation of anterior nasal mucosa, 276
cooling sensation of inspired air, 276

[Ethmoid nerve]
response to chemical stimuli, 35, 294
capsaicin desensitization, 35
unmyelinated axons, 33, 35

F

Facial manipulation, 230
source identification of personality variable, 218, 219, 221-225
Facial nerves, 104
Fluorescence, distribution of irritants in oral cavity, 214
Forced-choice method, threshold lowering, 119
Formaldehyde, 49, 299, 329
irritation inurement, 53
perceived "odor" intensity, 51
persistence, 50, 51
single-fiber response, 22
Formic acid:
chemically-provoked ethmoid responses, 25, 26, 28, 29
single-fiber responses, 23
Free nerve endings, 1, 51, 202
epithelial impediment to stimulus, 52
intercellular tight junctions, 3
stimulus access, 4
Fungiform papillae, oral cavity irritation, 202

G

Gasserian ganglion, *see* Trigeminal ganglion
Ginger, 236, 246, 247
Gingerol, 309
differences among oral irritants, 198
Glossopharyngeal nerve, 104
excitation of olfactory regions, 72
trigeminal versus olfactory input, 72

Goblet cells, 34
collateralization of peptidergic fibers, 12
Golgi apparatus, effects of capsaicin, 158
Gopher tortoise, 22
comparison of olfactory and trigeminal sensitivities, 97
amyl acetate, 97
benzylamine, 97

H

Heptanoic acid, trigeminal electrophysiologic threshold, chemically-provoked ethmoid responses, 25, 26, 28, 29
Heptanol, trigeminal electrophysiollogic threshold, chemically-provoked ethmoid responses, 25, 26, 28, 29
Hexanoic acid, trigeminal electrophysiologic threshold, chemically-provoked ethmoid responses, 25, 26, 28, 29
Hexanol, trigeminal electrophysiologic threshold, chemically-provoked ethmoid responses, 25, 26, 28, 29
Histamine, chemically-induced burning, 149
Homocapsaicin, 241, 245
Homodihydrocapsaicin, 241, 245
Horny epidermis, restriction of common chemical sense, 22
Horseradish, 46, 246, 247
Hydrochloric acid, 336
Hydrogen sulfide, 306
cerebral-evoked potentials to chemical stimuli, 124
evoked potential mapping, 129
Hyposmia:
laryngectomy-induced, 71, 72, 80
loss of nasal air flow, 83
trigeminal versus olfactory input, 72
severing of recurrent laryngeal nerve, 72

## I


Inflammation, neurogenic, 159-161, 334-335
Intensity:
  of physiological response, nasal, 24-31, 96-104
  of sensation
    nasal, 47-55, 106-111
    oral, 173-190, 203-208
Intercellular tight junctions, 3
Ion channel, modulation by peptides, 11
Ionine, single-fiber response, 23
Irritants, nasal cavity:
  chemical somesthesis, 46-53, 131
  classes of, 31
    breakage of S-S linkages, 31, 33
    lipid solvents, 31
    protein precipitants, 31
    reactions with nucleophilic groups, 31, 33
    reactions with SH groups, 31
  lipophilicity, 31
  loss of sensitivity to, 53
  response stages of capsaicin-sensitive primary afferents to, 161
  role of capsaicin, 141-162
  versus common chemical sense, 46
Irritants, oral cavity:
  avian consumption of, 290-320
  chemical stimulation, 171-190
    sensitization and cross-sensitization, 187-190
  decay time, 187-190
  differences among, 197-201
  distribution in, 214
    fluorescence, 214
  interactions of, 205-210
    cross-adaptation studies, 205-209, 215
    mixture experiments, 205-208
  mechanical effects, 179-186
    ethanol, 179, 183-186
    perceived pain reduction, 181, 183
    pressure, 179, 183-186
    vibration, 179-183
[Irritants, oral cavity]
  physiological coding, 201, 202
    receptor mechanisms, 201
    routes of access of irritant molecules, 202
  psychophysical studies, 171-190
  sensory discrimination between capsaicin and piperine, 199, 213
    evoked-responses, physical, 204
    peripheral physiology, 205
    potency, 202
    spatial pattern of responsiveness, 199, 202, 203
    tastant sensation inhibition, 205
    time course of sensation decay, 204, 205
  sip-and-spit procedure, 172-186
  source identification of personality variables, 217-227
    private body consciousness tests, 219-227
  thermal effects, 171-190
    lability, 176-179
    temperature-dependent phenomenon, 172-179
    time course of adaptation, 180-190
  stimulation mechanism, 297-302
Irritants, sensory:
  setting "acceptable" exposure limits, 298
  specificity of receptors, 300-302
    capsaicin, 300, 301
    *cis*-4-cyclohexylmethylcyclohexylamine, 301
    nicotine, 301
  worker/consumer protection, 297-302
Irritation, nasal cavity, 106, 108, 111, 120
  arousal by noxious chemicals, 21
  avoidance behavior, 21
  capsaicin and desensitization, 141-162
    chemonociception, 142-145
    mechanonociception, 145, 146
    noxious heat threshold, 143, 144

[Irritation, nasal cavity]
chemical somesthesis, 467-53
spatial summation, 47
temporal summation, 47
defining, 328, 329
epithelial injury, 43, 44
interaction between common chemical sense
and irritants, 46
and olfaction, 53-55
interspecies differences, 55
nasal reflex as a psychophysical adjunct, 55, 56
results of chemical irritation, 56
odor reduction via, 53-55
perceived "odor" intensity, 44-46
origins, 44, 45
perceptual "contamination," 44
psychophysical functions of, 44-46
response magnitude, 29
protection system, 29
stimulation
mechanism of, 59
via physical interactions, 43, 44
via carbon dioxide, 59
voltage clamp studies, 155-159
Irritation, ocular, 105-115, 119
Irritation, oral cavity:
avian consumption of irritants, 290-320
chemical effects, 187, 190
sodium chloride, 187-190
classical skin senses, 171-190
effects of tastants, 29-227
mechanical effects, 179-187
personality variable in perception, 217-227, 229
facial manipulation, 221-225
self-perception theory, 218, 227
source identification, 217-219
types of, 218
sensation-seeking, 229, 230
thermal effects, 172-179
capsaicin, 173-176
sodium chloride, 176-178
Isoamyl acetate, 63
nasal stimulation via physical interaction, 43, 44
suppression of receptor cell responses, 64
interaction between trigeminal afferents and olfactory mucosa, 64
Isomenthol, 283
Itch, 148

**K**

Kallman syndrome, 86
Kapsizimus, *see also* Capsaicin, toxicity of

**L**

Lacramatories, chemical trigeminal stimuli, 31
Lachrymators, 301
Lamina propria, 62, 202
substance P (SP)-like immunoreactivity, 61
Laryngectomy, air flow generation without larynx bypass, 74, 80
anosmia, inducement of, 71, 72
changes in mucosal secretion characteristics, 72
enhanced air flow, 83
improved olfactory ability, 83
patient predisposition for hyposmia, 81
trigeminal versus olfactory input, 71-91
odorant detection threshold, 72, 75-91
odorant identification, 86-91
two-interval forced-choice tracking procedure, 75
Larynx bypass, 77-80, 87-91, 93
odorants
detection threshold, 76, 82
identification, 86-91
restored air flow, 76, 81
sniffing behavior, 84, 85

Lavender, stimulation of olfactory nerve, 96
*l*-Carvone, trigeminal electrophysiologic threshold, chemically-provoked ethmoid responses, 25, 26, 28, 29
Licorice, trigeminal versus olfactory input, 71-91
  odorant identification, 86-91
Limonene, 243
  chemically-provoked ethmoid responses, 25, 26, 28, 29
  chemosensory-evoked potentials, 126
    sensory response differentiation, 126
  single-fiber response, 23
Linalool:
  nasal stimulation via physical interactions, 43, 44
  trigeminal electrophysiologic threshold
    chemically-provoked ethmoid responses, 25, 26, 28, 29
Lingual nerve, responses to capsaicin, 213
Lipid solubility, effects on nasal trigeminal nerve distribution, 40
Lipophilicity, 31
Liposomes, detection of chemical stimuli, 33
Locatable stimuli, somatosensory distribution, 124

**M**

Magnetometer, 128
Magnetoencephalography, 124, 127, 131
  CSSEP generation, 133
Maple syrup, 219
Maxillary nerve, 104
  olfactory and trigeminal control of reflexive responses to inhaled odorants, 96
    stimulatory effectiveness, 96
[Maxillary nerve]
  respiratory mucosa innervation
    reflex nasal secretions, 276, 277
Mechanonociception, capsaicin, 145, 146
  noxious stimulation, 145, 146
Mechanoreceptors, 39
Melanophores, trigeminal nerve ending stimulation, 31
Membrane potential, 33
  chemical stimuli detection, 33
*Mentha arvensis*, 281, 282
*Mentha piperita*, 281, 282
Methanol, trigeminal electrophysiologic threshold, chemically-provoked ethmoid responses, 25, 26, 28, 29
Menthol, 236, 246, 247, 305
  calcium conductance, 276, 288, 289, 294
  cerebral-evoked potentials to chemical stimuli, 124
    evoked potential mapping, 129
  chemically-induced burning, 149
  cold fiber stimulation, 328
  common chemical sense, 327, 335, 336
  medicinal uses, 276, 281, 283, 388, 289, 294
  nasal air flow, 283-288, 293-295
  production and chemistry, 281-283
    isomers of, 279, 283
  response in anosmics, 139
  sensory response differentiation, 124-134
  trigeminal electrophysiologic threshold, chemically-provoked ethmoid responses, 25, 26, 28, 29
Menthone, synthesis of menthol, 282
Mercaptide formations, 31
Mercaptobenzoic acid, 309
Methyl acetate, single-fiber responses, 23
Mitochondria, 156
Monomodal chemical stimulation, 125
Morning-after effect, *see* Shower shock

Mouth feel, 320
improvement of, 320
Mucous secretion, SP release, 65
Muscarinic inhibition, by SP, 63
Musk, single-fiber response, 23
Musk ketone, odorant stimulation of olfactory nerve, 97
Mustard, 194, 246
Mustard oil, chemically-induced burning, 149
Myelinated axons, 33

N

NaCl flux, collateralization of peptidergic fibers, 9
NaCl, *see* Sodium chloride
Nasal cavity, 1, 3, 98
air flow, 71, 72, 81, 91, 275-289
alteration of, 13
comparison of trigeminal and olfactory inputs, 71, 72
control of upper airway accessory respiratory muscles, 281, 289
dermal papillae, role of, 279
effects of nasal anesthesia, 287, 288
flow receptor stimulation, 39, 113
laryngectomy-induced dysfunction, 71, 83, 88
larynx bypass, 73-75
menthol, role of, 283-288
nasal congestion, 275, 276, 281
patency, 81
respiratory reflexes, 275, 280, 281
sensory receptors and nerve pathways, 275-280
sniffing behavior, 84, 85
anatomy of, 276-280
asymmetrical cycle, complications in stimulus interpretation, 59, 60
effects on trigeminal nerve
description of epithelia, 39
lipid solubility, 40
water solubility, 40
[Nasal cavity]
epithelial impediments to stimulation, 51, 52
innervation of trigeminal nerve by, 22, 31, 276, 279, 288, 289
antibodies, 19
collateral, 13
ethmoid (ophthalmic), 22-25
nasopalatine (maxillary), 22
odorants, 61-65
peptidergic, 1-14
monomodal chemical stimulation, 124
mucous ciliary movement, 20
antidromic stimulation, 20
patency, 72
peptidergic trigeminal fibers
distribution of, 3, 4
sparsity of, 19
presence of CGRP-immunoreactive fibers, 8, 13
reflexes
irritation-induced, 55, 56
spatial summation, 55, 56
stimulation by ammonia, 11
thermal stimulation, 26
transepithelial fibers, 7
similarity to CGRP-immunoreactive fibers, 7
vasodilation of, 13
Nasociliary nerves, 96
olfactory and trigeminal control of reflexive responses to inhaled odorants, 96
stimulatory effectiveness, 96
Nasopalatine nerve, 104
single-fiber responses, 22
Naxolone, opponent-endorphin response, 262
*n*-butyl acetate, psychophysical function, 45
time course of adaptation, 45
*n*-butyl alcohol, odor and pungency of, 52
Neoisomenthol, 283
Neomenthol, 283

Neuroblastoma cells, trigeminal nerve ending stimulation, 31
Neurokinin A, capsaicin-sensitive efferent response mediator, 160
Neuropeptides
  depletion by capsaicin, 35
  presence in nasal trigeminal unmyelinated nerve fibers, 2
    colocalization, 2
  release of, 11
    modulation of olfactory receptors, 11
Nicotine, 111, 301, 306
  response to odorants, 106
  trigeminal electrophysiologic threshold
    chemically-provoked ethmoid responses, 25, 26, 28, 29
Nociceptors, 39, 133, 172, 183, 186, 190, 332, 334, 335
  A-delta, 171, 202
  chemical irritation, 171-190
  chemo-, 142-146
  C-polymodal, 52, 142, 171
    stimulation by capsaicin, 172-176
    temperature sensitive afferent, 172-176
  effects of capsaicin, 235
  mechano-, 145, 146
  site establishment of cortical generators of CSSEP, 128
  stimulation of, 128, 132
Nocifensive reaction, 143, 146-148
Non-locatable stimuli, olfactory distribution, 124
Nonpeptidergic trigeminal sensory fibers, heat-pain stimulation, 19
Nonverbal response measures, 109
Nordihydrocapsaicin, 241, 245
Noxious heat threshold, 143, 144, 146, 169
  dose-dependent acute thermal hyperalgesia, 144, 145
  single unit recordings of cutaneous nerves, 153-155
Noxious stimuli, 8, 131, 171, 328-330, 333-335
  arousal of sense of irritation, 21
  CSSEP generation, 133
Nowlis-Green adjective checklist, facial manipulation, 221-227
  source identification of personality variables, 217-227
N-terminal peptide, interaction between trigeminal afferents and olfactory mucosa, amphibia, 62
NV ob, 64
  antidromic excitation of, 64
    SP depletion, 64
    transepithelial slow potentials, 64

## O

Octanoic acid, trigeminal electrophysiologic threshold, chemically-provoked ethmoid responses, 25, 26, 28, 29
Octanol, trigeminal electrophysiologic threshold, chemically-provoked ethmoid responses, 25, 26, 28, 29
Odorants:
  classical conditioning of cardiac acceleration, 102
  CSSEP generation, 133
  maxillary response to, 95
  neural bases of olfactory-trigeminal interaction, 96
  neural mediation of psychophysical and/or physiological responses to, 104-115
  non-olfactory pathways, 95-98
    ophthalmic, 95
    psychophysical, 96, 98, 104-115, 119
    psychophysiological, 98-100
  non-olfactory stimulation, anosmics
    changes in respiratory behavior, 104-115
    ocular, 96, 104-115

[Odorants]
olfactory nerve stimulation, 96
blast injection technique, 97
stream injection technique, 97
responses of normal and anosmic subjects, 95-121
animal studies, 98-104
historical overview, 95-98
human studies, 104-115
relative humidity, 108
role of septal organ of Masera, 95
trigeminal nerve stimulation
access of odorants to olfactory receptors, 97
greater superficial petrosal nerve, 97
psychophysical studies, 98, 99
response determination, 97
vomeronasal response to, 95
Odor-evoked electro-olfactograms (EOG), 64
latency of, 69
similarity with SP-induced transepithelial potentials, 62
OEP, *see* Olfactory-evoked potentials
Oil of Cloves, stimulation of olfactory nerves, 96
Oil-water partition coefficient, 32
receptor activation, 31
lipophilicity, 31
Olfactory bulb, 104
inhibition or facilitation of trigeminal input, 98
substance P release, 13
Olfactory-evoked potentials (OEP), cerebral-evoked potentials to chemical stimuli, 124
in anosmics, 124
Olfactory nerve, 21, 108, 121, 336, 337
alteration of prereceptor and perireceptor events, 11, 12
cerebral-evoked potentials to chemical stimuli, 123-134
common chemical sense, interactions with, 53-55
[Olfactory nerve]
directional olfactory orientation, 124
experimental use of highly selected panels, 229
input deprivation, 72
odorant detection, 72
laryngectomy-induced decrements, 71, 72
localization, 334
mediation by chemical stimuli, 139
odorant stimulation, 96, 97
trigeminal nerve
cohabitation with olfactory cells, 61
comparisons, 97
interactions with, 61-65, 96-115
response magnitude for irritants, 29
stimulation, 12, 33
support, 134
versus olfactory input, 71-91
Opponent-endorphin responses, mechanisms for preference development of chili peppers, 259-262
secretion of brain endorphins, 261-262
Ophthalmic nerve, 104, 110, 276
Oral cavity:
irritation, 171-190
role in respiration, 275, 276
Orange, stimulation of olfactory nerve, 96
Ozone, single-fiber response, 22

**P**

Pain, 123, 127, 131
carbon dioxide, 123, 127, 131
chemically-induced, 149
indirect activation of peptidergic fibers, 3
bradykinin production, 3
receptors and oral irritants, 215, 216
Paracellular shunts, 11, 12

Parker, G. H., 326, 327, 336
  common chemical sense, 21, 37
    sense of irritation, 21
PBC, *see* Private body consciousness
Pentanol:
  single-fiber response, 22
  trigeminal electrophysiologic threshold, chemically-provoked ethmoid responses, 25, 26, 28, 29
Penta-SP, interaction between trigeminal afferents and olfactory mucosa, amphibia, 62
Pentazocine, cerebral-evoked stimulation to chemical stimuli, 133
Pepper, red and black:
  components of, 197, 209
  black, 232, 234, 236, 238, 246
Peppermint:
  oil of, synthesis of menthol, 281, 282
  single-fiber responses, 23
Peptidergic trigeminal fibers, 5
  basal lamina, 5
  blood flow, 12, 13
  CGRP-SP
    location of, 5
  classes of, 2
  collateralization, 9-14
  effector endings, 20
  effects of stimulation, 8, 9
    alterations in peripheral tissue physiology, 9
    elicitation of sneeze, cough, and choke reflex, 1, 9
    modulation of preganglionic autonomic neurons, 9
    noxious and non-noxious stimuli, 8
    transmission of sensory information to CNS, 8, 9
  immunoreactivity
    to antisera, 2
    free nerve endings, 3
    proximity to epithelial surface, 2, 3
[Peptidergic trigeminal fibers]
  innervation of,
    afferent and effector functions, 1-14
    basal lamina, 5
    density in olfactory epithelium, 5
    epithelial maturation zone, 5, 6
    nasal glands, 5
    subepithelial CGRP-SP peptidergic fibers, 5, 6
  intranasal distribution of CGRP-SP fibers, 3, 4
    interspecies differences, 3, 5-8
    location in lamina propria, 6
    non-homogeneous distribution of intraepithelial fibers, 6
    transepithelial peptidergic fibers, 6
  mast cell activation, 20
  modulation of shunt permeability, 11
  neural plexus, 12
  origin in trigeminal nerve, 2, 3
  parasympathetic peptides, 11-13
  release of bioactive peptides, 11-13
  urodelean, 5
    density of itnraepithelia peptidergic fibers, 5
Perceptual "contamination," 44
Perigeminal fibers:
  association of sensitive and secretory functions in trigeminal nerve fibers, 65
  location of, 65
Perivascular neural plexus, 12
Phase boundary potential, chemical stimuli detection, 33
Phenethyl alcohol, 22
  chemosensory-evoked potentials, 126
    sensory response differentiation, 71-91
  in anosmics, 123, 139
  odorant stimulation of olfactory nerve, 97

[Phenethyl alcohol]
trigeminal electrophysiologic threshold
chemically-provoked ethmoid responses, 25, 26, 28, 29
Phenylthiocarbamide, 209
Piperine, 197, 201-206, 208, 209, 213, 214, 215, 246, 309
chemically-induced burning, 149
irritation, oral, 171-190, 198, 199
thermal effects, 171-190
private body consciousness tests, 219, 220
source identification of personality variables, 218-219
*Piper nigrum*, 198
Piquancy, 46, 237, 240, 246, 256, 262
Plasma extravasation:
SA mechanoreceptor activation, 152
capsaicin-induced, 152
SP release, 65
Plexus, 61
subepithelial, 6-8, 12
Polyethylene glycol, common chemical sense, 332
Polymodal nociceptors, 35, 150, 160, 169
capsaicin
dual sensory efferent function, 160
effects on C fibers, 152
noxious stimulation, 145, 146
role in common chemical sense, 326, 337, 338
Potassium currents, effects of capsaicin, 156, 157
Primary (SI) somatosensory cortices, CSSEP generation, 133
Private body consciousness (PBC)
source identification of personality variables, 218-227
capsaicin, 219-221
piperine, 219-221
quinine hydrochloride, 219, 220
Propanol, trigeminal electrophysiologic threshold, chemically-provoked ethmoid responses, 25, 26, 28, 29
Proprionic acid, 119, 120
response to odorants, 106, 108
changes in respiration, 113
nonolfactory receptor response, 111, 112
trigeminal electrophysiologic threshold, chemically-provoked ethmoid responses, 25, 26, 28, 29
trigeminal receptor stimulation, 104, 111
Proprioception, 22, 33, 275, 331
Protein receptors, specific:
chemical stimuli detection, 33
change in surface potential, 33
chili peppers, 233
Pungency, 43, 44, 46-48, 50, 54, 69
nasal pharyngeal sensations, 330
spices, 31
Pyridine, stimulation via physical interaction, 43, 44

## Q

Quality cooling, *see* Sensation quality
Quinine, 246, 247
hydrochloride, private body consciousness tests, 219, 220
sulfate, chemical-induced burning, 149
tastant sensation inhibition, 205, 209

## R

Receptor proteins, activation by physical adsorption, 31
oil-water partition coefficient, 31
Recurrent laryngeal nerve, trigeminal versus olfactory input, 72
RH, *see* Relative humidity
Relative humidity (RH), response to odorants, 108
Ruthenium red, capsaicin, 155

## S


Sal de dulce, 253
Sal de piquante, 253
Schwann cells, resistance to capsaicin, 142
Secondary (SII) somatosensory cortices, CSSEP generation, 133
Self-perception, oral irritation, 218-227
  psychophysical judgment, 217-227
  self-produced cues, 218-224, 227, 229
  situational cues, 218-224, 227
Sensation quality, 148, 199, 204, 330, 340
Sensory ganglion cells, presence of tachykinins, 2
Sensory innervation, non-olfactory:
  classical view, 1
    brainstem routes, 1, 2
    termination sites, 1
  reactions with trigeminal stimuli, 1
  recent evidence, 1, 2
    brainstem routes, 2
Sensory irritants, 329, 331
  concentration calculation of inspired material, 307
  trigeminal chemoreception, 297-302
Sensory nerve fibers, collaterals of, antidromic activation, 9
Sensory neuron blocking effect, 143
  capsaicin application, 143
Sensory response differentiation:
  localization after dichotic stimulation, 124-127
    directional smelling, 124
  monomodal chemical stimulation, 123-125
  perceptibility by anosmics, 123, 124
  topography of chemosensory-evoked potentials, 124-134
    brain mapping, 128, 129
    dependence on stimulus intensity, 131
    evoked potential mappings, 124
    site establishment of CSSEP, 124
Septal organ of Masera, 95
Shower shock (Morning-after effect), capsaicin irritation, thermal effects, 195
Silica, common chemical sense, 332
Slime molds, trigeminal nerve ending stimulation, 33
Smokers, 46
  inurement to airborne irritants, 53
  irritation of nasal mucosa, 55, 56
  single-fiber response, 23
Sniffing behavior, 83-85
  effect on trigeminal versus olfactory input, 85
Sodium chloride (NaCl), 221
  irritation, oral, 171-190
    chemical effects, 187-190
    thermal effects, 187-190
  private body consciousness tests, 219, 220
    facial manipulation, 223-225
  tastant sensation inhibition, 205
Somatosensory system, *see also* Trigeminal nerve
SP, *see* Substance P
Sphenopalatine ganglion, 8, 276
  collateralization of peptidergic fibers, 9
  synapse of collateral peptidergic fibers, 13
  trigeminal nerve collaterals, 8
Spices, 197, 198, 229, 230, 232, 236, 240
  inhibition by mastication, 180
  source identification of personality variables, 225-227
Spike activity, 64
Spinal trigeminal nucleus, 1, 8, 9
Steep stimulus onsets, cerebral-evoked potentials to chemical stimuli, 123
*Sternus vulgaris* (Starlings), response of trigeminal nerve to irritant consumption, 309-320
Sternutatories, as chemical trigeminal stimuli, 31

Substance P (SP), 19, 69
collateralization of peptidergic fibers, 12
growth modification of basal cells, 12
neurotrophic activity, 12
-containing fibers, 35
depletion by capsaicin, 35, 64
desensitization of neural response, 35
effect on olfactory neurons, 11
-immunoreactive nerve fibers
nasal epithelium, 2
oral cavity, 202
origin in trigeminal nerve, 2
-induced vasodilation, 63
intraepithelial cells, stimulation of, 12
-like immunoreactivity, 61-64
opponent-endorphin responses, 261
paracellular shunts
increase of anion permeability, 11
release of, 64
activation of trigeminal sensory fibers, 13
mucous secretion, 65
olfactory bulb, 13
plasma extravasation, 65
role as capsaicin-sensitive efferent response mediator, 160
secretagog action of, 63
SP-induced vasodilation, 63
topical administration, 12
alteration of prereceptor and perireceptor events, 12
trigeminal afferent interaction with olfactory mucosa, 64
alteration of spontaneous receptor cell discharge, 62
decrease in K+ conductance, 63
muscarinic inhibition, 63
secretory activity of Bowman glands, 63
SP 1-7, 62
transepithelial slow potentials, 62
[Substance P]
vasodilation via trigeminal nerve, 13
edema, 13
plasma extravasation, 13
Substance P antisera, 6
Sucrose:
private body consciousness tests, 219, 220
tastant sensation inhibition, 205
Sudden Infant Death, 280
Suffocants, as chemical trigeminal stimuli, 31
Sulfur dioxide, 306, 307
cerebral-evoked potential to chemical stimuli, 124
irritation of nasal mucosa, 55
Surface potential, detection of chemical stimuli, 33
Sustentacular cells, collateralization of peptidergic fibers, 12

**T**

Tachykinins:
locating SP receptors, 63
role as capsaicin-sensitive efferent response mediators, 160
substance P, 2
presence in sensory ganglion cells, 2
Taste buds, 65, 202
Taste receptors, stimulation by aliphatic alcohols, 31
Temporal summation, 56
Tetrodotoxin, inhibition of nerve endings, 142
Thaler illusion, mechanoreceptor response to cooling, 331
Thermal lability, 176, 179
Thiol alkylating agents, 31
Threshold limit value (TLV), 298
Time course of adaptation, 48, 49
common chemical sensations, 53
irritation, oral, 180-190
role of olfactory and irritant sensations, 45

[Time course of adaptation]
airborne irritants, 45, 46
sensation decay, 204, 205
TLV, *see* Threshold limit value
Tobacco, 236, 246, 246, 264
two stages of preference acquisition for, 250-251
Toluene, trigeminal electrophysiologic threshold chemically-provoked ethmoid responses, 25, 26, 28, 29
Trachea, paracellular shunts, anion shunts, 11
Transepithelial fibers, peptidergic innervation, 6, 7
Trigeminal (Gasserian) ganglion, 1, 22, 62
Trigeminal chemoreception, 1, 2, 9, 33, 76, 108, 121, 123, 124, 279, 202, 282
airborne irritants, 297-302
specificity of sensory irritant receptor, 300-302
worker/consumer protection, 298-300
antidromic stimulation
of ophthalmic component, 64
release of SP, 64, 65
vasodilation, 13
avian irritant consumption, 309-320
brain responses, 123-134
localization, 129-133
calcium conductance, 288, 289
capsaicin
desensitization and irritation, 35, 141-161
inhalation, 329
cerebral-evoked potentials to chemical stimuli, 123, 134
evoked potentials at subliminal levels of perception, 139
changes in olfactory stimuli perception, 14
olfactory sensitivity, comparisons, 97, 102
characterization, 31
[Trigeminal chemoreception]
collaterals of, 13
changes in prereceptor and perireceptor events in olfactory transduction, 12
common chemical sense, 326-328
CSSEP versus OEP, 123-134
distribution in nasal cavity, 39
electrophysiological threshold, 32
alcohol and partition coefficients, 31
chemically provoked ethmoid responses, 25, 26, 28, 29
fiber characteristics, 33-35
indirect activation of peptidergic fibers, 13
bradykinin production, 3
stimulus effectiveness of bradykinin antagonists, 3
interaction with olfactory receptor cells, 61-65, 96-115
cohabitation with olfactory cells, 61
decrease in K+ conductance, 63
discharge frequency, 63
electrophysiological evidence, 22, 64, 65
eye, 96, 104-115
immunohistological evidence, 61, 62
observation of SP-reactive cell bodies within trigeminal ganglion, 62
pharmacologic evidence, 62, 63
physiological implications, 65
receptor cell increase in discharge frequency, 63
regulation of cell environment and activity, 62
SP-induced mucosal secretory activity, 62
SP-induced transepithelial potentials, 62
versus olfactory input in laryngectomy, 71-91
localization, 334

[Trigeminal chemoreception]
location of nerve endings, 35
mediation of quality discrimination, 329, 340
menthol, 288, 289
effects on nasal sensation of air flow, 275
modulation of preganglionic autonomic neurons, 9
nasal cavity, physiology, 21-35
nasal reflexes, 276, 281
non-olfactory mediation of response to vapor phase odorants, 95-121
NV ob, 62
odorants, 96-100
changes in nasal secretion, 97
detection thresholds, 75-91
perception, 104
psychophysical study, 98-100
response determination, 97, 98
sniffing behavior, 85
stimulation of, 95-115
olfactory nerve, comparisons, 71-91, 97, 102
CSSEP generation, 134
laryngectomy-induced olfactory decrements, 71, 72
larynx bypass, 72-75
odorant detection threshold, 75-86
odorant identification, 86-91
ophthalmic branch, 110
oral irritants, 197-210
peptidergic fibers
substance P mediation, 13
receptor stimulation, cornea, 113
receptor proteins, 31
similarity with cholinergic receptors, 31
role as a protection system, 29
sensory irritant stimulation, 297-302
single-fiber responses, 22, 23
stimulus and receptor characteristics, 28, 29, 31, 33
spinal reflex system, 9
VPM, 9
Trigeminal (Gasserian) ganglion, 22
Trigeminectomy, 104, 108, 111, 330, 339
Turbinate, 7

U

Unmyelinated fibers, 33
role as trigeminal receptor, 35
Upper respiratory airway, 114

V

Vagus nerve, 104
excitation of accessory olfactory regions, 72
trigeminal versus olfactory input, 72
Valeric acid, trigeminal electrophysiologic threshold, chemically-provoked ethmoid responses, 25, 26, 28, 29
Vanilla, nasal air flow, 284, 286
Vanillin, 123
addition with carbon dioxide, 139
cerebral-evoked potentials to chemical stimuli, 124, 126, 137, 139
evoked potential mapping, 129
changes in airway patency, 81
laryngectomy
-induced hyposmia, 81
loss of nasal air flow, 81
mixture with carbon dioxide, 127, 134
trigeminal versus olfactory input
detection thresholds, 75-86
Vanillylamide, 233
Vanillyl nonamide, difference among oral irritants, 198-201, 202
Vasoactive intestinal peptide-containing peptidergic fibers, 2
Vasodilatation, capsaicin, 159, 160
Ventral posteromedial nucleus (VPM), trigeminal nasal chemoreception, 9
integration site for somatosensory and visceral sensory input, 9

Vibrissae, 277, 278
Vinegar, trigeminal versus olfactory input, 71-91
  odorant identification, 86-91
Voltage clamp, 155-159
Voltage-sensitive gating molecules, chemical stimuli detection, 33
Vomeronasal nerve, 277
  peptidergic innervation, 6, 7
    air flow, 13
    density of CGRP-SP fibers
  response of subjects of odorants, 98
Vomeronasal Pump, 7
  role of vomeronasal pump, 8
VPM, *see* Ventral posteromedial nucleus

**W**

Warm fibers, 176, 328
Water solubility, effects on nasal trigeminal nerve distribution, 40

**X**

Xylene, noxious heat threshold, 145, 149
Xylol, anosmic odorant detection, 97

**Z**

Zingerone:
  chemically-induced burning, 149
  response to odorant consumption, 309-320